The Essentials of
Supply Chain Management

The Essentials of Supply Chain Management

New Business Concepts and Applications

Hokey Min

*James R. Good Chair in Global Supply Chain Strategy at
Bowling Green State University*

Publisher: Paul Boger
Editor-in-Chief: Amy Neidlinger
Executive Editor: Jeanne Glasser Levine
Development Editor: Natasha Wolmers
Operations Specialist: Jodi Kemper
Cover Designer: Chuti Prasertsith
Managing Editor: Kristy Hart
Project Editor: Elaine Wiley
Copy Editor: Bart Reed
Proofreader: Laura Hernandez
Indexer: Tim Wright
Senior Compositor: Gloria Schurick
Manufacturing Buyer: Dan Uhrig

© 2015 by Hokey Min

For information about buying this title in bulk quantities, or for special sales opportunities (which may include electronic versions; custom cover designs; and content particular to your business, training goals, marketing focus, or branding interests), please contact our corporate sales department at corpsales@pearsoned.com or (800) 382-3419.

For government sales inquiries, please contact governmentsales@pearsoned.com.

For questions about sales outside the U.S., please contact international@pearsoned.com.

Company and product names mentioned herein are the trademarks or registered trademarks of their respective owners.

All rights reserved. No part of this book may be reproduced, in any form or by any means, without permission in writing from the publisher.

Printed in the United States of America

First Printing June 2015

ISBN-10: 0-13-403623-9
ISBN-13: 978-0-13-403623-6

Pearson Education LTD.
Pearson Education Australia PTY, Limited.
Pearson Education Singapore, Pte. Ltd.
Pearson Education Asia, Ltd.
Pearson Education Canada, Ltd.
Pearson Educación de Mexico, S.A. de C.V.
Pearson Education—Japan
Pearson Education Malaysia, Pte. Ltd.

Library of Congress Control Number: 2015935040

This book is dedicated to my late father B.J. Min, my mother H.W. Seo, my wife Christine, and my son Alexander Snow.

Contents

Acknowledgments xiii

About the Author xv

Chapter 1 **Principles of Supply Chain Management 1**
Learning Objectives 1
Evolution of the Supply Chain Concept 1
Total Systems Approach and Boundary Spanning 6
Conceptual Foundations of Demand Chain, Value Chain,
and Supply Chain 8
Strategic Alliances and Partnerships 10
Organizational Learning from Strategic Alliances 17
Interfaces among Purchasing, Production, Logistics, and Marketing 19
Theory of Constraints (TOC) for Supply Chain Management 20
Change Management for Supply Chain Management 22
Chapter Summary 25
Study Questions 26
Zara's Rapid Rise as a Cool Supply Chain Icon 27
Bibliography 31

Chapter 2 **Supply Chain Strategy: The Big Picture 37**
Learning Objectives 37
Strategic Dimensions 37
Red Ocean versus Blue Ocean Strategy 39
Strategic Supply Chain Planning Processes 40
Strategic Integration of Supply Chain Processes 42
The "Victory" (Winning Strategy) Model 44
Push versus Pull Strategy 46
Typology of Supply Chain Strategy 49
Internal Supply Chain Strategy Audits 51
External Supply Chain Strategy Audits 52

    Chapter Summary 53
    Study Questions 54
    Case: Dell, Inc.—Push or Pull? 54
    Bibliography 58

Chapter 3   **Customer Service: The Ultimate Goal of Supply Chain Management 61**
    Learning Objectives 61
    Understanding Customer Expectations and Perceptions 61
    Customer Service Elements 62
    Building Customer Relationships 67
    Service Delivery Performance 73
    Formulating a Winning Customer Service Strategy in a Supply Chain 79
    Chapter Summary 81
    Study Questions 82
    Case: Shiny Glass, Inc. 83
    Bibliography 86

Chapter 4   **Demand Planning and Forecasting 89**
    Learning Objectives 89
    Demand Management 89
    Demand Forecasting 94
    Sales and Operational Planning 105
    Collaborative Commerce 107
    The Bullwhip Effect 113
    Chapter Summary 116
    Study Questions 118
    Case: Seven Star Electronics: Demand Planning 119
    Bibliography 123

Chapter 5   **Inventory Control and Planning 127**
    Learning Objectives 127
    The Principles of Inventory Management 128
    Functions of Inventory 129
    Types of Inventory 130
    Inventory Classification 130
    Independent Demand Inventory Control and Planning 134
    Dependent Demand Inventory Control and Planning 147
    Distribution Resource Planning 155

    Just-In-Time Inventory Principles 160
    Basics of Cycle Counting 164
    Vendor Managed Inventory 167
    Chapter Summary 168
    Study Questions 169
    Case: Sandusky Winery 170
    Bibliography 172

Chapter 6   **Warehousing 175**
    Learning Objectives 175
    Warehouses in Transition 176
    Types of Warehouses 178
    Types of Warehouse Leases 180
    Warehousing Costs 181
    Warehouse Network Design 183
    Warehouse Layout 186
    Warehouse Asset Management 189
    Material Handling 190
    Order Picking 192
    Warehouse Productivity 193
    Warehouse Security and Safety 195
    Warehouse Automation 196
    Warehouse Workforce Planning 197
    Warehouse Management Systems 198
    Handling Returned Products and Reverse Logistics 205
    Chapter Summary 209
    Study Questions 210
    Case: One Bad Apple and Thousands of Headaches 211
    Bibliography 216

Chapter 7   **Transportation Planning 221**
    Learning Objectives 221
    Transportation as a Vital Link in the Supply Chain 222
    Central Place Theory 224
    Transportation Regulations and Deregulations 225
    Legal Forms of Transportation 233
    Carrier Management 235
    Surface Transportation 239
    Water Carriers 242

    Air Carriers 244
    Intermodalism 245
    Transportation Documentation 247
    Transportation Pricing 249
    Freight Rate Negotiation 250
    Revenue/Yield Management 252
    Transport Management Systems 253
    Terminal Operations 254
    Chapter Summary 256
    Study Questions 257
    Case: Louis Cab On Demand 259
    Bibliography 262

Chapter 8   **Sourcing** **265**
    Learning Objectives...265
    In-Housing versus Outsourcing 266
    Principles of Outsourcing 267
    Cost Analysis 270
    Value Analysis 277
    What Should Be Sourced...279
    Who Can Be Supply Sources 280
    Supplier Relationship Management 291
    Intermediaries for Sourcing 293
    Supply Risk Management..295
    Competitive Bidding versus Negotiation 299
    Global Sourcing 302
    E-purchasing and Auctions..305
    Chapter Summary 308
    Study Questions 310
    Case: Lucas Construction, Inc. 311
    Bibliography 313

Chapter 9   **Logistics Intermediaries** **319**
    Learning Objectives...319
    The Role of Intermediaries 320
    Types of Intermediaries 321
    Potential Challenges for Using Logistics Intermediaries 327
    3PL Market Trends 329
    Chapter Summary 335

	Study Questions 336
	Case: Falcon Supply Chain Solutions 336
	Bibliography 339
Chapter 10	**Global Supply Chain Management 343**
	Learning Objectives 343
	The Impact of the Free Trade Movement on Global Supply Chain Management. 344
	Global Market Penetration Strategy of Multinational Firms 347
	Strategic Alliances among Multinational Firms 351
	Global Outsourcing Trends 354
	Hidden Inhibitors Affecting Global Supply Chain Operations 356
	Managing International Distribution Channels 364
	Foreign Trade Zones and Free Trade Zones 366
	Import and Export Documentation 367
	Incoterms and International Payments 369
	Countertrade 373
	Transfer Pricing 375
	Cross-Cultural Negotiations 376
	Chapter Summary 378
	Study Questions 379
	Case: Aurora Jewelers 380
	Bibliography 385
Chapter 11	**Legally, Ethically, and Socially Responsible Supply Chain Practices 389**
	Learning Objectives 389
	Triple Bottom Line 389
	Types of Laws 390
	Laws Applicable and Relevant to Supply Chain Activities 392
	Disputes and Claim Resolutions 395
	Supply Chain Ethics 396
	Green Supply Chain Management 403
	Chapter Summary 412
	Study Questions 413
	Case: Jumping Footwear, Inc. 414
	Bibliography 416

Chapter 12 Measuring the Supply Chain Performance 419
    Learning Objectives 419
    Supply Chain Performance and Its Impact on the Bottom Line 419
    Supply Chain Performance Metrics 421
    Key Performance Indicators (KPIs) 425
    A Balanced Scorecard for Supply Chain Performance Measurement 426
    The Supply Chain Operations Reference (SCOR) Model 429
    Other Performance Tools 432
    Chapter Summary 433
    Study Questions 435
    Case: La Bamba Bakeries 436
    Bibliography 439

Chapter 13 Emerging Technology in Supply Chain Management 441
    Learning Objectives 441
    The Emergence of E-commerce 442
    Enterprise Resource Planning (ERP) 443
    Geographic Information System (GIS) 445
    Intelligent Transportation Systems 447
    Barcoding Systems 449
    Radio Frequency Identification (RFID) 450
    Artificial Intelligence 453
    Information Technology (IT) Project Management 454
    Future Trends of IT in Global Commerce 457
    Chapter Summary 461
    Study Questions 462
    Case: Red-Kitchen.com 463
    Bibliography 466

Index 469

Acknowledgments

This book project is ten years behind schedule. While working on this project, I often wondered if I would ever finish it. However, writing this book has been one of the most rewarding experiences I have ever had because it helped me educate myself about the dynamic evolution of supply chain principles and created a discussion forum with my beloved students. In particular, without the valuable feedback from my former executive and professional MBA students, I might not even have a chance to introduce this book to prospective readers and enthusiastic students who are curious about one of the hottest business topics around. Although a majority of the content of this book originated from the education/training materials I used for a series of executive education/training programs I organized at the UPS Center for Worldwide Supply Chain Management (UPSi) at the University of Louisville, I should give a ton of credit to my former mentors who initially guided me into the exciting field of supply chain management. Some of these mentors include Drs. Martha Copper, John Current, Bernard LaLonde, and David Schilling of the Ohio State University, Dr. Robert Markland of University of South Carolina, and Dr. Lori Franz of University of Missouri-Columbia. These professors taught me how to become a consummate scholar.

In addition, I would be remiss if I did not acknowledge a number of my longtime professional colleagues and friends who were willing to share their in-depth knowledge of supply chain management with me during my academic career. These individuals include Dr. Emanuel Melachrinoudis of Northeastern University, Dr. Michael LaTour of Cornell University, Dr. Thomas Speh of Miami University, Dr. Tom Goldsby of Ohio State University, Dr. Robert Lieb of Northeastern University, Dr. John Langley, Jr., of Georgia Tech, Dr. Arnie Maltz of Arizona State University, Dr. Paul Hong of University of Toledo, Dr. Seong-Jong Joo of Central Washington University, Dr. Ik-Whan Kwon of Saint Louis University, Dr. Angappa Gunasekaran of University of Massachusetts-Dartmouth, Dr. Thomas Lambert of Northern Kentucky University, Dr. Soon-Hong Min of Yonsei University (Korea), Dr. Byung-in Park of Chonnam National University (Korea), Dr. Yonghae Lee of Hanyang University (Korea), Dr. Mitsuo Gen of Waseda University (Japan) and POSTECH (Korea), Dr. Gengui Zhou of Zhejiang University of Technology (China), Mr. John Caltagirone of the Revere Group, Mr. Ken Ackerman of Ackerman Associates, Mr. Bob Prugar of UPS Supply Chain Solutions, Mr. Charles Harret of Northern Continental Logistics, Mr. Paul Kovarovic of DoubleDay Acquisitions, Inc., Mr. Steven Mulaik of the Progress Group, Mr. Jonathan Whitaker of Accenture, Ms. Catherine Cooper of Ozburn-Hessey Logistics, Ms. Alice Houston of Houston-Johnson, Inc., Mr. Mike Nye of The Andersons, Inc., Mr. Sam Yildirim of Stride Rite, Ms. Amy Thorn of the Distribution Business Management Association, Mr. David Johnson of GSI Commerce, Mr. Joe Cappel of Toledo-Lucas County Port Authority, Mr. Arthur Potter of Pegasus Transportation, Dr. Chris Norek of Chain Connectors, Mr. Joseph Kutka of Exel Logistics, Mr. Thomas Freese of Freese & Associates, Mr. Ned Sheehy of the Kentucky Motor Transport Association, Mr. Jeffrey Leep of the Genlyte Thomas Group, Mr. Victor Deyglio of the Logistics Institute (Canada), Mr. Ken Jamie of Lynnco Supply Chain Solutions, Mr. Ed Nagle of Nagle Trucking, and Mr. James Kim of Logos Logistics.

Last, but not least, I am very grateful to my family for their continuous support and encouragement, while putting up with my frequent retreats into my study room. Especially, I am greatly indebted to my late father, B.J. Min, for his inspiration. He valiantly fought against terminal cancer with his strong will, but he could not overcome God's will and passed away without seeing my book get published. Also, I owe a great debt of gratitude to my former students who tolerated the incomplete draft of this book and offered constructive comments for improving its quality and readability. Finally, I would like to thank Ms. Jeanne Levine of Pearson Education and her supporting cast for their professional job in editing this book.

Hokey Min
Bowling Green, Ohio

About the Author

Dr. Hokey Min is the James R. Good Chair in Global Supply Chain Strategy in the Department of Management at the Bowling Green State University. He was Professor of Supply Chain Management, Distinguished University Scholar, and Founding Executive Director of the Logistics and Distribution Institute (LoDI), the UPS Center for World-wide Supply Chain Management, and the Center for Supply Chain Workforce Development at the University of Louisville. He earned his Ph.D. degree in Management Sciences and Logistics from Ohio State University. His expertise includes global logistics strategy, healthcare supply chains, closed-loop supply chains, e-synchronized supply chains, service benchmarking, and supply chain modeling. He has published more than 170 scholarly articles in various refereed journals including *European Journal of Operational Research*, *Journal of Business Logistics*, *International Journal of Physical Distribution and Logistics Management*, *Journal of Supply Chain Management*, *Supply Chain Management: An International Journal*, *Journal of the Operational Research Society*, *International Journal of Production Research*, *International Journal of Production Economics*, *Transportation Journal*, and *Transportation Research*. He recently authored a book titled *Healthcare Supply Chain Management: Basic Concept and Principles*. He has also engaged in numerous consulting projects with more than 50 different organizations including UPS, Brown-Forman Beverage World-Wide, Syntel Inc., Nationwide Insurance, National Tobacco Company, Time-It Transportation, Pegasus Transportation Inc., Usher Transport Inc., Nagle Trucking, Houston-Johnson Inc., Master Halco, Briggs and Stratten, West-Point Stevens, ScanSteel Inc., Dixie Warehouse Services, GenLyte Thomas Industries, Owens Corning, Buckeye Cable Systems, Andersons Inc., BioFit, Kentucky Motor Transport Association, Korea Maritime Institute (KMI), Korea Ocean Research Development Institute (KORDI), The Korea Research Institute of Ships & Ocean Engineering (KRISO), and The Chinese Rural Energy & Environment Agency.

1

Principles of Supply Chain Management

"Silos—and the turf wars they enable—devastate organizations."
—Patrick Lencioni, *Silos, Politics and Turf Wars*

Learning Objectives

After reading this chapter, you should be able to:

- Understand the rationale behind and fundamental principles of supply chain management.
- Comprehend the differences between supply chain perspectives and traditional business perspectives.
- Identify the main drivers of supply chain links.
- Recognize the managerial benefits and potential challenges of the supply chain practices.
- Analyze the impact of supply chain management on the bottom line and the competitiveness of the organization.
- Understand the necessary changes and transformations required for the successful implementation of the integrated supply chain perspectives.
- Find ways to leverage the supply chain for business success.

Evolution of the Supply Chain Concept

Over the years, most firms have focused their attention on the effectiveness and efficiency of separate business functions such as purchasing, production, marketing, financing, and logistics. The lack of connectivity among these functions, however, can lead to sub-optimal organizational goals and create inefficiency by duplicating organizational efforts and resources. To capture the synergy of interfunctional and interorganizational integration and coordination across the supply chain and to subsequently make better strategic decisions, a growing number of firms have begun to realize the strategic importance of planning, controlling, and designing a supply chain as a whole. In today's global marketplace, individual firms no longer compete as independent entities with unique brand names, but rather as integral parts of supply chain links. As such, the ultimate success of a firm will depend on its managerial ability to integrate and coordinate the intricate network of business relationships among supply chain partners (Drucker, 1998; Lambert and Cooper, 2000). A supply chain is referred to as an *integrated system* that synchronizes a series of interrelated business processes in order to: (1) create demand for products; (2) acquire raw materials and parts; (3) transform these raw materials and parts into finished products; (4) add value to these products; (5) distribute and promote these products to either retailers or customers; (6) facilitate information exchange among various business entities (e.g., suppliers, manufacturers, distributors, third-party logistics providers, and retailers). Its main objective is to enhance the operational efficiency, profitability, and competitive position of a firm and its supply chain partners. More concisely, supply chain management is defined as "the integration of key business processes from end-users through original suppliers that provide products, services, and information and add value for customers and other stakeholders" (Cooper et al., 1997b, p. 2). A supply chain is characterized by a forward flow of goods and a backward flow of information, as illustrated by Figure 1.1 (Min and Zhou, 2002, p. 232).

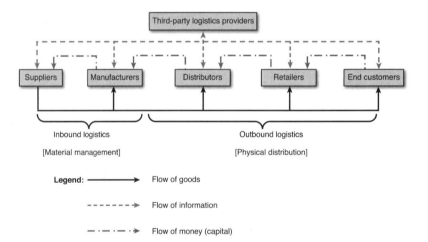

Figure 1.1. *The supply chain process*

Typically, a supply chain is composed of two main business processes:
- Material management (inbound logistics)
- Physical distribution (outbound logistics)

Material management is concerned with the acquisition and storage of raw materials, parts, and supplies. To elaborate, material management supports the complete cycle of material flow—from the purchase and internal control of production materials, to the planning and control of work-in-process, to the warehousing, shipping, and distribution of finished products (Johnson and Malucci, 1999). On the other hand, physical distribution encompasses all outbound logistics activities related to providing customer service. These activities include order receipt and processing, inventory deployment, storage and handling, outbound transportation, consolidation, pricing, promotional support, returned product handling, and life-cycle support (Bowersox and Closs, 1996).

Combining the activities of material management and physical distribution, a supply chain does not merely represent a linear chain of one-on-one business relationships, but a web of multiple business networks and relationships. Along a supply chain, there may be multiple stakeholders, composed of various suppliers, manufacturers, distributors, third-party logistics providers, retailers, and customers. For example, a supply chain for typical automobile seats linking suppliers, manufacturers, third-party logistics providers, and customers is graphically illustrated in Figure 1.2. As shown in this figure, the supply chain begins with customers such as Ford, General Motors, and Fiat-Chrysler, who need to use automobile seats as critical parts of their manufactured cars. At the next upstream stage of the supply chain, the car manufacturer often purchases automobile seats from the original equipment manufacturer (OEM). This OEM needs to acquire the parts and components of the automobile seats, including brackets, foam, fabric, and fasteners from tier-one suppliers fabricating those parts and components. Because these parts and components are made of metals, screws, bolts, plastics, and textiles, the tier-one suppliers should acquire some simple parts and raw materials from tier-two suppliers, who should obtains such parts and materials from tier-three suppliers such steel and yarn producers. These tier-three suppliers, in turn, obtain their sources of materials from ore mining and cotton plants at the furthest upstream of the supply chain. In case logistics activities involving the movement, handling, storage, and packaging of these materials, parts, components, and finished goods are outsourced from third-party logistics providers, the complexity of the supply chain network will be increased due to the possibility of both forward and reverse flow of products. As illustrated by this example, the typical supply chain cannot be explained by a linear linkage among the supply chain members.

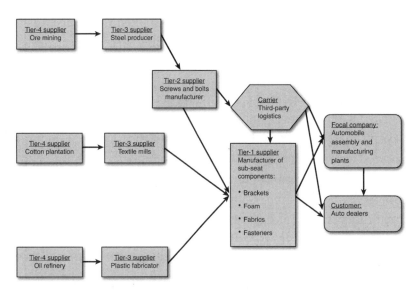

Figure 1.2. *The supply chain network for automobile seats*

In a nutshell, the concept of supply chain management has evolved around a customer-focused corporate vision, which drives changes throughout a firm's internal and external linkages and then captures the synergy of interfunctional, interorganizational integration and coordination. Herein, integration does not entail merger/acquisition or equity of the ownership of other organizations. The successful integration of the entire supply chain process can bring about a number of bottom-line benefits (Schlegel, 1999):

- **Improved customer service and value added**—Customer service can be improved through increased inventory availability, better on-time delivery performances, higher order fill rates, and lower post-sales costs.
- **Enhanced fixed capital**—Fixed capacity is maximized through a strategic partnership and joint planning that can increase overall capacity and throughput.
- **Utilized asset**—Asset utilization can be maximized by increasing inventory turns and closely aligning supply with demand.
- **Increased sales and profitability**—The ability to assess outcomes due to price changes, promotional events, and new product development can be enhanced through increased visibility resultant from information sharing among supply chain partners.

Financial benefits can be accrued from successful supply chain integration. For instance, thanks to streamlined supply chain integration, Dell's personal computer (PC) market share in the U.S. grew from 2.7% in 1995 to 24.1% in 2014 (Gartner, 2014). Similarly, Walmart, which happens to be another supply chain leader, enjoyed the rapid growth of its market share from 6.8% in 1992 to 17.1% in 2004 before declining to 11.4% in 2013 (Foster, 2006; Statistica, 2014). Despite these benefits of supply chain integration, firms engaged in this effort must be aware of the various challenges because of the unprecedented number and diversity of products and services available to customers in the era of mass customization. This variety will make it more difficult for a firm to predict customer needs and requirements. Therefore, the consequence of making forecasting errors will be more serious than ever before. Unfortunately, in a stretched supply chain with complex layers of suppliers and distributors, the severity of forecasting errors could be far beyond the level of compromise. Hardest hit by such forecasting errors are often upstream suppliers with little resources whose visibility of true demand is blindsided by distorted information passed by their immediate customers (e.g., manufacturers) and other downstream customers (e.g., distributors and retailers). This phenomenon was often explained by the so-called "bullwhip" effect.

The bullwhip effect is generally referred to as an inverse ripple effect of forecasting errors throughout the supply chain that leads to amplified supply and demand misalignment, where orders (perceived demand) to the upstream supply chain member tend to exaggerate the true patterns of end-customer demand because each chain member's view of true demand can be blocked by its immediate downstream supply chain member (Min, 2000; Lee et al., 1997a). The common symptoms of the bullwhip effect include delayed new product development, constant shortages and backorders, frequent order cancellations and returns, excessive pipeline inventory, erratic production scheduling, expedited shipments, and chronic overcapacity problems (Min, 2000; Lee et al., 1997b). The failure to mitigate or eliminate the bullwhip effect can disrupt the firm's revenue driver and adversely affect the firm's bottom line. According to Hendricks and Singhal (2005), supply chain disruptions led to:

- Significant reduction in stock returns relative to their benchmarks (e.g., 33% to 40% reduction over a three-year period)
- Increased share price volatility (e.g., 13.5% increase in share price volatility one year after supply chain disruptions)
- Decline in profitability (e.g., 107% drop in annual operating income, 7% decline in annual sales growth, and 11% annual total cost increase)
- Debilitating firm performances (e.g., at least two consecutive years of lower performances after supply chain disruptions)

Similarly, another worldwide survey of 602 financial executives conducted by FM Global and Harris Interactive indicates that supply chain disruptions are the biggest threat to a firm's revenue drivers (Yang and Gonzalez, 2006). Considering the enormous impact of supply chain disruptions on a firm's financial status, today's firms are increasingly pressured to manage their supply chain right. Thus, supply chain management has become the forefront of the firms' competitive strategy. The discipline of supply chain management, however, is still undergoing an evolutionary process. Table 1.1 summarizes the changes in the philosophy, focus, and performance metrics of supply chain management, from the earlier stages to the current era (see Martin and Towill, 2000).

Table 1.1. The Evolution of Supply Chain Management Disciplines

Evolution Stage	Time Period	Philosophy	Key Driver	Key Performance Metric
I	Early 1980s	Product driven	Quality	• Inventory turns • Production cost
II	Late 1980s	Volume driven	Cost	• Throughput • Production capacity
III	Early 1990s	Market driven	Product availability	• Market share • Order fill rate
IV	Late 1990s	Customer driven	Lead time	• Customer satisfaction • Value added • Response time
V	Early twenty-first century	Knowledge driven	Information	• Real-time communication • Business intelligence

Total Systems Approach and Boundary Spanning

A traditional business paradigm intends to react to unforeseen customer demand on a "push" basis by building buffers such as inventory that mitigate forecasting errors and hide distribution/production planning problems. The traditional business paradigm is also characterized by the sequential flow of information from one business function to another. Because the sequential information flow does not give an organization the opportunity to synchronize its functional activities and will impair its visibility throughout the planning processes, the same hidden problems will recur and the vicious cycle of inefficiency will continue without the problems ever being addressed. The best way to break this vicious cycle is to create a system that allows the organization to see the big picture of the business processes and then analyze the impact of the whole business processes on the organizational-wide goals rather than the departmental/functional goals. In other words, to continuously improve business processes, the traditional business paradigm should be replaced by the total system approach, which can create a whole greater than the sum of its parts. Therefore, the total systems approach is considered a major foundation for the supply chain concept.

The total systems approach regards the supply chain as an entity that is composed of interdependent or interrelated subsystems, each with its own provincial goals, but which integrates the activities of each segment so as to optimize the system-wide strategic objectives. To elaborate, the total systems approach is referred to as a "holistic, integrated approach" whereby all the business processes involving demand planning, purchasing, production, transportation, warehousing, and marketing are coordinated to make the best tradeoffs within them so as to achieve the optimal outcome for the whole system. For instance, the decision to increase inventory to make products more readily available to customers will help promote sales, but it would incur higher inventory carrying costs and warehousing costs. Without understanding such interdependence of the decision-making processes within the supply chain, the organization as a whole will continue to suffer from the downward spiral of declining productivity. That is to say, the total systems approach recognizes the fact that a decision made in one of the business functions can impact other functions of the organization. As such, the total systems approach enables the firm to assess how changes in business strategy and decisions affect the firm's across-the-board total costs and benefits.

The total systems approach to supply chain integration is often predicated on the five essential attributes displayed in Figure 1.3 (Miller and Berger, 2001, p. 13). As shown in this figure, collaboration is at the center of the total systems approach.

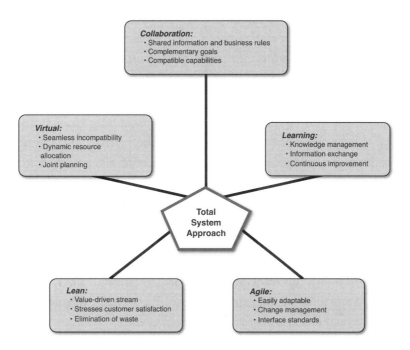

Figure 1.3. *The five essentials of the total systems approach to supply chain integration*

As the extended enterprise perspective brought by the total systems approach has become the important foundation of supply chain thinking, we have witnessed increasing boundary-spanning activities across the supply chain. Typically, these boundary-spanning activities have played three different roles:

- **Gatekeeping**—They single out potential suppliers and third-party logistics providers through a request for proposal (RFP) and then help the firm to make an informed decision as to who will be selected as the supply chain partner among a managerial list of candidates.
- **Transacting**—They develop all aspects of business trading opportunities with the potential supply chain partners on an equal footing.
- **Protecting**—They ensure conformance with contract terms and conditions, delivery schedules, product/service quality, and other partnership agreements (see Davis and Spekman, 2004, for details of boundary spanning roles).

Conceptual Foundations of Demand Chain, Value Chain, and Supply Chain

Although supply chain management has been hailed as an innovative way to compete in today's business world, its concept created a lot of confusion, as evidenced by the presence of more than 2,000 different definitions of supply chain management (see Gibson et al., 2005). Adding to the confusion, the term *supply chain* was often interchangeably used with *demand chain* and *value chain*. Therefore, it is important for us to synthesize these terms and differentiate among them when appropriate.

Because the ultimate goal of supply chain management is to serve the customer better, supply chain management begins with the understanding of customer values and requirements. Indeed, Poirier (1999) argued that the primary objective of supply chain improvements was to serve ultimate customers more effectively and therefore an analysis of the supply chain should focus on the "finish line" (demand), not the "starting point" (supply). To enhance the customer values and meet customer requirement, careful planning of demand-creation and -fulfillment activities is critical to the success of the whole organization. This planning cannot be articulated without understanding the dynamics of interrelated business activities and jointly developing ideas for business process improvement among the intra- and inter-organizational units. Therefore, any efforts geared toward the customer-centric and "pull" approach throughout the entire business processes are considered part of the demand chain.

In a context similar to the demand chain, a *value chain* is referred to as a series of interrelated business processes that create and add value for customers. Its intent is to disaggregate all of a firm's business processes into discrete activities to evaluate their level of contributions to the firm's value and then discern value-adding activities from non-value-adding activities. Herein, "value is the amount buyers are willing to pay for what a firm provides them and thus is measured by total revenue, a reflection of the price a firm's product commands and the units it can sell" (Porter, 1985, p. 38). Thus, the extent of value created and added by the firm often dictates its level of business success, because the higher the value, the greater the profit margin and competitive advantages.

As shown in Figure 1.3, the value chain focuses on the customer's value by connecting the customer's needs to every aspect of the value-adding business activities encompassing sourcing, manufacturing, logistics, and marketing across the organizational boundary. The value chain is often driven by four key imperatives (Bovet, 1999):

- **Reduced uncertainty**, which minimizes asset intensity through the reduction and elimination of inventory
- **Increased speed**, which minimizes the risk of obsolescence
- **Increased revenue** resultant from the maximization of customization and the subsequent customer loyalty
- **Increased productivity** through multiple asset productivity

Although Table 1.2 shows that the strategic focus and perspectives of the demand chain, the value chain, and the supply chain are somewhat different from one another as described by Sherman (1998, p 2), their fundamental concepts and ultimate goals are not distinctively different in that all these concepts are customer-centric and stress the importance of coordinated linkage between business activities to the firm's competitiveness. Therefore, these three terms can be regarded as synonyms. To put it simply, the supply chain originates at the sources of supply and flows toward the customer, whereas the demand chain flows backward from the customer and ends up with the enterprise. The value chain is created when the supply chain is in sync with the demand chain. Regardless of the semantic differences, the supply chain benefit may be maximized by following the seven principles outlined by Copacino (1997):

- *Understand the customer values and requirements* so that the firm can identify how to align its operations to meet its customers' requirements and needs.
- *Manage logistics assets* such as warehouses, terminals, transportation equipment, and pipeline inventory across the supply chain with the help of both the downstream and upstream supply chain partners.
- *Organize the customer management* in such a way that the firm can provide a single point of contact to the customer for information and post-sales follow-ups.

- *Formulate joint sales and operations plans* as the basis for a more responsive supply chain by sharing real-time demand and forecast information within and across the supply chain.
- *Leverage manufacturing and sourcing flexibility* by utilizing postponement strategies.
- *Focus on the synergies of strategic alliances and relationship management* by building the sense of mutual trust and the spirit of gain sharing.
- *Develop customer-driven performance measures* to drive the behavior of all supply chain members across the supply chain.

Table 1.2. The Comparison of Demand Chain, Value Chain, and Supply Chain

	Demand Chain	**Value Chain**	**Supply Chain**
The Role in Demand Planning	Demand creation	Demand performance	Demand fulfillment
Key Strategy	Product strategy	Financial strategy	Channel strategy
Primary Activities	New product development Market research Sales and promotion Forecasting	Cost accounting Pricing Revenue management Capital investment Value-added services	Procurement Manufacturing Logistics Payment
Key Stakeholders	End customers	Shareholders	Supply chain partners

Strategic Alliances and Partnerships

To create synergy resultant from the integration of business functions and activities, a growing number of firms have made a conscious effort to forge strategic alliances (or partnerships) with other firms. Generally speaking, a *strategic alliance* is a voluntary relationship between two or more organizations that is formed based on the mutual need of these independent organizations (e.g., suppliers, manufacturers, distributors, retailers) without being constrained by ownership, control, and equity investment (Devlin and Bleackley, 1988; Varadarajan and Cunningham, 1995). In a strategic alliance, business decisions are made jointly to achieve the agreed goals of aligned organizations that share resources, information, profits, knowledge, and risk. Those goals are to: (1) gain access to new customer bases; (2) offer more product/service offerings for the customers; (3) pool resources in light of the large outlay required; (4) to learn new know-how from the alliance partners; (5) utilize the existing personal network to reach new supplier bases; (6) enhance the market position in present markets; (7) add credibility to the business (Pucik, 1988; Doz, 1996; Das and Teng, 1998). For example, Daimler AG, the world's biggest truck maker, aligned with the Chinese truck manufacturer Foton to build Mercedes-Benz trucks in China and sell the Chinese brand models of its partner internationally because of rising demand for low-cost commercial vehicles in emerging markets. In exchange for its favor, Daimler AG would offer Foton's Auman brand trucks in emerging markets such as Africa and the Middle East to save on design and labor costs to build new vehicles specifically for those markets. As such, both Daimler AG and Foton would have a chance to penetrate into new market bases in emerging economies through this strategic alliance.

Similarly, Nissan and Renault once formed an alliance to expand their presence in both Asian and U.S. markets. Although these examples illustrate a *horizontal alliance* between competitors in the same industry, a strategic alliance can be formed vertically between trading partners across the supply chain. An example of a *vertical alliance* includes an earlier supplier involvement that brings the expertise and collaborative synergy of parts/components suppliers into the new product design process of a manufacturer. For example, in the 1990s, Chrysler established an alliance with its supplier to develop new antilock brake technology for Neon with the promise of a long-term purchase contract for that supplier. This alliance saved Chrysler a substantial amount of R&D costs.

Despite numerous benefit potentials, inter-organizational alliances have often been plagued by ambiguities in the level of commitments (or responsibilities) from each partner; a lack of mutual trust between partners and the subsequent lack of information sharing; imbalances in the channel power and risk sharing; and differences in organizational culture and strategic focus. For example, GM and the now-defunct Korean automaker Daewoo signed a deal in 1999 that would help GM's effort to double its share of sales in the Asian market to 10% in five years, while helping Daewoo bring in much-needed capital and radically streamlining its sprawling global production operations. In addition, this alliance was intended to help GM gain access to "cheap" labor in Korea, while it allowed Daewoo to gain access to GM's "superior" engineering skills and an entry into the U.S. auto market. Unfortunately, this alliance failed mainly due to corporate cultural differences and a lack of strategic fit between the two companies—Daewoo was known for its aggressive market expansion strategy whereas GM was in the process of downsizing its production capacity. To make matters worse, labor in Korea was no longer cheap, owing to rapid economic growth and the presence of a strong labor union, thus causing an unexpected cost increase in this alliance's Korean operations.

To cope with these risks and challenges, firms entering into a strategic alliance should understand what kind of partnership they are forming, what kind of role each partner should play, and how they should manage the partnership. Such understanding will be enhanced by defining the structures of their partnership. Following the conceptual framework suggested by Cooper et al. (1997b), these structures are classified as follows:

- The type of a supply chain partnership
- The structural dimensions of a supply chain network
- The characteristics of process links among supply chain partners

The Type of a Supply Chain Partnership

When a supply chain network is being structured, it is necessary to identify who the partners of the supply chain are. Yet, an inclusion of all potential partners may complicate the total network because it may explode the number of partners added from one tier level to another (Cooper et al., 1997a). The key is to identify the type of the partners who are critical to the value-added activities and determine a manageable number of the supply chain partners given resources.

Lambert et al. (1998) classify supply chain partners into two distinctive types: primary and secondary partners. In general, *primary partners* (focal companies) are autonomous channel captains or strategic business units that actually perform operational and/or managerial activities designed to create a specific product or service for a particular customer or market. These primary partners can be manufacturers such as Dell or mass-merchants such as Walmart and Target. In contrast, *supporting partners* are companies that simply provide resources (e.g., assets, application software, real-estate property), knowledge, and utility for the supply chain. These supporting partners can be transportation carriers, consulting firms, third-party logistics providers, IT service providers, online brokers, and educational institutions. The categories are not exclusive, however, because a firm can be both a primary partner and a supportive partner of the supply chain, performing primary activities related to one process and supportive activities related to another process.

Although the distinction between primary and supporting supply chain partners is not obvious in all cases, it allows the firm to define who the furthest upstream and downstream members of the supply chain are and identify where customer demand actually starts. The furthest upstream members of the supply chain typically represent supporting partners, whereas the furthest downstream represent the end of the supply chain where no further value is added and the product and/or service is consumed. The furthest downstream (or point of consumption) may coincide with a *value-offering point*

(VOP), where a customer allocates demand to his or her upstream supply chain partner (e.g., retailer, distributor, or manufacturer). According to Holmstrom et al. (1999), a VOP determines how and when customer demand is triggered to upstream supply chain partners and defines the economics of the customer (i.e., the tradeoff between value creation and transaction costs). Such a clarification of the supply chain network eventually helps supply chain partners understand what scope of the supply chain problem should be addressed.

The Structural Dimensions of the Supply Chain

Understanding the structural dimensions of a supply chain is a prerequisite for building the supply chain link. In general, there are two structural dimensions: horizontal structure and vertical structure. The *horizontal structure* refers to the number of tiers across the supply chain. The supply chain may be lengthy, with numerous tiers, or it may be short, with just a few tiers. The *vertical structure* refers to the number of suppliers and customers represented within each tier, as illustrated by Figure 1.4 (Lambert et al., 1998, p. 3).

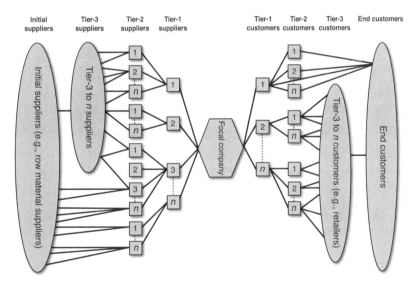

Figure 1.4. *Supply chain dimensions*

As such, an increase or reduction in the number of suppliers and/or customers will alter the dimension of the supply chain. For example, as some companies make strategic moves toward either supply base reduction or customer selectivity, the supply chain becomes narrower. Outsourcing (inclusion of third-party logistics providers) or functional spin-offs will also alter the supply chain dimension by lengthening and widening the supply chain.

The Characteristics of Supply Chain Links

Porter (1985) stresses the strategic importance of linkages among supply chain activities, because the linkages could lead to competitive advantages. To fully exploit the benefits of such linkages, a firm should understand the specific characteristics of the linkages (or links) to which it is connected. Lambert et al. (1998) have identified four distinctive characteristics of supply chain links:

- Managed business process links
- Monitored business process links
- Unmanaged business process links
- Non-member business links

Managed process links are the ones where the firm (typically a primary supply chain partner or a channel captain) integrates a supply chain process with one or more customers/suppliers. These links may connect multi-tier supply chain partners as the firm is actively involved in the management of tier one and a number of other links beyond tier one. Due to its direct involvement, the firm may allocate resources (e.g., manpower, equipment, technology, and know-how) to its partners and share information with them. *Monitored process links* are not fully controlled by a firm (typically a primary supply chain partner), but the firm is involved in monitoring or auditing how the links are integrated and managed. *Unmanaged process links* are the ones that the firm neither actively manages nor monitors. With these links, the firm fully trusts its partners' ability to manage the process links appropriately and consequently leaves the management responsibility up to them. *Non-member process links* are the ones between both partners and non-members of the company's supply chain. Such links are not integral parts of the firm's supply chain structure, but can dictate the performance of the firm. Additionally, different characteristics of supply chain links affect the firm's allocation of resources and the subsequent supply chain planning. Therefore, those characteristics should be factored into the supply chain partnership process.

Supply Chain Drivers

Because the establishment of common goals is critical for successful strategic alliances, goal setting will be the first step of the supply chain partnership. To set common goals, supply chain partners need to figure out what will be the major driving forces (drivers) behind the supply chain linkages. These drivers include customer service initiatives, monetary value, information/knowledge transactions, and risk elements (Min and Zhou, 2002).

Customer Service Initiatives

Though difficult to quantify, the ultimate goal of a supply chain is customer satisfaction. Put simply, *customer satisfaction* is the degree to which customers are satisfied with the product and/or service received. The following list represents typical service elements in a supply chain:

- **Product availability**—Due to random fluctuations in the demand pattern, downstream supply chain partners often fail to meet the real-time needs of customers.
- **Response time**—Response time represents an important indicator of the supply chain flexibility. Examples of response time include time-to-market, on-time delivery (a percentage of a match between the promised product delivery date and the actual product delivery date), order processing time (the amount of time from when an order is placed until the order is received by the customer), transit time (duration between the time of shipment and the time of receipt), cash-to-cash cycle time (the amount of time from when a product has begun its manufacturing until it is completely sold; this metric is an indicator of how quickly customers pay their bills), and downtime (a percentage of time resources that are not operational due to maintenance and repair).

Monetary Value

The monetary value is generally defined as a ratio of revenue to total cost. A supply chain can enhance its monetary value through increasing sales revenue, market share, and labor productivity, while reducing expenditures, defects, and duplication. More specifically, the monetary value is categorized as follows:

- **Asset utilization**—Asset utilization can be estimated by several different metrics, such as net asset turns (a ratio of total gross revenue to working capital), inventory turns (a ratio of annual cost of goods sold to average inventory investment), and cube utilization (a ratio of space occupied to space available).
- **Return on investment (ROI)**—This is a typical financial measure determining the true value of an investment. Its measure includes the ratio of net profit to capital that was employed to produce that profit, or the ratio of earnings in direct proportion to an investment.
- **Cost behavior**—In the supply chain framework, cost management requires a broad focus, external to the firm. Thus, cost may be viewed as a function of strategic choices of the firm's competitive position, rather than a function of output volume (Shank and Govindarajan, 1993). In other words, a traditional cost classification (fixed versus variable cost), which works at the single firm level, may not make sense for the supply chain network affected by multiple cost drivers (e.g., scope and scale). An alternative cost management principle for a

supply chain framework includes activity-based costing (ABC), target costing, and cost of quality (COQ). Because the application of the aforementioned cost management principles to the supply chain is still at the evolutionary stage, most of the firms engaged in the supply chain still use traditional cost measures such as inventory carrying cost, inventory ordering cost, transportation cost, and product return cost.

Information/Knowledge Transactions

Information serves as the connection between the various phases of a supply chain, allowing supply chain partners to coordinate their actions and increase inventory visibility (Chopra and Meindl, 2004). Therefore, successful supply chain integration depends on the supply chain partners' ability to synchronize and share "real-time" information. Such information encompasses data, technology, know-how, designs, specifications, samples, client lists, prices, customer profiles, sales forecasts, and order history.

- **Real-time communication**—The establishment of collaborative relationships among supply chain partners is a prerequisite to information sharing. Collaborative relationships cannot be built without mutual trust among supply chain partners and technical platforms (e.g., the Internet, electronic data interchange, extensible markup language, enterprise resource planning, and warehouse management systems) for information transactions. The effectiveness of real-time communication hinges on the supply chain partners' organizational compatibility, which facilitates mutual trust, and technical compatibility, which solidifies electronic links among supply chain partners. Because organizational and technical compatibilities are hard to measure, some surrogate metrics such as the rate of electronic data interchange (EDI) transactions (a percentage of orders received via EDI) and the percent of suppliers accepting electronic orders/payment can be used.
- **Technology transfers**—The collaboration fostered by supply chain partners can be a catalyst for the research and development (R&D) process throughout the supply chain. The rationale is that a firm, which initiated technology development, can pass its technology or innovative know-how to its supply chain partners, thereby saving R&D cost and time. Therefore, a successful transfer of technology can help supply chain partners enhance their overall profitability.

Risk Elements

The important leverage gained from the supply chain integration is the mitigation of risk. In the supply chain framework, a single supply chain member does not have to stretch beyond its core competency because it can pool the resources shared with other supply chain partners. On the other hand, a supply chain can pose greater risk of failure due to its inherent complexity and volatility. Braithwaite and Hall (1999) note that a supply chain would be a veritable hive of risks unless information is synchronized, time is compressed, and tensions among supply chain members are recognized. They also observe that supply chain risks (emanating from sources external to the firm) will always be greater than risks that arise internally, because less is known about them. Thus, supply chain partners need to profile the potential risks involved in supply chain activities. The following list summarizes such profiles:

- **Risk of quality failure**—As illustrated by the recall in May of 2000 of 6.5 million Firestone tires that were susceptible to tread separation in harsh driving conditions, the consequences of failing to ensure quality failure at the upstream supply chain can be enormous. This is due to the interdependence of supply chain partners. Similarly, order-picking errors and failures to schedule adherence should be prevented at the furthest upstream supply chain (if possible, at the initial source of supply).
- **Risk of information failure**—One of the well-known consequences of information failure in the supply chain is the bullwhip effect, where orders at the upstream supply chain members tend to exaggerate the true consumption of end customers (e.g., Lee et al. 1997; Min 2000). Because the bullwhip effect will create phantom demand and subsequent overproduction and overstock, its risks should be assessed prior to the development of the supply chain network. One way of reducing such risks is to postpone the final assembly, branding, purchasing, packaging, and shipment of products until they are needed.

Organizational Learning from Strategic Alliances

Because today's competition is increasingly knowledge based, knowledge can be a key productive resource of the firm. Recognizing the opportunity to leverage knowledge as a competitive differentiator, many firms that are strategically aligned with other firms across the supply chain may be interested in learning about customer needs, market forces, industry dynamics, business acumen, and technological know-how from their supply chain partners. Thus, one of the incentives for forming strategic alliances among supply chain partners can be organizational learning. For example, according to the 15th annual Third-Party Logistics Survey conducted by Langley in 2010, many third-party logistics providers (or 3PLs) gave their clients new and innovative ways ("knowledge") to improve logistics productivity by creating long-term 3PL-client relationships with an open and collaborative spirit. In general, *organizational learning* is defined as unconscious or deliberate development of new knowledge that has the potential to improve organizational capability and performance in the short and long run (Fiol and Lyles, 1985; Slater and Narver, 1995; Bell et al., 2002). Organizational learning consists of knowledge acquisition, dissemination, and shared interpretation of knowledge across the organizations (Sinkula, 1994; Min, 2001). The knowledge that can be gained from other organizations has two types:

- **Explicit knowledge**—Knowledge that can be codified and easily transmitted to partnering organizations. Its examples are knowledge often captured in the forms of documented texts, tables, charts, figures, diagrams, scientific formula, mathematical expressions, and written performance metrics (Nonaka, 1991).
- **Tacit knowledge**—Knowledge that cannot be easily communicated to partnering organizations due to its implicit, noncodified content. Examples include technical expertise and job-specific insights possessed by vehicle dispatchers, airplane pilots, ocean vessel navigators, insurance underwriters, and logistics engineers.

Organizational learning can occur at two different levels of the decision-making hierarchy (see Fiol and Lyles, 1985; Sinkula, 1994):

- **Operational learning**—Allows partnering organizations to improve their day-to-day practices and policies through the mutual detection and correction of errors and inefficient operations
- **Strategic learning**—Allows partnering organizations to redefine their overall missions, strategies, goals, and philosophies through the development of innovative thoughts and ideas

Regardless of the type of organizational learning, the extent and effectiveness of organizational learning may be greatly influenced by the level of trust between partnering organizations. Without mutual trust, organizations in the supply chain may be unwilling to share their information and consequently reluctant to diffuse their innovations and technology to their supply chain partners. Therefore, it is important to build trust among supply chain partners before forming strategic alliances and then exploiting the opportunity to learn from each other. Indeed, a study conducted by Kwon and Suh (2004) indicated that the presence of trust greatly improves supply chain performances. On the other hand, a lack of trust among supply chain partners often results in inefficient and ineffective performance as the transaction costs increase. Also, in order to build trust among the partners, there should be an opportunity for the firm to interact with its partners through an *open communication* channel. Once trust is established, organization learning necessitates *experimentation,* which involves unlearning of failing routines, and encouraging knowledge sharing between the partners to develop new or innovative practices (Kwon and Suh, 2004; Krikke and Caniëls, 2010). The last stage of organizational learning, as shown in Figure 1.5, involves *alignment* that changes routines and eliminates bad habits, while embracing a better way of doing things.

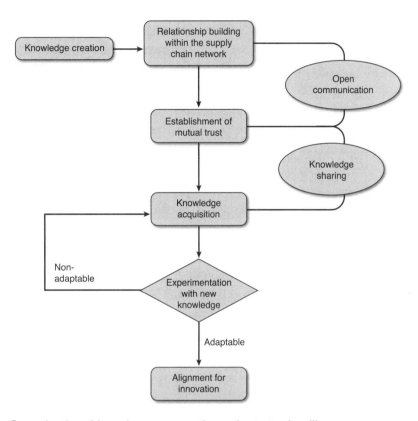

Figure 1.5. *Organizational learning process through strategic alliance*

Interfaces among Purchasing, Production, Logistics, and Marketing

According to Ballou et al. (2000), there are three dimensions of supply chain management:

- **Intrafunctional coordination**, which administers the activities and processes *within* the particular function (e.g., logistics) of a firm
- **Interfunctional coordination** between logistics and purchasing, logistics and production, and logistics and marketing *among* the functional areas of the firm
- **Interorganizational coordination**, which takes place *between* legally separated firms such as manufacturers and their suppliers.

This section emphasizes the interfunctional coordination between several business functions. For instance, the production of finished goods requires acquisition of raw materials, parts, and components, which triggers a set of purchasing activities. Also, the quality of incoming materials and parts obtained through purchasing often determines the quality of finished goods for sales and distribution. Thus, purchasing is tied to production, marketing, and logistics.

From cost perspectives, an effort to lower logistics costs by consolidating smaller shipments and using a slower mode of transportation would increase lead time and consequently increase inventory carrying costs, while hurting customer services through the delayed deliveries. That is to say, a change in logistics activities directly influences the effectiveness and efficiency of production and marketing. In particular, logistics is closely linked to marketing through their roles in customer services. As graphically displayed in Figure 1.6, logistics contributes to the "place" dimension of the marketing mix by creating both "spatial" (moving products to the customer's location) and "temporal" (holding inventory until the customer demands it) utilities for the customer (Lambert, 1976, p. 7). By the same token, real-time point-of-sales information fed by the retailer would prepare for the manufacturer to produce and keep the appropriate stock level and requires less need for large warehousing space. This example illustrates the influence of marketing on production and logistics. Considering these interdependences among purchasing, production, logistics, and marketing, a balancing act is needed to achieve the desired benefits of the firm as a whole. Such a balancing act can be put together effectively through supply chain integration.

Theory of Constraints (TOC) for Supply Chain Management

The strength of the supply chain link can dictate the effectiveness and efficiency of the supply chain partnership and the ultimate success of the supply chain. To maximize the supply chain benefit, supply chain partners should uncover weak links and prevent variations in supply chain capacity (e.g., production/distribution capacity and inventory) and supply chain performances. Perhaps one of the most effective ways of doing so is to apply the *theory of constraints (TOC)* to supply chain management. The core idea in TOC is that every system such as profit-making firms must have at least one constraint that limits the system from getting more of whatever it strives for and consequently determines the output of the system (Noreen et al., 1995). A *constraint* is anything in an organization that hampers the organization's progress or increased throughput. Thus, the firm's failure to manage this constraint leads to the significant decline in its productivity. The same TOC analogy can be made to the supply chain, where the weak supply chain link can limit the effectiveness and efficiency of the entire supply chain. In other words, the supply chain will fail at the weakest link.

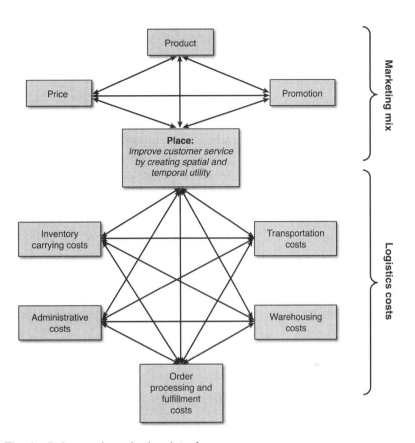

Figure 1.6. *The logistics and marketing interface*

For example, the part production slowdown and the subsequent delivery delays caused by the upstream supplier would increase the lead time for the downstream manufacturer and distributor and then result in product shortages at the retailer. These product shortages would not allow the retailer to meet customer needs and consequently would deteriorate customer services. In this example, the supplier's production capacity will become the system's (supply chain's) constraint. In TOC terms, the supplier production capacity will be regarded as the "drum" that sets the beat for the entire supply chain. The size of the inventory held by the supplier will be viewed as the "buffer," because it buys time needed to recover from the anticipated disruptions occurring in the upstream supply chain. The "rope" is symbolic of the link between the upstream and downstream supply chains, where the rate of the final sales or distribution does not exceed the supplier's production capacity.

This drum-buffer-rope (DBR) logic of TOC thinking would protect against variability at the constraint and ensure the continuous improvement of the supply chain processes. Considering the usefulness of TOC thinking to supply chain management, the supply chain partners may consider the following TOC focusing steps to optimize the supply chain benefits (see Dettmer, 1997 for the detailed discussion of the five TOC focusing steps).

1. Identify the weakest link in the supply chain.
2. Decide what to do to get the most out of the weakest link (constraint) without committing to potentially expensive changes.
3. Adjust the rest of the supply chain processes to a "setting" that would enable the constraint to operate at the maximum effectiveness.
4. Take whatever action is required to eliminate the constraint.
5. Once the current constraint is broken, keep on looking for other constraints to continuously improve the supply chain performances.

Change Management for Supply Chain Management

The transition from the traditional silo-based business paradigm to supply chain management requires traumatic changes in organizational structures, cultures, and business strategy. Unless such changes are properly managed, the firm may suffer from degrading employee morale, frequent bottlenecks, and increased resistance to supply chain transformations. Therefore, supply chain transformation without prepared change management may defeat the purpose of supply chain initiatives.

According to Fries (2005), change management is generally composed of three steps:

1. **Unfreezing**, which involves tossing out old ideals and processes to break old habits
2. **Changing**, which initiates a new system or major process transformation and creates a team to facilitate the needed transformation throughout the organization
3. **Refreezing**, which involves practicing what was implemented, learned, and accepted by the change management team and the impacted organizational units

To elaborate, change management for supply chain transformations may follow the detailed steps described in Table 1.3 (see Hughes et al., 1998; NAPM InfoEdge, 2000; Dittmann, 2013; Min, 2014 for supply chain transformations). Because a vision sets the tone for change management, the supply chain vision should be created with input and feedback from all stakeholders, including affected employees. In general, a vision refers to "a picture of the future with some implicit or explicit commentary on what people should strive for to create that future" (Kotter, 1996, p. 68). The vision goes beyond authoritarian decrees and micromanagement to break through all the forces that support the status quo (see Figure 1.7). To bring about successful supply chain transformations, the change management team may develop the checklist summarized in Table 1.4 and then make sure that they do not get trapped in common mistakes. Per Sambukumar and Vijayan (2011), these mistakes include the following:

- **Falling into the "leading supply chain practices" trap**—Each organization is unique in that it serves different customer bases and therefore needs to develop a different business strategy. Although so-called known leading practices of the supply chain leaders (e.g., Walmart and Dell) are helpful for supply chain transformations, unfiltered adoption of other companies' successful supply chain practices that are not coherent with the company's business strategy can create unexpected challenges.
- **Falling into the "do-it-all" technology trap**—There is no doubt that a state-of-the-art information and communication technology (ICT) helps the company enhance its supply chain visibility and thus increases the chance for successful supply chain transformations. However, many companies may not realize that ICT is nothing more than a catalyst for supply chain transformations and is not a solution for such transformations.
- **Falling into the organizational cultural trap**—Supply chain transformations are predicated on the elimination of functional or organizational silos. Due to the remnants of internal politics and organizational cultural barriers, the company may run into difficulty in integrating and synchronizing business activities across the different units and organizations.

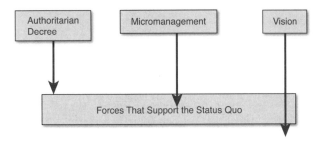

Figure 1.7. *Breaking through resistance with vision*

Table 1.3. Change Management Steps for Supply Chain Transformations

Step	Action Plans	Key Questions to Ask
Unfreezing	• Create visions and define goals for stakeholders and supply chain partners. • Assess the readiness for change. • Develop the change management team and designate change leaders (agents). • Create the detailed plan for change and then communicate visions and expected outcomes to stakeholders and supply chain partners.	• What are we going to accomplish? Why do we have to change the current practices? Who will be affected by a change? • What challenges and opportunities do we face during the change process? What resources are available for change? Who will sponsor the change? What is the scope and scale of a change? What type of resistance can be expected? • Who will plan the change? Who will lead the change? Who will coach the change? Whom are we partnering with? • How do we build awareness around the need for change? How do we communicate with stakeholders about the risk of not changing? Who will be responsible for what? How do we synchronize supply chain transformations with the company's overall business strategy? How can we make supply chain transformations as simple as possible?
Changing	• Map key processes for change. • Identify financial needs and cost drivers. • Define performance metrics. • Develop incentive schemes and manage resistance.	• What will it take to achieve the stated goals through change? Where do we start? What are our next steps for bringing about successful change? How do we deploy new initiatives? • How much do we have to invest in new initiatives? Is it worth investing? How do we measure and control costs? • How do we assess our progress? What are our desirables? • How do we share benefits and risks with our supply chain partners? How do we identify, understand, and manage resistance throughout the organization?
Unfreezing	• Collect the performance data and monitor the current progress. • Analyze feedback from stakeholders and supply chain partners and take correction actions needed to reengineer the processes. • Build in new ways of doing business in the corporate culture and create new visions. • Celebrate and recognize success.	• Do we meet or exceed the benchmark standard we set? Is there a sign for a successful change? • Can we do better? What needs to be done to improve the current practices? • Do we increase the customer value? Are there noticeable cultural changes? How do we stack up against our competitors? • How do we reward the change agents and those involved in change initiatives to cement and reinforce the change?

Table 1.4. The Checklist for Supply Chain Transformations

Item	Checklist
1	Articulate reasons for change brought up by supply chain principles.
2	Locate change agents, sponsors, and supporters of supply chain transformations at an early stage. Mobilize commitment through cross-functional and interorganizational diagnosis of the challenges and opportunities for supply chain transformations.
3	Develop training sessions (e.g., workshops, seminars, and briefings) for affected stakeholders (e.g., employees) to raise the level of thinking about available initiatives.
4	Make the benefits of the changes tangible by focusing on quick deliverables and immediate successes (achievement of short-tern goals). Also, set performance targets.
5	Concentrate the initial wave of changes on breaking down silos between functional departments and organizational boundaries.
6	Identify an appropriate and structured change management methodology.
7	Set the supply chain agenda straight and orchestrate the overall impact of such agenda on the organization's competitiveness.
8	Involve top management directly rather than delegating the implementation of strategic supply chain initiatives to lower management.
9	Prioritize the business missions critical to the supply chain success (e.g., the ones with the biggest financial impact).
10	Destabilize the status quo by shaking up the power structure inside.
11	Ensure the continuous reinforcement of supply chain transformations.

Data sources:

Hughes, J., Ralif, M., and Michels, B. (1998), *Transform Your Supply Chain: Releasing Value in Business*, London, UK: Thomson Business Press, p. 146

NAPM InfoEdge (2000), "Setting Up the Relationship to Succeed," *NAPM InfoEdge*, 5(2), p. 13.

Chapter Summary

The following summarizes key takeaways from this chapter:

- The supply chain concept has emerged as the cornerstone of the twenty-first century's competitive business strategy, although similar concepts such as integrated logistics management, value chain, and demand chain have existed in the past. A key to the supply chain success is how efficiently and effectively multiple business functions and firms across the traditional organizational boundaries break their barriers to create synergies among themselves.
- Although supply chain management is more than a fad, organizations that are interested in supply chain transformations should examine whether their strategic fits, structures, and cultures are right for such transformations. Prior to taking

advantage of enormous benefit potentials of supply chain integration, one should be aware of potential challenges and risks associated with such integration.

- To induce the business partners into supply chain transformations, the firm should provide those partners with measurable (quantifiable) benefits and risks accrued from supply chain integration. Thus, the firm should clearly understand the specific impacts of supply chain integration on its business bottom lines, such as return on assets, profitability, revenue growth, and market shares.
- Once supply chain partnerships are built, each supply chain partner (including the focal company) needs to understand its specific role and mission to avoid duplicated efforts that would defeat the purpose of supply chain integration.
- The level of supply chain potency is often dictated by the performances of the weakest supply chain links. Therefore, the theory of constraints can be a useful tool for supply chain management.
- Before jumping onto the supply chain bandwagon, the firm should prepare for a radical change brought about by supply chain transformations. That is to say, change management should always precede supply chain management.

Study Questions

The following are important thought starters for class discussions and exercises:

1. What are the key differences between the traditional business paradigm and the integrated supply chain concept?
2. What is the total systems approach?
3. How is supply chain management different from logistics management?
4. Are there more than semantic differences among the demand chain, value chain, and supply chain?
5. What is the role of supply chain management in creating and adding value for customers?
6. Why has supply chain management emerged as a field of important study and practice? What is the driving force behind the supply chain concept? How does supply chain management impact the bottom line of the business organization?
7. Think of any organizations you are familiar with (e.g., Kroger, Target, Procter & Gamble, IBM, Exxon Mobil, Pizza Hut) and then graphically display their supply chain networks (or develop a detailed supply chain map for your organization).
8. What are the major challenges for integrating business functions and interorganizational efforts across the supply chain? For example, how can the firm overcome the differences in organizational culture, channel power and

leadership structures, and compatibility to make the supply chain integration successful?

9. On what level should business processes be integrated among supply chain partners?
10. How would you link between purchasing and logistics? How would you link between manufacturing and logistics?
11. How would you identify and select the right supply chain partner?
12. What are the potential barriers to forming the supply chain partnership?
13. What are four distinctive types of business process links throughout the entire supply chain?
14. What are the expected payoffs from building the supply chain partnership?
15. What is the theory of constraints (TOC)? How is it related to supply chain management? Can it help to manage the supply chain better?
16. How do you prepare for the changes in business practices by taking the supply chain initiatives?

Zara's Rapid Rise as a Cool Supply Chain Icon

This case was written based on various published stories about Zara and inspiration from the toolbox of the Council of Logistics Management.

New Heights in the Fashion Industry

Founded in 1975, Zara is the world's largest clothing retail chain and is based in Galicia, Spain. Its parent company, the Inditex group, reached annual sales of 16.7 billion Euros in 2013, twice as much as its annual sales of 8.2 billion Euros in 2006 (http://www.inditex.com/investors/investors_relations/financial_data).

According to Investing.com, Zara's inventory turnover ratio is 29.58, as compared to an industry norm of 5.29 in 2013. Its gross profit margin (26.57%) is substantially higher than its major competitors. Thanks to its recent achievement, Zara is on the list of the *Forbes* world most valuable brands in 2013. Zara's remarkable success is attributed to the "fast fashion" philosophy in which the latest, trendy but inexpensive fashions are quickly distributed into its retail stores based on constant dialogues with the customers. Zara focuses on "putting the customer in control" in that it monitors customer reactions carefully by noting what they buy and do not buy. This differs from the traditional method of forecasting fashion trends and then promoting the clothing line through fashion shows, news media, and magazines.

Zara constantly records customer feedback given to the staff, and that feedback is relayed to its headquarters daily. Based on such feedback, its team of in-house designers quickly introduces new designs and then its factories manufacture clothes based on those designs (Levin, 2013). Zara's key customer bases are college-age kids and young females under the age of 30 who look for inexpensive but stylish fashions. Zara also sells men's clothing lines, aiming at the stylish and youthful. In Europe, Zara has attracted a cult following from many young people. Most U.S. young adults have rarely shopped at Zara, but that seems likely to change in the near future. In 2008, Zara overtook Gap as the biggest apparel retailer in the world. Zara opened an additional 360 stores in just 2012 alone. Zara has grown from 179 stores, mostly in Spain, to more than 6,100 stores with 116,110 employees in 86 countries, including the U.S., Canada, and Russia (Reuters, 2013). With its eyes on U.S. customers, Zara can offer these customers an option previously unavailable to them, thus competing head-to-head against established U.S. brand apparel stores such as Gap, Abercrombie & Fitch, and J. Crew. It has a unique, youth-oriented, no-advertisement marketing strategy in which logistics plays an important role. Zara's supply chain process defies conventional wisdom, as illustrated in Exhibit 1.1.

Exhibit 1.1. *Comparison of Zara's Supply Chain Process with Conventional Wisdom*

Sequence	Conventional Wisdom	Zara's Supply Chain Process
1	Develop designs based on prediction.	Reach customers.
2	Find supply sources or contract manufacturers.	Bring customers to the store and listen to them at the point of sale.
3	Contract manufacture.	Introduce designs based on customer feedback.
4	Promote and distribute.	Manufacture, while offshoring some.
5	Reach customers.	Sell online and offline through its own chain stores.

Zara's Cool Supply Chain Practices

Zara manufactures more than half of its clothing lines in Europe (primarily Spain and Portugal), with most of the remaining lines being produced in or sourced from other continents, including Morocco in Africa, Mexico in North America, and China in Asia. These worldwide manufacturing operations pose many supply chain challenges. In particular, Zara's expanding global reach may derail its focus on its fast fashion (speed-to-market) strategy. Zara's fast fashion strategy allows it to introduce a new style from concept to store shelf within 10 to 14 days in an industry where nine months is the norm. Typically, based on the fixed distribution schedule, Zara's store managers order new clothes twice a week and expect to receive their orders twice a week. This means that the inventory changes often, with most items staying on the shelf for only a month or less. As such, a short lead time is crucial for this strategy. However, unlike its primary European markets, where its stores are located close to each other and concentrated in the center of downtown, many retail stores are typically spread out throughout the country in the emerging U.S. and Chinese markets. Also, Zara has to ship via air from Europe to North America. Ocean shipping is not a viable shipping option for Zara because it would add at least another ten days to the time it takes to get the product into the customers' hands, thus undermining the speed-to-market ("quick-response" logistics) and "rapid-fire" fulfillment strategies. In an effort to reduce order cycle time and enhance supply chain efficiency, Zara's store managers utilize handheld electronic ordering devices that relay the order information to its distribution and manufacturing sites on a real-time basis. Such electronic ordering practices helped Zara's headquarters update and maintain spare transportation, warehousing, and production capacity (e.g., unfilled trucks, a mega distribution hub, and extra manufacturing capacity). This kind of strategy enhances Zara's supply chain visibility and allows it to introduce a new line of apparel following the latest fashion much faster than its rivals. For example, it takes less than two weeks for a new skirt to get from the Zara design team in Spain to a Zara store in Paris or Tokyo, as much as 12 times faster than its competitors, as of 2001 (Newsweek, 2001).

In addition to this practice, Zara's supply chain edge lies in the vertical coordination and integration of the supply chain by owning several layers in its global supply chain and performing flexible manufacturing practices. For instance, nearly 60% of Zara's merchandise and 40% of its fabric are produced in-house or purchased from its own subsidiaries, including subcontract manufacturers, to reduce the risk of supply disruptions and quality failures. In other words, Zara wants to control all the steps of supply chain operations involved in the sales of stylish clothes: from design and sourcing to fabrication, distribution, and sales. This controlled manufacturing environment allows Zara to experiment with new technologies and leverage them for its productivity. One of those technologies includes robotics that Zara uses in cutting and dyeing fabric in 23 highly automated factories. Another distinguished feature of Zara's manufacturing is a limited production run, which bring several benefits. First, limited runs allow the Zara-branded product to cultivate the exclusivity of its offerings and give its customers a sense of prestige because their clothes never look like someone else's. Second, limited runs pressure customers to buy instantly at full price, because the clothes on its store shelf will be quickly out of stock or removed from the shelf to make room for something new. Finally, limited production runs allow Zara to minimize a risk of stockpiling unwanted clothes. Artificial scarcity created by limited production runs mean that there is not much to be disposed of at discounted prices or written off as obsolete inventory when the season ends or the fashion changes. Indeed, Zara sells 85% of its stock at the regular price, compared to the industry norm of 50% (Booz & Company, 2007).

Supply Chain Vulnerability

Despite the proven success of the aforementioned supply chain strategy, Zara is not problem free. To elaborate, its production is still heavily concentrated in Spain, where average wages are relatively high as compared to low-cost sourcing countries such as Bangladesh, and thus the current manufacturing practice may undermine its philosophy of offering affordable clothing lines. Any weather, labor, or terrorist disruptions to the area will have a serious impact on its global distribution, because there are few alternative distribution centers in Europe. Because production is carried out mostly in a small radius of Northern Spain, Zara is also vulnerable to financial instabilities in Europe as most of its cost base is denominated in Euros, which has been devalued due to recent financial crisis in Europe. Finally, unstable oil prices resulting from the political unrest in Middle-Eastern countries will affect profits because twice-weekly air express deliveries mean higher transportation costs.

Another challenge stems from its diversified product lines with multiple channel sales: online sales through its website (www.zara.com) or brick-and-mortar retail store sales. In comparison to its competitors' 2,000–4,000 items, Zara puts 11,000–12,000 different items varying in color, size, and fabric on the store shelves in a single year (Ghemawat and Luis Nueno, 2006). Putting a variety of goods on the shelves in worldwide store locations requires an unusual (though not unique) logistics strategy for the company. Zara ships goods from its single distribution center (or hub) in Spain to other foreign markets via air express, usually in small batches. In the 1970s, The Limited used a similar logistics strategy to support its test marketing, air expressing small quantities of new styles from Asia to U.S. stores. In Zara's case, however, the speedy shipments (e.g., 24- to 48-hour delivery to its stores from the distribution center) are its core business/supply chain strategy of "fast fashion," not just test marketing. Under its Just-In-Time delivery and manufacturing principles, Zara also ships frequently, allowing lower inventories while serving its multinational market from a single distribution center in Spain. However, with rising fuel cost, this air express delivery strategy may backfire. To make this strategy more challenging, Zara's products have multicountry labels and can be redistributed to another store in another country where they may fare better, if any particular product line is not selling well in any particular country.

Although Zara used to trade higher transportation costs for lower warehousing and inventory carrying costs thanks to the compact density of Zara's store locations in Europe, it would not be feasible for the company to achieve the same level of logistics efficiency in less condensed store locations in North America. Notice that Zara's small, boutique-type store that specializes in clothing lines with short shelf lives needs much faster turnaround time than others. Without any doubt, Zara's fast transportation supports its speed-to-market strategy, which is the heart and soul of its new product development and market penetration strategy. However, the recent disruption of air express services resulting from unexpected events, such as a volcano eruption in Iceland in 2010 and 2014, has created another headache for this strategy.

New Directions

You are summoned to report to Zara's CEO's office to explain the reason for recent supply chain problems. Your boss has also made it clear he wants you to come up with immediate solutions to these problems. Zara's taskforces have been assembled and informed that "quick and sensible solutions" are needed as soon as possible. As one of Zara's management team, you should address the following questions:

- What are the most important prerequisites for sustaining Zara's competitive advantages?
- Can Zara's air express delivery strategy be sustained? If not, what would be the better alternatives?
- What is the extent of supply chain risks associated with Zara's dependence on a single distribution center and in-housing manufacturing operations in Spain?
- Should Zara reassess its supply chain strategy to successfully penetrate the U.S. market?
- Do you think that Zara should consider modifying its pull supply chain strategy to overcome current supply chain challenges? If so, why? Otherwise, why not?

Bibliography

Ballou, R.H., Gilbert, S.M. and Mukherjee, A. (2000), "New Managerial Challenges from Supply Chain Opportunities," *Industrial Marketing Management*, 29, 7–18.

Bell, S.J., Whitewall, G.J. and Lukas, B.A. (2002), "Schools of Thought in Organizational Learning," *Academy of Marketing Science*, 30(1), 70–86.

Booz & Company (2007), *Keeping Inventory—and Profits—Off the Discount Rack: Merchandise Strategies to Improve Apparel Margins*, Unpublished Report, San Francisco, CA: Booz Allen and Hamilton.

Bowersox, D.J. and Closs, D.J. (1996), *Logistical Management: The Integrated Supply Chain Process*, New York, NY: The McGraw-Hill Company, Inc.

Braithwaite, A. & Hall, D. (1999), Risky Business: Critical Decisions in Supply Chain Management, *Logistics Consulting Partners*, Hertfordshire, United Kingdom: LCP Ltd.

Bovet (1999), "Value Webs: The Next Business Revolution," Unpublished Report on Logistics, 3–9, Boston, MA: Mercer Management Consulting.

Chopra, S. and Meindl, P. (2004), *Supply Chain Management: Strategy, Planning and Operation*, Upper Saddle River, New Jersey: Prentice-Hall.

Cooper, M.C., Ellram, L.M., Gardner, J.T. and Hank, A.M. (1997a), "Meshing Multiple Alliances," *Journal of Business Logistics*, 18(1), 67–89.

Cooper, M.C., Lambert, D.M., and Pagh, J.D. (1997b), "Supply Chain Management: More Than a New Name for Logistics," *The International Journal of Logistics Management*, 8(1), 1–13.

Copacino, W.C. (1997), *Supply Chain Management: The Basics and Beyond*, Boca Raton, FL: St. Lucie Press.

Das, T.K. and Teng, B.S. (1998), "Resource and Risk Management in the Strategic Alliance Making Process," *Journal of Management*, 24(1), 21–42.

Davis, E.W. and Spekman, R.E. (2004), *The Extended Enterprise: Gaining Competitive Advantage Through Collaborative Supply Chains*, Upper Saddle River, NJ: Financial Times Prentice Hall.

Devlin, G. and Bleackley, M. (1988), "Strategic Alliances—Guidelines for Success," *Long Range Planning*, 21(5), 18–23.

Dettmer, H.W. (1997), *Goldratt's Theory of Constraints: A Systems Approach to Continuous Improvement*, Milwaukee, WI: ASQC Press.

Dittmann, P.J. (2013), *Supply Chain Transformation: Building and Executing an Integrated Supply Chain Strategy*, New York, NY: McGraw-Hill.

Doz, Y.L. (1996), "The Evolution of Cooperation in Strategic Alliances: Initial Conditions or Learning Processes?", *Strategic Management Journal*, 17(1), 55–83.

Drucker, P.F. (1998), "Management's New Paradigms," *Forbes*, October 5, 152–177.

Fiol, C.M. and Lyle, M.A. (1985), "Organizational Learning," *Academy of Management Review*, 10(4), 803–813.

Foster, T.A. (2006), "World's Best-Run Supply Chains Stay on Top Regardless of the Competition," *Global Logistics and Supply Chain Strategies*, 10(2), 27–35.

Fries, K.E. (2005), "Practical, Hands-on Change Management," Presented at the 28th Annual WERC Conference, Dallas, Texas.

Gartner (2014), "Gartner Says Worldwide PC Shipments in the Third Quarter of 2014 Declined 0.5 Percent," *Newsroom*, October 8, http://www.gartner.com/newsroom/id/2869019, retrieved on February 17, 2015.

Ghemawat, P. and Luis Nueno, J. (2006), "Zara: Fast Fashion," Harvard Business School Case: 9-703-497.

Gibson, B.J., Mentzer, J.T., and Cook, R.I. (2005), "Supply Chain Management: The Pursuit of a Consensus Definition," *Journal of Business Logistics*, 26(2), 17–25.

Hendricks, K. and Singhal, V.R. (2005), *The Effects of Supply Chain Disruptions on Long-term Shareholder Value, Profitability, and Share Price Volatility*, Unpublished Report, Atlanta, GA.

Holmstrom, J., Hoover Jr., W.E., Eloranta, E., and Vasara, A. (1999), "Using Value Engineering to Implement Breakthrough Solutions for Customers," *The International Journal of Logistics Management*, 10(2), 1–12.

Hughes, J., Ralif, M., and Michels, B. (1998), *Transform Your Supply Chain: Releasing Value in Business*, London, UK: Thomson Business Press.

Johnson, G.A. and Malucci, L. (1999), "Shift to Supply Chain Reflects More Strategic Approach," *APICS-The Performance Advantage*, 28–31.

Kotter, J.P. (1996), *Leading Change*, Boston, MA: Harvard Business School Press.

Krikke, H. and Caniëls, M.C.J. (2010), "How Can Inter-organizational Learning Support Re-designing Modular Supply Chains?: A Literature Synthesis," Presented at the International IPSERA Workshop on Customer Attractiveness, Supplier Satisfaction and Customer Value, November 2010, University of Twente, Netherlands.

Kwon, I.G. and Suh, T. (2004), "Factors Affecting the Level of Trust and Commitment in Supply Chain Relationships," *Journal of Supply Chain Management*, 40(2), 4–14.

Lambert, D.M. (1976), *The Development of an Inventory Costing Methodology: A Study of the Costs Associated with Holding Inventory*, Oak Brook, IL: National Council of Physical Distribution Management.

Lambert, D.M. and Cooper, M.C. (2000), "Issues in Supply Chain Management," *Industrial Marketing Management*, 29, 65–83.

Lambert, D.M., Emmelhainz, M.A., and Gardner, J.T. (1996), "Developing and Implementing Supply Chain Partnerships," *The International Journal of Logistics Management*, 7(2), 1–17.

Lambert, D.M., Cooper, M.C., and Pagh, J.D. (1998), "Supply Chain Management: Implementation Issues and Research Opportunities," *The International Journal of Logistics Management*, 9(2), 1–19.

Langley, Jr., C.J. (2010), *2010 Third-Party Logistics: Results and Findings of the 15th Annual Study*, Atlanta, GA: Capgemini Consulting and Georgia Institute of Technology.

Lee, H.L., Padmanabhan, V., and Whang, S. (1997a), "Information Distortion in a Supply Chain: The Bullwhip Effect," *Management Science*, 43(4), 546–558.

Lee, H.L., Padmanabhan, V., and Whang, S. (1997b), "The Bullwhip Effect in Supply Chains," *Sloan Management Review*, 93–102.

Levin, S. (2013), "How Zara Took Customer Focus to New Heights," *Credit Union Times,* April 9, http://www.cutimes.com/2013/04/09/how-zara-took-customer-focus-to-new-heights, retrieved on June 25, 2013.

Martin, C. and Towill, D. (2000), "Marrying the Lean and Agile Paradigms," *Proceedings of the EUROMA 2000 Conference,* 114–121.

Miller, T.E. and Berger, D.W. (2001), *Totally Integrated Enterprises: A Framework and Methodology for Business and Technology Improvement*, Boca Raton, FL: St. Lucie Press.

Min, H. (2000), "The Bullwhip Effect in Supply Chain Management," in *Encyclopedia of Production and Manufacturing Management*, edited by P. Swamidass, Boston, MA: Kluwer Academic Publishers, 66–70.

Min, H. (2014), *Healthcare Supply Chain Management: Basic Concepts and Principles*, New York, NY: Business Expert Press.

Min, H. and Zhou, G. (2002), "Supply Chain Modeling: Past, Present and Future," *Computers and Industrial Engineering*, 43(1), 231–249.

Min, S. (2001), "The Role of Marketing in Supply Chain Management," in *Supply Chain Management*, edited by J.T. Mentzer, Thousand Oaks, CA: Sage Publications, Inc., 77–100.

NAPM InfoEdge (2000), "Setting Up the Relationship to Succeed," *NAPM InfoEdge*, 5(2), 13–15.

Newsweek (2001), "Zara Clothing Retail Model Based on Lean Inventories and Market Flexibility Could Change the Tuture of Manufacturing," *Newsweek*, September 9, http://www.newsweek.msnbc.com, retrieved on October 10, 2005.

Noreen, E., Smith, D., and Mackey, J.T. (1995), *The Theory of Constraints and Its Implications for Management Accounting*, Montvale, NJ: The North River Press.

Porter, M.E. (1985), *Competitive Advantage: Creating and Sustaining Superior Performance*, New York, NY: The Free Press.

Poirier, C.C. (1999), *Advanced Supply Chain Management*, San Francisco, CA: Berrett-Koehler Publishers, Inc.

Pucik, V. (1988), "Strategic Alliances, Organizational Learning, and Competitive Advantage: The HRM Agenda," *Human Resource Management*, 27(1), 77–93.

Reuters (2013), "Has Zara Reached Saturation Point? Far from It, Investors Bet," *The Business of Fashion*, November 5, http://www.businessoffashion.com/2013/11/has-zara-reached-saturation-point-far-from-it-investors-bet.html, retrieved on December 22, 2013.

Sambukumar, R. and Vijayan, A. (2011), *The Top Three Reasons Supply Chain Transformations Fail*, Unpublished Report, Bangalore, India: WIPRO Consulting Services.

Schlegel, G.L. (1999), "Supply Chain Optimization: A Practitioner's Perspective," *Supply Chain Management Review*, 11(4), 50–57.

Shank, J.K. and Govindarajan, V. (1993), *Strategic Cost Management: The New Tool for Competitive Advantage*, New York, NY: The Free Press.

Sherman, R.J. (1998), *Supply Chain Management for the Millennium*, White Paper, Oak Brook, IL: Warehousing Education and Research Council.

Sinkula, J.M. (1994), "Market Information Processing and Organizational Learning," *Journal of Marketing*, 58, 35–45.

Slater, S.F. and Narver, J.C. (1995), "Market Orientation and the Organizational Learning," *Journal of Marketing*, 59(3), 63–74.

Statistica (2014), "Retail Market Share of Walmart Stores in the United States in 2012 and 2013, Based on Share of Retail Sales," http://www.statista.com/statistics/309250/walmart-stores-retail-market-share-in-the-us/, retrieved on February 17, 2015.

Varadarajan, P.R. and Cunningham, M.H. (1995), "Strategic Alliances: a Synthesis of Conceptual Foundations," *Journal of the Academy of Marketing Science*, 23(4), 282–296.

Yang, J. and Gonzalez (2006), "Top Threats to Revenue," *USA Today*, February 1, 1B.

2

Supply Chain Strategy: The Big Picture

"Leadership, by the best management thinking, requires a vision of what the future should be and the ability to influence others to achieve excellence. But is a passion for excellence enough? You have to have the passion, but you also need the system, the tools to achieve your vision. When asked what made him a better player than others with similar abilities, Wayne Gretzky replied that he skated to where the puck is going to be, not where it had been."
—Anonymous

Learning Objectives

After reading this chapter, you should be able to:
- Understand the strategic nature of supply chain thinking.
- Comprehend the differences among strategy, philosophy, and doctrine.
- Formulate the winning supply chain strategy and align it with the business strategy.
- Develop alternative supply chain strategies and select the "right" supply chain strategy from among them.
- Develop the framework to assess the efficiency and effectiveness of the supply chain strategy.
- Find ways to adjust the supply chain strategy to changing business environments.

Strategic Dimensions

Because supply chain management cuts across business functional and organizational boundaries, its impact is much broader and longer lasting. Therefore, supply chain management is inherently tied to strategic decision making. Put simply, the role of strategy is to plan the use of resources to meet objectives (Bracker, 1980). In other words, a strategy is a series of plans to integrate an organization's long-term objectives of supporting markets. It differs from a corporate philosophy, which concerns itself with a way of doing business. It also differs from a business doctrine, which represents a code of beliefs such as slogans. Depending on the extent of its scope, a strategy can be divided into three categories (Hill, 2000):

- **Corporate strategy**—This is concerned with decisions by the business as a whole in terms of the industry sectors (e.g., transportation, cosmetic, auto manufacturing, construction, public utility) in which it wishes to compete.
- **Business unit strategy**—This is concerned with decisions on the target markets (e.g., seniors, teenagers, and women) in which the business currently competes or wishes to compete in the future. These decisions involve the coordination and integration of new product/service development, branding, customer relationship management, quality assurance, and delivery scheduling. This strategy can be sub-classified into cost, differentiation, and focus. A *cost strategy* zeroes in on low cost to sustain the firm's competitive edge. A *differentiation strategy* stresses product/service innovations to attract new customers. A *focus strategy* develops a unique market niche to concentrate on the firm's current strength (Porter, 1998; Webster, 2008).
- **Functional strategy**—This focuses on the management and control of the range of tasks supporting each specific business function, such as marketing, operations, purchasing, logistics, and finance. It also determines the bases on which the function will support the desired competitive advantage (Wheelwright, 2004).

Regardless of the aforementioned strategic scope, the elevation of supply chain perspectives to a strategic position expands the responsibility of a supply chain manager in charting the direction of the organization and its partnering organizations. Thus, the supply chain manager should carefully formulate and select supply chain strategy. The formulation of the supply chain strategy may begin with answering the following fundamental business questions:

- What do we do best?
- How can we improve what we have been doing?
- Where do we go from here?

To elaborate, the question of what we do best is concerned with the identification of an organization's core competency in terms of its principal strengths, business focus, financial/production capacity, and managerial talents. This question also dictates the organization's make-or-buy (i.e., outsourcing) decisions. For example, the core competency of FedEx lies in areas such as express parcel delivery services and asset-based warehousing and transportation services. Therefore, to make an outsourcing decision as to whether it should acquire off-the-shelf warehousing and transportation technology, FedEx should decide whether such technology development is its main service prowess.

Once the organization identifies its areas of core competency, it needs to gain greater dominance in those areas by increasing competitiveness and the subsequent sales opportunities. As such, the question of how we improve what we have been doing is concerned with the enhancement of a competitive position in the market and the continuous improvement of product and service quality. For example, the use of promotional strategy through discount coupon offerings and interest-free deferred payment may answer such a question. Finally, the question of where we go from here is concerned with the adaptation of the organization to innovation and various environmental changes (e.g., regulations, economic fluctuations, globalization, customer preferences, and market shifts). The answer to such a question includes new product development, new service offerings, new market penetration, and diversification.

The selection of proper strategy is followed by the formulation of alternative strategies. In general, the strategy selection decision involves the analysis of the costs and benefits associated with alternative strategies and their probability of success (Kerin and Peterson, 1993).

Red Ocean versus Blue Ocean Strategy

Although alternative strategies can be classified into three different categories with respect to their decision hierarchy, they can be broken down into two types—red ocean strategy and blue ocean strategy—with respect to their language of competition and fundamental business model. Generally speaking, red ocean strategy is designed to capture a greater share of existing demand in the clearly defined market boundaries by outperforming industry rivals through low-cost offerings and product/service differentiation (Kim and Mauborgne, 2005b). This strategy, however, cannot provide a sustainable competitive edge over the company's rivals because prospects for profit and growth will be reduced as the existing market gets crowded and the competition gets tougher over time. In contrast, blue ocean strategy aims to target untapped market space so that the company can create new demand and opportunities for highly profitable growth in the unexplored market well beyond the existing industry boundaries (Kim and Mauborgne, 2005a). Table 2.1 summarizes the key differences between these two strategies.

For example, in an effort to create uncontested new market space, the Himalaya Drug Company pioneered the use of modern science to rediscover and validate the potency of herbal medicine, which was revered but not regarded as a proven treatment of ailments. By focusing on the development of safe, side-effect free, and natural remedies that were neglected by most of the traditional pharmaceutical companies, Himalaya Drug thrives in its new market and its products have been endorsed by many doctors around the globe (Chaganti, 2008). Likewise, the Korean third-party logistics service provider (3PL) CJ GLS has strengthened its competitive position by developing a relatively new uncontested market called "RFID-based electronic logistics business" and then leveraging its application service provider (APS) to connect manufacturing to distribution processes. CJ GLS has also targeted often-overlooked small and medium enterprises (SMEs) as their key customer bases whose technological capabilities were constrained due to their limited resources and technological know-how and has provided end-to-end supply chain solutions covering the entire logistics service, from tag installation, to ONS registration of the clients' products, to product history management, while playing a new role as the fourth-party logistics service provider (4PL) (Kim et al., 2008).

Table 2.1. A Comparison of the Red Ocean and Blue Ocean Strategies

Red Ocean	Blue Ocean
Industry boundaries are defined and accepted, and the competitive rules of the game are known. Therefore, prospects for growth and profit are dimmed and reduced.	A consistent pattern of strategic thinking behind the creation of new markets and industries where demand is created rather than fought for and the rule is irrelevant.
Red ocean represents all the industries in existence today. Red ocean strategy accepts constraints of the game—and pursues either differentiation or low-cost strategy. A majority of new ventures are known to target red ocean.	Blue ocean provides companies with guidelines on how to escape from intense competition from the same market place, where there are limited customers with an increasing number of competitors, by creating a *new* marketplace where there is less competition, if any. Pursue both differentiation and customer value.
Commoditization of product/service offerings. Cutthroat competition for the existing market and demand.	Superb business performance by creating a leap in value for customers. Innovative technological capabilities do *not* necessarily create the blue ocean.

Strategic Supply Chain Planning Processes

Regardless of strategic dimensions, strategy cannot be put forward without maintaining a delicate balance between learning from the past and shaping up new courses of action to lead the organization toward a future state, which may include a substantial departure from its past conduct (Hax and Majluf, 1988). Because the substantial departure from its past conduct can create organizational resistance to change, strategic planning requires a series of careful steps that are conducive to the purposeful management of change. These steps may follow three levels of processes suggested by Bower and Doz (1979):

- **Cognitive processes**—These are the processes of individuals on which the understanding of the environment of strategy are based. These processes include the identification of the market segmentation where the organization's growth opportunities exist.
- **Social and organizational processes**—These are the processes by which perceptions are channeled and commitments are developed. These processes involve conveying a sense of direction to the organization's primary stakeholders, such as shareholders, boards of directors, employees, and all other relevant constituencies, through direct and open communication and the building of consensus among them to move forward with the strategic plan.
- **Political processes**—These are the processes by which the power to influence purpose and resources is shifted. These processes involve the empowerment of the decision-making authority and the provision of incentives to key players who can make new strategic plans a reality.

From the supply chain perspective, the preceding processes can be further broken down into the following processes within the strategic framework proposed by Hax and Majluf (1984):

- **Creating the vision of the firm**—The vision of the firm specifies the current and future expected business scope, markets, geographical coverage of the business activities, key business partners sharing information, resources and risks, and corporate philosophy facilitating the integration of business activities into a supply chain.
- **Developing the supply chain planning guidelines**—The vision of the firm should be translated into pragmatic guidelines with respect to supply chain planning horizons, assessment of external environmental factors (e.g., technical, social, and political climates), assignment of managerial responsibilities, and corporate performance goals.
- **Formulating strategic action plans**—Once the supply chain planning guidelines are set, strategic action plans should be formulated to sustain a long-term competitive advantage. These action plans focus on the identification of core competencies residing in the firm, which will determine the unique potential for competitive leadership, the allocation of available resources (e.g., funds and people), and the development of the business portfolio.
- **Evaluating the success or failure of strategic action plans**—The specific progress of strategic action plans should be monitored based on clear performance measurements that could provide early warning signals for potential failures. These performance measurements include cost/benefit analyses, risk analyses, financial efficiency, coherence to planned schedules, and relative priorities of action plans.

- **Gaining support from the stakeholders, including top management**—To make strategic action plans successful, the strategic planner should garner organization-wide support and financial commitments. Therefore, this final stage of the strategic planning process should include efforts to ensure strategic and budgetary commitments from key stakeholders who will be affected by the strategic action plans.

Strategic Integration of Supply Chain Processes

LaLonde (1997) stresses that the goal of delivering enhanced customer service and adding economic value to the supply chain can only be achieved through the synchronized management of the flow of physical goods and associated information from sourcing to consumption. The synchronization of all channel activities throughout the supply chain starts at the upstream of the supply chain, sourcing. In other words, the integration of supply chain processes begins with the linkage of sourcing to manufacturing.

The role of sourcing in controlling manufacturing costs and dictating manufacturing schedules has dramatically increased over the last few decades, as a growing number of firms have outsourced their manufacturing activities, including new product development, quality control, and technological innovations to reduce capital investment, speed up product innovation, and enhance manufacturing flexibility. Reflecting such a trend, the average expenditure for purchased goods and services reached more than 50% of sales revenues (Giunipero and Brand, 1996). The increased role of sourcing in manufacturing necessitates new strategies. One such strategy is Segmented Purchasing Strategies (SPS), proposed by Copacino (1997). SPS manages purchasing activities based on the level of exposed supply risk (e.g., sourcing alternatives, product availability, and quality variances among suppliers) and the extent of economic opportunity (e.g., cost saving potentials, value-adding opportunity, and product innovations). To elaborate, SPS segments families of products to be manufactured in four distinctive categories and develops different sourcing strategies as summarized next. The four families of products are categorized as follows (Copacino, 1997):

- **Nuisance products**—These include maintenance, repair, and operating supplies (MROs) and other low-value, noncritical products whose sourcing activities can be repetitive and routine and therefore can be automated.
- **Bottleneck products**—These represent important but low-volume materials that do not require frequent purchases.
- **Commodities**—These are products that can be easily substituted for other alternatives and are therefore easier for the buyer to obtain potential bargains.

- **Critical products**—These represent "must-have" or unique items that are essential for meeting product quality requirements and necessary customer services.

Once the firm formulates a sourcing strategy based on manufacturing plans, then the logistics activities of moving and storing incoming materials, components, and parts from the chosen suppliers to the buying firm need to be orchestrated. For instance, the firm that embraces a just-in-time (JIT) sourcing strategy must make a conscious effort to deliver incoming materials, parts, and components to the manufacturer's locations without quality failures, in-transit damages and losses, and extensive delays. To be consistent with JIT sourcing strategy, the firm needs to develop a JIT logistics strategy, which often requires the frequent use of a prime (faster) mode of transportation and more reliable contract carriers. In addition, the firm should formulate a logistics strategy in such a way that supply disruptions resulting from unexpected traffic congestions, accidents, terrorist attacks, natural disasters, and labor strikes can be minimized. Here are some examples of logistics strategies that can mitigate potential supply disruptions:

- **Postponement**—Delay the formation of the finished product/service bundle as long as possible. The "last-minute" customization made available through postponement will reduce the customer response time without relying on the supplier's production capacity.
- **Co-location of manufacturers and their suppliers**—Proximity of supplier bases to the manufacturing facilities will help reduce lead time and thus enhance the responsiveness to unexpected supply disruptions.
- **Avoidance of choke points**—The use of intermodal transportation modes such as piggyback and fishyback that can bypass choke points and are therefore less prone to traffic congestion should be considered to increase delivery reliability.

As mentioned previously, the sourcing strategy should be linked to the logistics strategy. Whereas the inbound logistics strategy is tied to the sourcing strategy, the outbound logistics strategy is often linked to the marketing strategy. Given the increased homogenization of products and the rapid reduction of product-to-shelf cycles in today's business world, logistics leverage can help the firm sustain its competitive advantage and differentiate itself from its competitors because it is more difficult for other firms to copy logistics leverage than price and product features—the rationale being that logistics leverage requires a unique, well-coordinated relationship among multiple parties in the distribution channel that can only be built upon the development of mutual trust over time and the costly investment in the channel infrastructure (Mentzer and Williams, 2001).

A good example is the success of Dell Computers, which transformed commodity industry practices by selling its personal computers online and leveraging its logistics superiority to create customer value. For example, Dell allowed its customers to directly access technical advice about specific product configurations through its websites and guided customers to configure and order the personalized products they wanted online. Because the electronically ordered products were directly delivered to customers by Dell's third-party logistics service providers (UPS and Airborne Express) without going through the distribution layers of middlemen and/or retail-outlets, Dell successfully created "direct marketing channel" to its customers and subsequently reduced customer response time. As illustrated by the Dell's business model, logistics leverage can be a viable aspect of marketing strategy. Considering the interdependence of logistics, sourcing, and marketing, logistics strategy can be formulated by taking the following steps suggested by Murray (1980):

1. **A supply chain overview**—Describe the logistics strategy in general terms and its relationship to sourcing and marketing functions.
2. **A statement of the logistics objectives**—State and prioritize the specific long-term objectives to achieve over five or more years with respect to cost control and customer services.
3. **An outline of tactical and operational plans**—Describe and document individual transportation, warehousing, order processing, and inventory strategies in detail that are necessary to support the overall logistics plans.
4. **A forecast of the necessary resources to undertake the strategic logistics plans**—Ensure the human, capital, and financial resources required to undertake the strategic logistics plans successfully.
5. **An evaluation of the bottom-line impact of the logistics strategy**—Develop the metrics that can measure the extent of success brought by the logistics strategy in terms of corporate profits, shareholder value, customer service performance, and market share.

The "Victory" (Winning Strategy) Model

Porter (1980) argues that a company can increase its competitive advantage by developing a value-creating or value-adding strategy. Such a strategy often builds on the company's distinctive competencies, including its innovation capability. Considering that supply chain principles can be viewed as an innovative way of doing business, the company may leverage supply chain excellence to improve its competitive advantage, which stems from its ability to create or add customer value. As shown in Figure 2.1, several different strategies can help the company enhance its customer value and the subsequent bottom line.

- The manipulation of the company's production/purchasing/sales *volume*
- The adjustment of the company's product/service *velocity*
- The control of the company's product/service *variability*

For example, a company's increased production volume creates economies of scale and thus reduces its product's unit price, which in turn increases customer value due to a lower-priced product for the customer. Also, if the company can shorten its customer response time by reducing its lead time, its customer service will be improved and therefore its customer value will be enhanced. Furthermore, if the company diversifies its product lines and increases its customization options, it will give more choices to its customers and thus increase customer value. Although factors such as volume, velocity, and variability can positively affect the customer value and the company's bottom line, as illustrated in Figure 2.1, the simultaneous increase in volume, velocity, and variability is nearly impossible due to the conflicting nature of volume, velocity, and variability. For instance, the company's volume production takes more time and therefore reduces its velocity (speed). This is why just-in-time (JIT) production, which aims to increase the company's velocity, is designed to make a product one at a time in its ideal setting. On the other hand, increased variability will undermine the company's strategy of mass production and fast customer response due to the additional time needed for customization and differentiation. That is to say, volume is inversely related to velocity, whereas variability is inversely related to volume and velocity. Also, it should be noted that variability can hurt visibility because diversified product/service offerings tend to make demand forecasts more difficult. Visibility has a delicate relationship with volume and velocity in that increased volume, such as a large amount of buffer inventory, can reduce visibility, whereas increased velocity, such as faster production runs and quick deliveries, reduces the forecasting time horizon and thus increases visibility. Considering the aforementioned complex dynamics involving volume, velocity, variability, and visibility, the company should factor all these into its strategic decision and development of the supply chain framework.

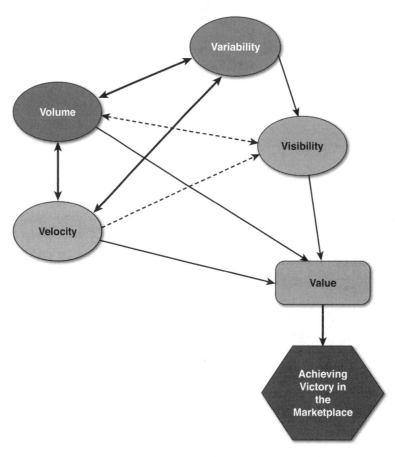

Figure 2.1. *A "Victory" model for supply chain strategy*

Push versus Pull Strategy

The decision as to how your company will source, produce, promote, and distribute its goods and services can shape the company's supply chain strategy. In a broad sense, this decision may be based on the two distinctive categories of business strategy summarized in Table 2.2 (Spearman and Zazanis, 1998; McNeil, 2011). These categories are push strategy and pull strategy. Generally, a push strategy involves building products and stockpiling these products well in advance of actual customer orders and purchases based on the forecast of anticipated customer demand. In other words, this strategy "pushes" the product to the end user or customer through sales promotions and other incentives. Because this strategy relies heavily on the long-term forecast of customer demand, it does not reflect the actual input from the marketplace and therefore can be error-prone. Although this strategy allows for level production and stable workforce/materials, it often fails to meet changing demand patterns of customers in volatile business environments and subsequently tends to build up unwanted inventory. As an alternative to this strategy, pull strategy is increasingly used in the supply chain. A pull strategy is a demand-driven strategy that determines the production, distribution, and service delivery schedules based on the actual customer demand. In other words, this strategy makes and distributes what is "pulled" from the end user or customer at the rate of his or her true demand requirements. As such, this strategy can better adapt to changes in customer tastes, preferences, and expectations without producing and distributing what is not wanted by the customer. The key benefits of this strategy are as follows (see Hagel and Brown, 2005; Simchi-Levi et al., 2008):

- Shorter lead times thanks to the better anticipation of customer orders and demand
- Decreased inventory levels throughout the supply chain, including retailers and manufacturers
- Decreased system variability
- Wider ranges of production and distribution options
- Better response to changing markets

The pull strategy, however, may pose the following problems:

- It's harder to leverage economies of scale due to small-lot production and distribution based on only what is needed by the end user or customer.
- It does not work in all cases (and is especially not suitable for standardized products with high sales/demand volume).
- It's difficult and more time consuming to implement due, in part, to its disciplined approach and the need for accurate, real-time demand information.

Table 2.2. Push versus Pull Strategy

Differentiator	Push Strategy	Pull Strategy
Prime business driver	• Maximize the utilization of physical and human resources or infrastructure at the lowest cost.	• High levels of customer service through quick responsiveness and flexibility to meet fluctuating and uncertain customer demand.
Supply chain strategy	• Plan and operate based on the anticipated demand. • Rely on demand forecasting and S&OP. • Maximize economies of scale and volume discounts.	• Operate in response to actual customer demand. • Explore postponement for customization. • Emphasize lean principles.
Lead time	• Long, but a prime focus should be to reduce lead time.	• Short.

Differentiator	Push Strategy	Pull Strategy
Pricing strategy	• Pricing is a key means for balancing supply and demand. • High-low pricing strategy may work better.	• Pricing does not normally impact short-term demand. • Everyday low pricing strategy may work better.
Manufacturing strategy	• Long and level production runs. • Production based on pre-planned manufacturing schedules or production quota. • Control throughput (or capacity).	• Short and flexible production runs. • Jobs are pulled by each work station from the previous station • Control work-in-process (WIP) and reduce WIP levels and variability.
Marketing strategy	• Put advertisements in front of someone who does not necessarily have any knowledge of the company or product/service. • Outbound marketing that devises ways to place products/services before prospective buyers. • Typically use paid advertising via TV, radio, billboard, catalog, etc.	• Motivate consumers to find the company and its product/service on their own. • Inbound marketing that aims to generate interest and responses from prospective buyers. • Use frequent web-based posts such as blogs, e-books, social media, etc.
Inventory	• Need a high level of safety stock. • Large, less frequent orders and shipments. • Larger warehouse space. • Emphasize prior inventory planning, cycle counting, and ABC inventory classification. • Push as close to customer locations as possible.	• Typically low (no more than 10% of the expected demand). • Keep inventory on as-needed basis. • Place smaller, more frequent orders and shipments. • Prevent or reduce the need for inventory write-offs and markdowns. • Higher out-of-stock risk.
Supply chain relationships	• Supplier relationships tend to be most critical because they help minimize the risk of supply disruptions. • Establish collaborative relationships with customers to minimize forecasting errors.	• Customer relationships are critical. • Leverage vendor-managed inventory built on a long-term partnership with suppliers.
Essential analytical tools and information technology applications	Sales forecasting techniques, EOQ, MRP systems, etc.	Point-of-sale (POS) systems, CRM (Customer Relationship Management) software, RFID, DRP systems, etc.

Considering the pros and cons of push and pull strategies, as described here, some companies may want to consider using the push-pull strategy, which is a hybrid version that exploits the best features of these contrasting strategies rather than opting for one or the other. A push-pull strategy determines the production, distribution, and service delivery levels based on a combination of forecasts and specific customer orders by integrating forecasted data into pull-based production and distribution schedules. In this strategy, the initial portion of supply chain activities (i.e., upstream supply chain activities such as the acquisition of raw materials or the replenishment of part inventories) will be scheduled based on long-term forecasts, whereas the later or final stages of the supply chain (i.e., downstream supply chain activities such as the assembly and packaging of products) will be scheduled based on actual customer orders. A postponement strategy is a good illustration of this push-pull strategy. In the postponement strategy, a company delays some supply chain activities such as assembly, packaging, labeling, and painting until true customer demand is revealed, while building products in semi-finished forms based on the projection of their future demand. Once actual orders come in, these semi-finished products will be customized immediately in production and distribution facilities close to customers and then distributed quickly to the customers. As such, this strategy can minimize waste (e.g., inventory) and the subsequent total supply chain cost, while reacting to customer demand quickly. Other benefits of this strategy include better reaction to demand variability, reduction in risks of obsolescence, reduction in logistics costs, and improved competitiveness due to cost savings and enhanced customer services.

Typology of Supply Chain Strategy

Here are some of the fundamental supply chain management questions we raise every day:

- What are we going to make and deliver?
- What materials and parts have to be acquired to make the product? How much, and when?
- What and how much inventory do we have now?
- Which resources are needed to make and deliver the product?

To answer these questions, a careful strategic plan has to be developed. This plan may vary depending on the type of products the company and its supply chain partners would like to produce and sell. Considering the need for a varying supply chain strategy for different products, Fisher (1997) proposes two distinctive supply chain strategies: (1) efficient and (2) responsive with respect to the type of product (either functional or innovative).

Further extending the strategic framework developed by Fisher (1997), Mason-Jones et al. (2000) categorize supply chain strategies into three different types: lean, agile, and "leagile." To elaborate, a lean supply chain strategy aims to develop a value stream from suppliers to final customers to eliminate all kinds of buffering costs in the supply chain and to ensure a level production schedule in order to improve process efficiency and then maintain the competitive advantage through economies of scale in a stable and predictable marketplace. On the other hand, an agile supply chain strategy aims to develop a flexible and reconfigurable network with partners to share competences and market knowledge in order to survive and prosper in a fluctuating market environment by responding rapidly to market changes. Finally, a leagile supply chain strategy, which combines some elements of the lean and agile strategies, utilizes make-to-stock/lean strategies for high-volume, stable-demand products while using make-to-order/agile strategies for everything else. Thus, a leagile supply chain strategy can have flexible production capacity to meet surges in demand (e.g., seasonal demand) or unexpected requirements while using postponement strategies for "platform" products that are made to forecast and then go through final assembly and configuration when the final customer order is placed (Christopher and Towill, 2001; Goldsby et al., 2006).

Going one step further, Lee (2002) and Vonderembse et al. (2006) put forth a demand and supply uncertainty framework that produces four different types of supply chain strategies—efficient, risk-hedging, responsive, and agile—as summarized in Table 2.3. A firm may pursue efficient supply chains when the market is mature and competitive advantage is achieved primarily through low cost and high productivity. Firms take an efficient supply chain strategy mainly to manufacture quality products efficiently and to provide customers with reliable services. A risk-hedging supply chain strategy can be used when a supply chain is fraught with the presence of uncertainty. In addition to the retail industry, hydro-electric power and some food producers are examples of this category (Lee, 2002). To leverage supply uncertainties, a firm would increase buffer stock for its core products and attempt to share the cost of the safety stock with its supply chain partners. In general, a responsive supply chain strategy is suitable for firms that offer a variety of innovative or customized products tailored for specific customer demands and flavors (Fisher, 1997). To accommodate customers' constantly changing demands, this supply chain may postpone making the final form of a product until the demand becomes known. Fashion apparel, computer, and popular music industries often use this strategy (Lee, 2002). Though somewhat similar to a responsive supply chain strategy, an agile supply chain strategy is the most flexible and the most market-driven strategy among four categories of supply chain strategies. An agile supply chain strategy aims at inducing flexibility and velocity to adjust promptly to volatile market conditions and to unpredictable sources of supply. Thus, an agile supply chain strategy focuses on fast market response time and quick product development time, while minimizing supply disruptions by streamlining information flows across the supply chain. Firms that exploit an agile supply chain strategy can be found in the high-end computer and semiconductor industries.

Table 2.3. Types of Supply Chain Strategies

Feature	Efficient Supply Chain Strategy	Risk-hedging Supply Chain Strategy	Responsive Supply Chain Strategy	Agile Supply Chain Strategy
Supply uncertainty	Low	High	Low	High
Demand uncertainty	Low	Low	High	High
Priority	Cost efficiency through waste reduction	Risk mitigation	Customer responsiveness	Time-sensitive customization
Product type	Functional or commoditized	Functional or commoditized	New or innovative	New or innovative
Strategic focus	Cost and quality	Cost, flexibility, quality	Speed, flexibility	Speed, flexibility, continuous improvement
Key driver for a supplier relationship	Transaction	Resource sharing	Adaptability	Knowledge sharing and technology transfer

Internal Supply Chain Strategy Audits

To reflect constantly changing customer needs and preferences, the supply chain strategy needs to be periodically evaluated, fine-tuned, and modified. One effective way of doing this is to conduct a supply chain strategy audit in which supply chain managers detail the strategic priorities and ensure that every unit within the organization pulls together to achieve shared organizational objectives without the less-than-optimal or inconsistent alignment of strategic priorities. Also, the audit should reveal a need for shift in strategic priorities and development of new supply chain strategies. When conducting an audit of the supply chain strategy, the company should take the following steps, which are analogous to the ones proposed by Stock and Lambert (2001):

1. Establish the task force to assist in the audit process.
2. Prepare a list of key questions that will be raised by the task force for audit interviews. These questions may be related to the firm's internal interface among different business units, its competitors' supply chain strategy, the extent of outsourcing, the level of internal resistance to organizational changes, competitive priorities, supply chain efficiency, and market niches and segments.
3. Develop specific performance metrics that enable the firm to gauge its competitive strengths and weaknesses relative to its competitors and then to identify the firm's challenges and opportunities in the chosen market.

4. Collect the firm's financial records and transaction data while soliciting the opinions of internal employees regarding the efficiency and effectiveness of the implemented supply chain strategy via personal interviews.
5. Identify alternative strategies that may be substituted for the current strategy for continuous improvement.
6. Recommend the required changes and improvement of the implemented strategy to top management.

External Supply Chain Strategy Audits

Whereas the internal supply chain strategy audit examines how congruent or harmonious the chosen supply chain strategy is with the firm's strategic missions and organizational culture, the external supply chain strategy audit examines how much the chosen supply chain strategy contributes itself to improvements in customer services and an increase in market share. Therefore, the external supply chain strategy audit requires the solicitation of feedback from the firm's ultimate customers and supply chain partners, such as suppliers and third-party logistics providers. The external audit is often composed of random customer surveys, forums with selected supply chain partners, and in-depth interviews with selected customers representing different market segments. Key questions to be addressed by the external audit include the following:

- What core competencies has your firm been building to compete for scarce resources and market share?
- Which segments of the market has your firm targeted for its niche and future growth?
- How does your firm stack up against its most effective competitors in terms of price, quality, delivery services, and product innovation? Can your firm's current strategy attract new customers?
- How does your firm identify its "order-winning" and "order-qualifying" criteria? How dynamically do those criteria change over time? How quickly and proactively can your firm respond to those changing criteria by reformulating supply chain strategies?
- Which chosen supply chain strategies are considered value-adding or non-value-adding strategies from the customer perspectives?
- How well is your firm's supply chain strategy in sync with its supply chain partners' corporate missions and organizational structures?
- How often and willingly can the supply chain–related information regarding demand forecasts, new product development, shipment status, and supply chain

network restructuring be shared among supply chain partners? How compatible are your firm's communication capabilities to those of its supply chain partners?
- What performance metrics should be used to evaluate the effectiveness and efficiency of your firm's entire supply chain? What value propositions does your firm prioritize?

Chapter Summary

The following summarizes key takeaways from this chapter:
- The supply chain thinking is rooted in a firm's strategy and leverages core competencies to create value for its customers, sustain a competitive advantage over its interfirm rivals, and protect the long-term interest of its stakeholders. The strategy can be classified into three different levels: corporate, business unit, and functional. A corporate strategy represents a pattern of action plans and decisions that help identify the firm's unique strengths and leverage those to stay in the market and then expand the firm's customer bases. A business unit strategy is a long-range plan to enable the firm to win its business continuously, set competitive priorities, and integrate functional activities such as sourcing, production, sales, promotion, and distribution. The formulation of a specific business unit strategy is often the responsibility of top executives. A functional strategy is a detailed plan of actions and decisions that aim to control and utilize given resources at the department level, such as operations, marketing, logistics, finance, and human resources.
- In general, many companies tend to decide on their competitive priorities based on several different categories, such as price, quality, time, customization, and innovation. This decision often hinges on so-called "order-qualifying" criteria and "order-winning" criteria. According to Hill (2000), order-qualifying criteria includes minimum price, prompt delivery services, and product warranty competitive enough to gain the consideration of the targeted customers. However, to attract new customers and beat the competition, the firm should meet its order-winning criteria by exceeding the industry norm and the common expectation of customers. This can often be accomplished by introducing more innovative product/service bundles.
- A supply chain strategy that is not aligned with the supply chain members' competitive priorities, core competencies, and organizational culture is likely to fail. Therefore, solicitation of feedback from supply chain partners, customers, and other stakeholders is critical to the successful implementation and the necessary transformation of a supply chain strategy. Such feedback can be channeled to policy makers through periodic internal and external audits. To make these audits more meaningful, the firm should develop measurable (quantifiable) standards

that help its stakeholders understand and evaluate the specific impacts of a supply chain strategy on its business bottom lines, such as return on investment, profitability, revenue growth, and market share.

Study Questions

The following are important thought starters for class discussions and exercises:

1. Explain how supply chain thinking is tied to strategy.
2. Compare and contrast strategy, philosophy, and doctrine.
3. Think of an organization that you are familiar with and identify its strategy, philosophy, and doctrine, if any.
4. What are three levels of strategy? How are they different from each other?
5. What are the steps to take to develop a strategic plan?
6. What is a segmented purchasing strategy?
7. How is a sourcing strategy linked to a logistics strategy?
8. What are the differences between the push and pull strategies? How do you choose the decoupling point for a push-pull strategy?
9. What are the four types of supply chain strategy? How would you select the right supply chain strategy?
10. How would you conduct supply chain strategy audits?
11. What are the key issues to be addressed for the external supply chain strategy audit?
12. What are the differences between "order-winning" and "order-qualifying" criteria?
13. What are core competencies? How would you identify the core competencies of a company?
14. Who is primarily responsible for developing the supply chain strategy?

Case: Dell, Inc.—Push or Pull?

This case was written based on various published stories about Dell and inspiration from the toolbox of the Council of Logistics Management.

The Brief History of Dell's Postponement Strategy

Postponement is the act of delaying or putting off products or service, which seems counterintuitive when a company wants to achieve high levels of customer service. Yet, postponement is a key element in the business model that helped Dell, Inc., thrive in the increasingly competitive personal computer (PC) market in the U.S., with more than $56.9 billion in 2013 annual revenue (http://www.sec.gov/Archives/edgar/data/826083/000082608313000014/dell10-kafy2013.htm, Dell Inc. 2013 Form 10K). Dell, Inc., is the namesake and brainchild of Michael Dell, who started his computer business in 1984. Dell's years of success are attributed to its once innovative business model that pioneered the idea of selling and distributing products directly to its customers without relying on brick-and-mortar store sales. When it first entered the PC market, Dell chose to sell directly to end customers as well as businesses primarily over the Internet or by telephone. Dell's website allows the customer to select specific models and options as well as the preferred method of payment and shipment without direct personal contact. After the order is placed, the customer can access the Dell website and check the status of his or her order at any time. Once the order is confirmed, Dell acquires the parts/components and builds the PC to each customer's specific needs/options, shipping the finished goods directly to the customer. This order fulfillment process is different from the traditional practice of building PCs and stocking them with the anticipation of future orders. Thus, Dell does not have to deal with inventory headaches and offers customers the convenience of shopping online with the paperless online order form. In this business model, Dell lets the customer decide exactly what he or she wants, how soon he or she needs it, how he or she wants the order shipped, and how he or she wants to pay, rather than "pushing" their pre-made products into the customer's hands without many options.

Supply Chain Strategy

A key element of Dell's business strategy is the use of "voice of the customer," which allows the customer to play "what-if" scenarios with prices and features until he or she is satisfied with the ultimate value—a combination of product configuration, price, and service (ordering, delivery, and post-sales). Also, to add value to the PC that a customer wants, Dell engages its entire organization in transforming "end-to-end" supply chain practices, from the traditional retails sales/distribution to customer-direct sales/distribution, as displayed by Exhibit 2.1 (Gilmore, 2011).

Voice of the Customer

I am looking for customized PC	I want to buy PC without visiting the store.	My order is paid and processed.	My order is in production.	My order is shipped and is in transit.	My order is delivered and inspected.	I need post-sales support.
Pre-configuration	Online	Credit checking and credit and payment	Build (assemble)-to-order ("stapling yourself to an order") and product testing	Multiple shipping options and shipment tracking	Order confirmation	Warranty
Customer configurations	Telephone	Electronic ordering system				Technical support

Exhibit 2.1. *Dell's end-to-end supply chain process*

As Exhibit 2.1 illustrates, Dell only needs inventory in partially completed PCs (parts/components) during the preconfiguration stage, but not in finished products throughout its supply chain. To further reduce the burden of holding parts/components inventory, Dell requires its suppliers to keep inventory on hand in its small warehouses (called "revolvers") or supplier logistics centers (SLCs) that are located within a few miles of Dell's assembly plants. Each of the SLCs is shared by a group of suppliers who pay rent for using this facility. Because each PC is assembled after the sale, there is little dead stock—that is, inventory that is outdated or poorly configured and therefore is never sold to customers. More importantly, before the delivery of the ordered PC, Dell has already been paid via credit card, so its cash flow has been dramatically improved. Dell's gains from this non-traditional business model have been impressive: lower inventories and the subsequent high inventory turnover; elimination of the reseller's markups; quicker customer response; and faster cash-to-cash cycle time. According to CSIMarket.com, Dell turned its inventory 97.54 times at its peak in 2004 and averaged 54.49 during the past decade. Its inventory turnover ratio is the second highest in the computer hardware industry as of 2013. Its average inventory turnover is five times higher than its biggest rivals, such as Hewlett-Packard (HP), which registers an average inventory turnover ratio of 12.15. There is no magic or secret behind Dell's success: Instead of stretching its supply chain through a retail distribution channel in finished form, its inventory consists of parts/components whose assemblies will be postponed until the customer orders a PC with optional features he or she really wants. This postponement strategy allows Dell to establish pricing points for each model and type of computer. With lower supply chain costs, Dell can offer competitive prices to its customers while still using high-end components such as the Phobus gaming/home theatre system.

Changing Times

In the fourth quarter of 2008, Dell still enjoyed success with a reported quarterly revenue of $16 billion, but fell short of the expected earnings for a company that was once at the pinnacle of the industry. Ever since, Dell has continued to slide in its market share and profit margin. For example, after losing its market leadership to Hewlett Packard in 2007, Dell's share of the worldwide PC market fell from 14.2% to 13.2% between 2007 and 2008 and its share of an already contracting PC market slipped to just 10.7% in 2012, from 18.1% in 2001 (Gartner Group, 2009; De La Merced and Hardy, 2013). Although Dell's recent struggles were affected by a multitude of factors, including the acquisition of EqualLogic, Alienware, Everdream, and Perot Systems; severance payments; and a worldwide economic downturn, Dell has been subject to a litany of cost- and productivity-related issues that most of us never saw coming over the last few years. To address these issues, Dell continues to lay off workers amid global restructuring and recently announced that it would cut 2,000 to 3,000 people from its payroll. It also closed its desktop manufacturing facilities with the hopes of reducing expenses as much as $3 billion a year. In addition, Dell is looking to update its product design, compress manufacturing cycles, increase investment in cloud computing technologies, get new product to market quicker, and develop new high-value client-focused software, peripherals, and services. Leveraging a 36% rise in combined sales in Brazil, Russia, India, and China (BRIC), Dell tries to exploit emerging markets and focus on notebook sales to boost both revenue and profit. Despite these efforts, Dell is still concerned about its shrinking market share in the worldwide PC market. As of the third quarter of 2014, Dell's archrival, HP, is the world leader with 19.8% of the market share, whereas Dell ranks third with 12.8% market share, behind Lenovo's 17.9% market share (Statistica, 2014).

So, what caused the recent trouble for Dell? Did it stem from its rising research and development (R&D) costs for rapidly changing technology? Was it because Dell was unaware of market changes and relied too heavily on its old trick of selling direct to customers in changing times? Or was it simply that Dell's popularity has run its course and HP realized what consumers want today before Dell did? In fact, Dell's rivals (HP, Apple, and Acer) have already duplicated some of Dell's postponement ("push-pull") strategies and became more competitive than ever before. As such, Dell's old trick of selling direct to customers no longer works well, and its market share has dwindled in the last several years. Dell began to realize that its iconic but old business model should be changed because consumers suddenly are more likely to buy a PC at their local retail outlet (e.g., Best Buy, Sam's, or Costco) than ever before—the rationale being that online purchase and call-center support might leave some customers uncertain of what they will actually get.

To regain its strong foothold in the dynamic PC market, Dell was forced to start selling its PCs in retail outlets (such as Walmart) again after abandoning such a marketing strategy in 1994. Dell is now selling its PCs in about 10,000 retail stores worldwide, including Carrefour and Tesco. However, some management experts believe that the company was too late—HP already owned that brick-and-mortar market with its long ties to discount retail giants such as Sam's and Costco. Another challenge faced by Dell is the fact that the computer industry has quickly become a commoditized business, where a Dell PC is virtually the same as an equally equipped HP product. Therefore, Dell was often forced to engage in price wars against its rivals. With a dwindling profit margin in the PC industry, Dell is fighting an uphill battle to stay competitive and remain profitable. As a professional consultant for Dell, you would like to help the company avoid any more trouble on the horizon and develop innovative ways to entice both past and new customers. List and justify your specific recommendations for Dell with its pros and cons. Prior to making these recommendations, you should consider the following dilemmas first:

- Why does the direct sales model that worked so well in the past no longer work these days?
- Does the simultaneous use of multiple sales/distribution channels (e.g., direct online sales and sales through retail stores) create a channel conflict?
- What are the potential problems associated with traditional retail sales?
- What managerial changes have to be made to regain market share and become a global leader of the PC industry?
- How can you make Dell's supply chain more resilient than ever before?

Bibliography

Bracker, J. (1980), "The Historical Development of the Strategic Management Concept," *Academy of Management Review*, 5(2), 219–224.

Bower, J.L. and Doz, Y. (1979), "Strategy Formulation: A Social and Political Process," *Strategic Management: A New View of Business Policy and Planning*, edited by C.W. Hofer and D. Schendel, Boston, MA: Little, Brown and Company.

Chaganti, S.R. (2008), *Pharmaceutical Marketing in India*, New Delhi, India: Excel Books Private Limited.

Christopher, M. and Towill, D. (2001). "An Integrated Model for the Design of Agile Supply Chains," *International Journal of Physical Distribution and Logistics Management*, 31(4), 235–246.

Copacino, W.C. (1997), *Supply Chain Management: The Basics and Beyond*, Baca Raton, FL: The St. Lucie Press.

De La Merced, M.J. and Hardy, Q. (2013), "Dell in $24 billion Deal to Go Private," *New York Times*, February 5, 2013, http://dealbook.nytimes.com/2013/02/05/dell-sets-23-8-billion-deal-to-go-private/?_r=0, retrieved on March 15, 2013.

Fisher, M. L. (1997), "What Is the Right Supply Chain for Your Product," *Harvard Business Review*, 75(2), 105–116.

Gartner Group (2009), "Gartner Says in the Fourth Quarter of 2008 the PC Industry Suffered Its Worst Shipment Growth Rate Since 2002," *Newsroom Press Release*, January 15, http://www.gartner.com/newsroom/id/856712, retrieved on May 27, 2010.

Gilmore, D. (2011), "The Lessons from Dell's Supply Chain Transformation," *Supply Chain Digest*, May 18, 2011, http://www.scdigest.com/assets/FirstThoughts/11-03-18.php?cid=4330, retrieved on October 7, 2012.

Giunipero, L.C. and Brand, R.R. (1996), "Purchasing's Role in Supply Chain Management," *The International Journal of Logistics Management*, 7(1), 29–38.

Goldsby, T. J., Griffis, S. E., and Roath, A. S. (2006), "Modeling Lean, Agile, and Leagile Supply Chain Strategies," *Journal of Business Logistics*, 27(1), 57–80.

Hagel III, J. and Brown, J.S. (2005), *The Only Sustainable Edge: Why Business Strategy Depends on Productive Friction and Dynamic Specialization*, Boston, MA: Harvard Business School Publishing.

Hax, A.C. and Majluf, N.S. (1984), "The Corporate Strategic Planning Process," *Interfaces*, 14(1), 47–60.

Hax, A.C. and Majluf, N.S. (1988), "The Concept of Strategy and the Strategy Formation Process," *Interfaces*, 18(3), 99–109.

Hill, T. (2000), *Operations Management: Strategic Context and Managerial Analysis*, New York, NY: Palgrave.

Kerin, R.A. and Peterson, R.A (1993), *Strategic Marketing Problems: Cases and Comments*, 6th Edition, Boston, MA: Allyn and Bacon.

Kim, C., Yang, K.H., and Kim, J. (2008), "A Strategy for Third-party Logistics Systems: A Case Analysis Using the Blue Ocean Strategy," *Omega*, 36(4), 522–534.

Kim, W.C. and Mauborgne, R. (2005a), "Blue-ocean Strategy: From Theory to Practice," California Management Review, 47(3), 105–121.

Kim, W.C. and Mauborgne, R. (2005b), *Blue Ocean Strategy: How to Create Uncontested Market Space and Make Competition Irrelevant*, Cambridge, MA: Harvard Business Press.

LaLonde, B.J. (1997), "Supply Chain Management: Myth or Reality," *Supply Chain Management Review*, 1(1), 6–7.

Lee, H. L. (2002), "Aligning Supply Chain Strategies with Product Uncertainties," *California Management Review*, 44(3), 105–119.

Mason-Jones, R., Naylor, B., and Towill, D. R. (2000), "Engineering the Leagile Supply Chain," *International Journal of Agile Management Systems*, 2(1), 54–61.

McNeil, K. (2011), "The Importance of Differentiating Push and Pull Supply Chains for Creating Unique Business Models," http://www.advantageinternational.com.au/SUPPLY-CHAIN-MANAGEMENT/Getting-the-most-from-your-Supply-Chain/PUSH-VERSUS-PULL-SUPPLY-CHAINS.asp, retrieved on August 25, 2011.

Mentzer, J.T. and Williams, L.R. (2001), "The Role of Logistics Leverage in Marketing Strategy," *Journal of Marketing Channels*, 8(3/4), 29–47.

Murray, R.E. (1980), "Strategic Distribution Planning: Structuring the Plan," *Proceedings of the 8th Annual Conference of the National Council of Physical Distribution Management*, 220–221.

Porter, M.E. (1998), *Competitive Advantage: Creating and Sustaining Superior Performance*, New York, NY: Free Press.

Simchi-Levi, D., Kaminsky, P., and Simchi-Levi, E. (2008), *Designing and Managing the Supply Chain: Concepts, Strategies and Case Studies*, 3rd edition, New York, NY: McGraw-Hill.

Spearman, M.L. and Zazanis, M.A. (1992), "Push and Pull Production Systems: Issues and Comparisons," *Operations Research*, 40(3), 521–532.

Statistica (2014), "Market Share of PC Vendors of PC Shipments Worldwide from 2011 to 2014, by Quarter," *The Statistics Portal*, http://www.statista.com/statistics/269703/global-market-share-held-by-pc-vendors-since-the-1st-quarter-2009/, retrieved on October, 31, 2014.

Stock, J.R. and Lambert, D.M. (2001), *Strategic Logistics Management*, 4th edition, New York, NY: McGraw-Hill.

Vonderembse, M. A., Uppal, M., Huang, S. H., and Dismukes, J. P. (2006), "Designing Supply Chains: Towards Theory Development," *International Journal of Production Economics*, 100(2), 223–238.

Webster, S. (2008), *Principles and Tools for Supply Chain Management*, Boston, MA: McGraw-Hill Irwin.

Wheelwright, S.C. (1984), "Strategy, Management, and Strategic Planning Approaches," *Interfaces*, 14(1), 19–33.

3

Customer Service:
The Ultimate Goal of Supply Chain Management

"A service is not something that is built in a factory, shipped to a store, put on a shelf, and then taken home by a consumer. In short, a service is not a thing."
—G. Lynn Shostack, "Designing Services That Deliver," *Harvard Business Review*

Learning Objectives

After reading this chapter, you should be able to:

- Understand the basic elements of customer service and the role of customer service in strengthening the supply chain.
- Comprehend the differences between products and services.
- Build trust and subsequent long-term relationships with customers to win future business from loyal customers continuously.
- Develop the tools to measure service performance.
- Identify potential sources of service failures and prevent those failures.
- Find ways to improve customer service to stay competitive in the market.
- Understand the secrets behind a winning customer service strategy.

Understanding Customer Expectations and Perceptions

Without customers, businesses cease to exist. Therefore, a firm's ability to attract and retain customers often dictates its business success. More specifically, customer service has a direct impact on the firm's market share, revenue, profitability, and competitiveness. Customer service also has affected inventory carrying cost, because it sets the level of inventory that the firm should maintain to make a product available to customers whenever they want to buy it. Considering the profound impact of customer service on the firm's bottom line, concerted efforts throughout the supply chain should be made to improve the level of customer satisfaction. The improvement of customer satisfaction begins with the understanding of customer needs, preferences, and expectations. However, such understanding is very difficult to gain due to the elusive and changing nature of the customer mindset that shapes customer service. That is to say, customer service can be viewed differently by different customers during different time periods. Regardless, there are some similarities among all customers in conceptualizing customer service.

In a broad sense, customer service is referred to as all the activities occurring at the interface between the customer and the corporation that enhance or facilitate the sale and use of the corporation's products and services (La Londe and Zinszer, 1976). Put simply, customer service represents a chain of events that is in the business of keeping customers (Davis, 1971). Those events include the firm's efforts to make a product available to the customer when he or she needs it, to deliver the product in a timely manner, to inform the customer of the current order status, and to respond to the customer as quickly as possible if he or she complains. Therefore, customer service can be viewed as performance standards that meet customer expectations. Because customer expectations may change over time, customer service in the form of performance standards cannot provide an accurate picture of what it really is. Rather than listing merely what the firm should accomplish to keep its customers, customer service should reflect corporate philosophy that involves the order process and links the upstream supplier to the downstream customer with the goal of increasing future sales. From this perspective, customer service can be defined as "a customer-oriented corporate philosophy which integrates and manages all of the elements of the customer interface within a predetermined optimum cost-service mix" (La Londe and Zinszer, 1976). In the next section, such elements will be identified and explained in detail.

Customer Service Elements

As discussed earlier, different firms and industries attach varying degrees of significance to customer service and consequently include different lists of categories in conceptualizing customer service. For example, Simon (1965) has identified five elements essential to customer service: anticipation, accuracy in problem identification, completeness, responsiveness, and problem-solving efficiency. These elements merely represent part of the firm's capability to predict and handle customer complaints, but not ways to enhance product sales, customer loyalty, and product value. To broaden the scope of customer service, Hopkins and Bailey (1970) have incorporated the physical distribution perspective into a wide range of marketing perspectives. Their categories of customer service include the following:

- Pre-sales and post-sales distribution services
- Technical service
- Product maintenance and repair services
- Service support for distributors and dealers
- Management aids to industrial/business customers

Similarly, Levy (1981) has identified seven elements perceived to be important to customer service. These are general financial aspects of physical distribution management, the salesperson's responsibilities, promotion, general product information, package identification, inventory management, and general physical distribution management.

From a more comprehensive viewpoint that regards customer service as the combination of the firm's activities, performance standards, and philosophy, La Londe and Zinszer (1976) group the elements of customer service into three distinctive phases of the interface between the firm and its customers, as displayed by Figure 3.1:

- Pre-transaction elements
- Transaction elements
- Post-transaction elements

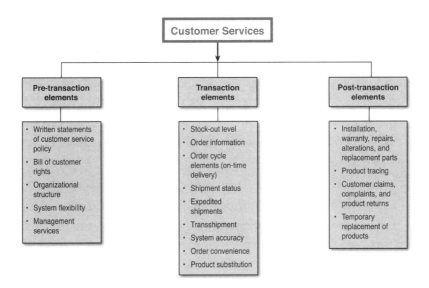

Figure 3.1. *Elements of customer service*

To elaborate, pre-transaction elements are characterized by nonroutine, policy-related activities that require management input due to their significant impact on product sales. The specific elements of pre-transaction customer service include the following:

- **A written statement of customer service policy**—This statement would be based on customer needs, service standards, and report channels in case of service failures. In the textile industry where one third of the firms surveyed by Davis (1971) did not have a written statement of customer service policy, neither management nor the customer had specific customer service guidelines to follow and therefore service problems were difficult to manage.
- **Provision of a written statement of customer service policy to the customer (bill of customer rights)**—Without allowing customers to be familiar with a firm's customer service policy, it would be difficult for customers to find recourse when the firm's service performance levels are not met. In other words, the customers should be notified of the detailed procedures of how they can communicate with the firm if service fails. Regardless, it was found that fewer than 10% of the firms made such statements available to their customers (Davis, 1971). Similarly, La Londe and Zinszer (1976) discovered that 16% to 25% of the firms across the industry had a written customer service statement and less than 12% of those firms provided such a statement to their customers.
- **Organizational structure**—Though no organizational structure is immune from the potential of service failure, an organizational structure must allow for proper communication and interfaces among functions involved in developing the customer service policy. In addition, both responsibility and authority should be assigned to those individuals who manage those functions. The names, contact numbers, and email addresses of those individuals should be given to customers for points of contact.
- **System flexibility**—The flexibility is necessary to react to customer complaints in case unplanned events such as inclement weather, labor strikes, earthquakes, fire, floods, and power failure abruptly disrupt the service process.
- **Management services**—Educational services for dealers and customers, such as the development of training manuals and seminars/workshops designed to improve customer service, can be considered pre-transaction elements of customer service.

Transaction elements are those activities that are most commonly associated with customer service. Therefore, when emphasis has been placed on customer service, transaction elements receive the most attention from management. They include the following:

- **Stockout level**—The stockout level is a measure of lost sales due to a low level of product availability. However, a single measure of stockouts can result in a distorted priority or misleading picture of the actual availability of critical inventory items (Ronen, 1982). Therefore, stockouts should be recorded by individual product and by customer and then should be monitored based on the priority of the product and customer.

- **Order information**—Order information involves the ability to provide the customer with fast and accurate information about inventory status, order status, expected shipping and delivery dates, and backorder status (Stock and Lambert, 2001). This information became increasingly important because firms have begun to substitute information for inventory. For example, grocery stores such as Walmart use a point-of-sale system to accurately predict future demand and keep the inventory minimum. Online communication with the customer under an electronic ordering system can further speed up the flow of order information to the customer and keep the customer informed in a timely manner.

- **Order cycle elements**—The order cycle is the total elapsed time from initiation of the order by the customer until the completion of customer delivery. The order cycle is typically composed of order communication, order entry, order processing, order picking and packing, and delivery. Although it is total order cycle time that concerns the customer, there may be elements of the order cycle that are more unpredictable but critical to the firm. For example, the early arrival of the product as a result of prompt delivery may lead to unnecessary inventory build-up for the customer. Therefore, reliable service is more appreciated by the customer than quick service, due to the necessity of building inventory to compensate for the unexpected variations of the order cycle time (Bowersox et al., 1968).

- **Expedited shipment**—In case of emergency, expedited shipments are needed to reduce transit time and the subsequent order cycle time. Because expedited shipments often require a premium (faster but more expensive) mode of transportation, such as air freight, it is important for the firm to determine which customers qualify for expedited status and which ones do not. Therefore, the service policy regarding the use of expedited shipments should be developed based on the customer's contribution to the firm's profitability.

- **Transshipment**—Transshipment can take place when one of the regional warehouses lacks certain stock that the customer wants to buy while other regional warehouses keep that stock at a high level. Also, in anticipation of customer demand, which varies from one market to another, transshipments can be made from one regional warehouse to another.

- **System accuracy**—System (or order) accuracy refers to the accuracy of quantities, prices, and product specifications ordered by the customer as well as billing. System error, such as order fulfillment error, can be costly because it results in reorder, reshipment, and refund. Therefore, the automation of system activities or electronic ordering systems should be considered to minimize the chance of system error.

- **Order convenience**—Order convenience refers to the degree of difficulty that a customer experiences when placing an order. Order inconvenience can be caused by administrative red tape, confusing order forms/procedures, and limited

payment options. With the advent of e-commerce, online order has been widely used by today's customers.
- **Product substitution**—When either a stockout or a product quality problem is experienced, the customer should be given an option of replacing the product ordered with an alternate product of comparable features or quality.

In contrast with pre-transaction and transaction elements, post-transaction elements of customer service focus on post-sales support. In general, as the cost of the product increases, the value placed on post-sales increases. For example, luxurious goods such as fancy cars often come with greater post-sales service. The post-sales service may be included as part of the purchase price. If post-sales service is unbundled, a customer can get any package of post-sales services as long as he or she is willing to pay for it. Service elements that can provide customers with peace of mind while they are using the products include the following:

- **Installation, warranty, alterations, repairs, and replacement parts**—Because these elements are crucial for product functions and maintenance, they can be a significant factor in the customer's purchase decision. The successful rendering of these elements of customer service rests on the following (La Londe and Zinszer, 1976):
    - Assistance in seeing that the product is functioning as expected upon initiation of usage by the customer
    - Availability of parts and repair equipment and personnel
    - Documentation support for the field force and accessibility to a supply of parts
    - An administrative function that validates warranties
- **Product tracing**—Product tracing is needed to keep track of products that can potentially harm their users or create potential environmental hazards. In the wake of a series of food poisoning incidents that killed some humans and pets, there is a growing need to trace and recall potentially dangerous products from the marketplace. The use of radio-frequency identification (RFID) tags can simplify the product-tracing process.
- **Customer claims, complaints, and product returns**—When customers are dissatisfied with a product they bought or with their post-sales service, they would like an open communication channel through which they can express their concerns about the product and return it for a substitution or refund. To handle customer claims, complaints, and returns in a timely manner, a specific corporate policy should be developed. Such a policy can help to reduce the recurrence of customer claims, complaints, and returns.
- **Temporary product replacement**—The temporary replacement of a product during repairs or while awaiting the arrival of a new purchase will enhance

post-sales service. In particular, if such a product happens to be a daily necessity, such as an automobile, washer, dryer, dishwasher, and cookware, its temporary replacement (e.g., a loaner car during auto repair) can significantly reduce inconvenience caused by product failure.

Building Customer Relationships

To keep customers, the firm should make customers trust its service capabilities and make them believe it can deliver its services as promised. Customer trust cannot be built without providing the customers with a satisfying service experience in their first transaction. Because customers are more likely to be satisfied when their needs and preferences are met, the firm should learn more about its customers' needs and preferences by communicating with them over the long term. Imagine walking into a local restaurant and the waitress says hello and calls you by your first name. The waitress remembers exactly what your favorite drink is, how your steak should be cooked, and which dessert you prefer. You would always appreciate such personalized service and are highly likely to return to the same restaurant in the future. Likewise, customer relationship management (CRM) is at the core of customer service.

In general, CRM is considered a business practice that is intended to improve service delivery, build social bonds with customers, and secure customer loyalty by nurturing long-term, mutually beneficial relationships with valued customers selected from a pool of more than a few customers (Min, 2006). As such, CRM often focuses on valued customers who have repeatedly purchased a great amount of products and remain committed to the particular company over an extended period of time. In general, the longer customers remained with a particular company, the more profitable they became to the company (Reichheld and Sasser, 1990; Lovelock and Wright, 2002). Because these valued customers are more profitable to serve than others, the company's customer retention efforts should be geared toward them. The first step in identifying these valued customers is to segment the customer population into loyal patrons and disloyal customers.

Customer Segmentation

Customer segmentation refers to the categorization of a heterogeneous customer base into a number of smaller, more homogeneous subgroups with similar needs, behaviors, and values (Zikmund et al., 2003). As summarized in Table 3.1, there are several ways to segment the customer base, including geographic, demographic, lifestyle, and behavioral. Geographic segmentation is based on the national, regional, and political boundaries that separate certain groups of potential customers from the others. For example, customers living in densely populated areas with high fuel prices and narrow streets such as Japan, Korea, and Taiwan tend to prefer small compact cars, whereas their counterparts living in sparsely populated areas with low fuel prices and wide roads such as the United States tend to prefer large gas-guzzlers. As such, different marketing strategies intended for different foreign markets should be developed to meet different customer needs. In addition to the geographical characteristics, demographic profiles are often used to segment the customer base. For example, high-income customers may prefer to drive high-end, luxurious cars such as Lexus, Acura, BMW, Mercedes, and Cadillac brands, whereas low-income customers cannot afford to buy those cars. Therefore, the sales and promotional efforts for luxury cars should be geared toward high-income customers. Also, the age of the customer may affect his or her preference. For instance, the Cadillac brand is known to be popular among senior citizens, but not among younger generations. The common demographic characteristics that could be the basis for customer segmentation are gender, age, income, educational level, marital status, and ethnicity/race.

Individual lifestyles and shopping behaviors may also be used to segment markets. For example, young married couples with dual incomes and no kid tend to spend more on furniture and nesting, whereas empty nesters tend to spend more on travel and continuing educational programs (Zikmund et al., 2003).

Table 3.1. Typical Ways to Segment Customer Markets

Segmenting Variable	Category
Geographic	• National • Regional • Climate • ZIP Code
Demographic	• Gender • Age • Income • Educational level • Marital status • Ethnicity/race

Segmenting Variable	Category
Lifestyle	• Activities (e.g., outdoor activities, sports, entertainment) • Interests (e.g., shopping, travel) • Opinions (e.g., political, social) • Values (e.g., religion, education) • Family life cycle (e.g., young single, young married, married with kids, empty nester, widow/widower)
Individual behavior	• Shopping channel (e.g., catalog, television, online) • Prestige (e.g., brand, monetary value, fashion) • Loyalty (e.g., repeat patronage) • Value (e.g., lowest price, advanced high technology, resale value) • Communication media (e.g., telephone, email, in-person)

Customer Acquisition

Establishing customer relationships begins with customer acquisition. Customer acquisition aims to attract new customers, build their trust in the company's product offerings and service capabilities, and bring back those customers for future business opportunities. Because it takes two to tango, the success of customer acquisition rests on the firm's ability to provide loyal customers with economic incentives and social bonds strong enough to keep them satisfied and to recover the cost of winning new customers. Examples of economic incentives that can be offered to new customers include affinity programs (e.g., university alumni association with member benefits), payment of commissions to loyal customers for their referrals, and promotional offers (e.g., discounts, coupons, gifts) for customers who have switched from the competition. For instance, long-distance call carriers such as AT&T often redeemed the first several months of telephone bills to attract their new customers. Similarly, H&R Block recently offered a 50% discount on their tax preparation fee for new customers who defected from other tax accounting firms. Many local banks offer goody bags with nominal gift items to attract new customers. Because the offering of such incentives can be costly, the firm should carefully discern high-value, prospective customers who are worthy of the investment. Although the extent of investment required to attract new customers may vary depending on the industry, the level of competition, and the contribution of customers to profitability, it is necessary for the firm to weigh the investment risk against the value of new customers.

Customer Retention

Whereas customer acquisition represents the firm's effort to establish customer relationships, customer retention represents the effort to enhance customer relationships.

In other words, customer retention aims to strengthen relationships with existing customers by increasing the value of customer loyalty to the firm. It is known to be true that relationships become more profitable to the firm over time because customers are less likely to switch to other firms once their bonds with the particular firm grow over time (Reichheld and Sasser, 1990; Bhattacharya, 1998). For example, a customer who has already accumulated substantial frequent flyer mileage with a particular airliner would be reluctant to buy a new airline ticket from another company. Similarly, as the length of the relationship increases, the customer loyalty to the firm is less likely to be affected by diminishing customer satisfaction (Gruen, 1995).

As discussed previously, a key to successful customer retention is the firm's ability to increase the level of social bonding or psychological attachment with the firm, thus leading to the customer's long-term commitment to the firm. This commitment can take three different forms: continuance, affective, and normative (Allen and Meyer, 1990). *Continuance* commitment is motivated by self-interest based on a customer's assessment of the cost of leaving the firm (Gruen, 2000). *Affective* commitment is based on the customer's overall positive feeling toward the firm (Allen and Meyer, 1990), and *normative* commitment is based on the customer's sense of felt obligation to the firm (Gruen, 2000). In addition to social bonding, offering financial incentives such as discounts, free product/service upgrades, and priority services to loyal customers is a popular means of retaining customers. Casinos providing free drinks and lodging to high rollers is considered such an example.

Another way to retain customers is to build structural, interactive relationships with loyal customers by creating an interactive information system that can help customers solve service problems, reinforce purchases, and obtain up-to-date information about new promotional offers and personalized benefits through email and websites (Zikmund et al., 2003).

Customer Communication and Education

Having the right information in a timely fashion, in the appropriate amount, and delivering it in the right style and at the right tempo are essential for maintaining healthy relationships with customers (Sisodia and Wolfe, 2000). The rationale being that a bond with customers can emerge as a result of exchanging knowledge or emotion with them (Storbacka and Lehtinen, 2001). This knowledge may involve contractual obligations (e.g., legal bonds), technological connections (e.g., the Internet network), and unique service offerings (e.g., ultra-high definition TV signals). Emotional bonds can be created when customers attach themselves to certain service providers by exclusively doing business with them for a long period of time. To effectively manage the exchange of knowledge and emotion with customers in the service encounter, the firm should establish a channel of communication with customers and inform them of ongoing service progress, new service offerings, and changes in customer service policy. For example, it has become a service norm that both FedEx and United Parcel Service (UPS) provide customers with information about the shipment status of parcels via their websites. L.L. Bean, which is regarded as one of the leading catalog/online sales outlets, ensures that 85% to 90% of phone calls made by its customers will be answered within 20 seconds. L.L. Bean also follows up with customers on product inquiries, delivery confirmation, product returns, and exchanges not long after selling products to its customers so that it can continue to interact with them and maintain its service credibility.

In particular, the availability of interactive communication media such as the Internet and email allows the firm to deliver real-time, personalized information to a large number of customers at a low cost. Leveraging such media, the firm can educate its current and prospective customers about the comparative value of its product/service offerings, product/service bundles, customer loyalty, and customer referrals. On the other hand, loyal customers can educate the firm about consumer behavior, future market trends, and the level of competition through their feedback to the firm. To summarize, the success of customer relationship management rests with how well the firm can maintain two-way communication channels with its customers.

Customer Selection

The strength and chemistry of customer relationships can vary from one customer to another due to differences in the customers' perceived value and finances. Relationships with some customers may be stronger and smoother than others. Therefore, certain customers' contribution to the company's profitability may be larger than others. If the value of a relationship with a certain customer is outweighed by the estimated loss of ending that relationship, the company may be better off by terminating the relationship and investing its energy and resources to strengthen the relationship with valued customers. The rule of thumb indicates that if the turnover rate of the customer base is above 20% (i.e., customers change every five years), a company would rather invest more heavily in generating new relationships while terminating some relationships than sustaining a relationship with every customer (Storbacka and Lehtinen, 2001). However, the company should realize that the time and cost involved in changing from one customer to another can be ten times as high as that involved in keeping a current customer (Hakansson and Snehota, 2000). Similarly, Gosney and Boehm (2000) have noted that the advertising cost for getting a new customer is five times more than that of advertising to retain existing customers. Therefore, a portfolio of customer relationships should be developed based on the careful estimation of customer profitability that weighs the asset aspect of customer retention against the liability aspect of customer retention while considering the opportunity cost of losing a current customer. In estimating customer profitability, the company must answer the following questions:

- How does the company determine profitability?
- What is the appropriate time horizon for assessing the long-term customer profitability?
- Which customers have the biggest profitability potential as compared to others?
- How much is the company dependent on valued customers who account for a majority of the company's profits? Or how is the profitability of the company distributed among customers?
- Which customer bases are eroding the company's profitability over the last few years?
- How should the company's resources be allocated among different customer groups (valued customers versus unimportant customers)?
- Which customer relationships should be terminated to prevent the recurring loss?

Notice that in certain industrial markets, 20% of the customers accounted for 225% of the total customer-base profitability (Cooper and Kaplan, 1991). For instance, in the banking industry, 20% of the bank customers accounted for between 130% and 200% of the total profits (Storbacka, 1994). Considering that the disproportional number of customers represents the major profit base for the company, customer selectivity makes economic sense.

A Measure of Customer Relationship Management (CRM) Success

To manage customer relationships properly, the company needs to know whether or not its relationships with its customers are on the right track and worthy of the further investment. The sound customer relationship would show a healthy return on investment (ROI) for both the customer and the service provider. Such ROI may stem from the increased revenue, increased customer participation in future promotions and sales through increased customer referrals, reduced customer turnover and defection rates, reduced cost of maintaining the relationship, or reduced cost of maintaining the relationship for the service provider. From the customer perspective, such ROI may include increased service support, more personalized benefits, and enhanced customer satisfaction. The inter-dynamics and potential benefits/costs of the customer relationship are summarized in Figure 3.2 (Zikmund et al.).

Once the costs and benefits of the customer relationship are identified and estimated, the ROI of the customer relationship can be measured, which can be a basis for justifying future investment in the customer relationship. Although the ROI is useful for measuring the financial success of CRM initiatives, it may not reflect the nonfinancial customer value, such as positive customer experience with service encounters. Therefore, a growing number of companies have begun to use the Customer Management Assessment Tool (CMAT), which is designed to measure the nonfinancial customer value created by CRM and then to compare the company's CRM progress with that of its competitors. CMAT is the leading CRM assessment tool for the company that wants to understand how well it is managing its customer relationships in comparison to its rivals and figure out how it can create more customer value from CRM (Woodcock et al., 2003).

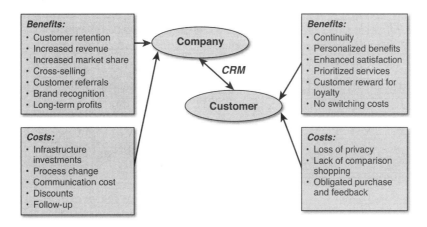

Figure 3.2. *Potential costs and benefits of the customer relationship*

Service Delivery Performance

An analysis of the gaps between the customer experience and the company's service process outcome is the key to monitoring the company's service delivery performance and managing customer service successfully (Czarnecki, 1999). Such gaps can be the main sources of customer dissatisfaction and characterized as service failures. To reduce or eliminate those gaps, the company should measure them relative to the company's own service standards and the performance of its competitors. In general, service performance measurement begins with a customer service audit and then the development of specific service standards, as displayed in Figure 3.3 (La Londe and Zinszer, 1976). This section will elaborate on service audit and service standard establishment processes that become the basis for defining service quality and developing the benchmarks.

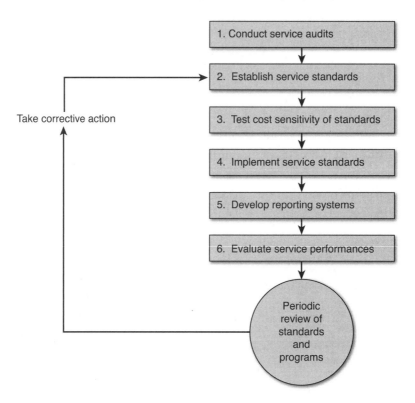

Figure 3.3. *The customer service measurement process*

Service Audits

The service audits may involve the evaluation of the current level of customer service from internal perspectives, the appraisal of the impact of customer service on the firm's market share and profitability, and the benchmark of customer service performance relative to the firm's competitors. Prior to evaluating the existing customer service, a customer service audit should begin with the development of clear customer service objectives, the formulation of service plans, and the establishment of reporting channels (Lambert and Lewis, 1980). A key to the success of service audits is the accuracy and timeliness of information obtained from customer feedback, employee suggestions, and external benchmarks. To ensure the accuracy and timeliness of such information, service auditors need to develop communication flow from customers to the company, within the company, and across the company on a periodic basis.

Service Standards

Customer service standards should reflect what the customer needs rather than what management thinks customers need. They should also provide an objective measure of service performance as well as management areas for corrective action (La Londe and Zinszer, 1976). Customer service standards can be established based on three perspectives, as shown in Figure 3.4 (Lambert and Lewis, 1980):

- In-stock level
- Transit time
- Order cycle consistency

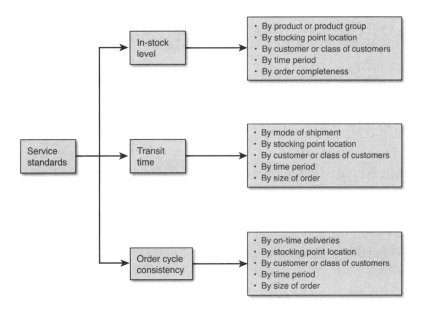

Figure 3.4. *The examples of customer service standards*

Service Quality

Put simply, service quality is serving the customer right the first time. More formally, service quality is defined as "error-free" service that consistently meets or exceeds the expectations and needs of the customers. The service quality comes from conformance to customer expectations that are cognitive descriptions of the quality in perception, economic quality ("affordability"), and availability. Because quality in perception represents the subjective feeling of customers, service quality may change over time and therefore is a very elusive concept. As such, some service providers may not be aware of the common needs of their customers. If they are, some customer needs may be outside the scope of service due to the economic burden of expanding the scope of service. Considering this dilemma, the definition of service quality should address the following questions:

- What will make customers happier?
- What shapes the customers' needs and expectations?
- What elements constitute service quality?
- How and when does the service provider ask for customer feedback?
- What does it take to win repeat customer business for the services?
- What is considered the industry norm for service quality?
- Where is the service failure most likely to occur?
- In what media and how often does the service provider communicate with customers to detect possible service problems?
- How can service quality be measured?

As shown in Figure 3.5, service quality is typically composed of three elements: service product, service environment, and service delivery, which may or may not involve a physical product (Rust and Oliver, 1994). The service product is the service as it is designed to be delivered (Rust and Oliver, 1994). For example, high-speed broadband Internet service that allows customers to browse and search through a variety of websites and communicate with their friends and families is a service product. The service environment includes the ambience, space, function, surrounding, and facility to facilitate and enrich the service delivery process. An example is a restaurant's dining tables, internal layout, design, and atmosphere, all aiming to enhance the dining experiences for customers. The service delivery is the way the service is presented to customers. For instance, the transportation of airline passengers to the destination of their choice, on time, and in a professional manner that conforms to their expectations illustrates service delivery.

Service Gaps

Due to the inherent difficulty in conceptualizing and evaluating service quality, it can fail. Service quality is often the indicator of differences between expectations and perceptions on the part of management, employees, and customers. Such differences are called "service gaps." According to Parasuraman et al. (1985), there are four possible service gaps (see Figure 3.6):

- **Gap 1**—The difference between what customers expect of a service and what management perceives customers expect

- **Gap 2**—The difference between what management perceives and customers expect and the quality specifications set for service delivery
- **Gap 3**—The difference between the quality specifications set for service delivery and the actual quality of that service delivery
- **Gap 4**—The difference between the actual quality of service delivery and the quality of that service delivery as described in the company's external communications

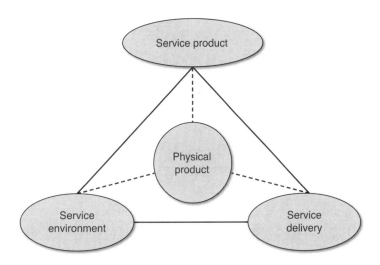

Figure 3.5. *Key components of service quality*

Gap 1 is the most common difference between what customers actually want and what managers think their customers want. For example, hotel guests may prefer a clean and quiet room over plush furniture and a fast broadband Internet connection, which hotel managers think hotel guests would appreciate (see, e.g., Min and Min, 1996). Gap 2 may occur when managers fail to develop the appropriate method of measuring service quality, because they do not know which specific elements service quality is composed of. Also, Gap 2 may be the result of lack of management commitment to service quality or lack of investment in service improvement efforts. For example, if transportation carriers do not specify what they mean by "on-time" delivery services to their customers, the assurance of on-time delivery to their customers would be meaningless. Gap 3 often occurs when there are variances in the skills, experiences, and aptitudes among the employees of service providers that lead to inconsistency. For example, a waitress in a restaurant may be very courteous to you on a certain day, but on your next visit another waitress may be somewhat rude and blunt. This is a reason why the amount of tip left by the restaurant's customers could vary from one visit to another. Gap 4 is often termed the "promise gap." It lies between what the service provider promises to deliver in its communications and what it actually does deliver to the customer (Bateson, 1992). For example, if the local pizza restaurant promises to deliver its freshly baked pizza to its customer within 15 minutes of ordering but ends up being half an hour late, the pizza restaurant has created gap 4. Another example of gap 4 is a broken promise made to the airline passenger whose seat was reserved in advance but who is bumped from the scheduled flight due to overbooking.

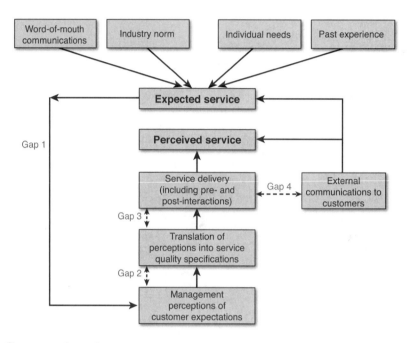

Figure 3.6. *Sources of service gaps*

Service Benchmarks

Customer service cannot be improved without being measured. Measurements are intended for comparisons that provide a snapshot of where the service provider stands vis-à-vis its competitors and industry norms. Therefore, the measurement of customer satisfaction with the service performance of the company can tell whether there is room for service improvement. Prior to measuring customer satisfaction, the company must answer the following questions (Myers, 1999):

- What are the objectives of the customer satisfaction measurement?
- Who will be surveyed? For instance, are both customers and employees of the service provider surveyed?
- What tools are needed for measurement?

- Where do service attributes come from? Do they come from focus groups of customers or the service provider's own marketing research?
- What type of measurement scale should be used?
- Should a customer satisfaction index be used? A customer satisfaction index (CSI) is a single number that management believes will tell how the service provider is doing as far as customer satisfaction is concerned (Chakrapani, 1998).

Because the ultimate goal of measuring customer satisfaction is to develop a service standard that can be the basis for continuous service improvement, the service provider needs to set a reliable service standard. A service benchmark represents such a standard.

In general, a service benchmark is the yardstick needed for a continuous service improvement process by which a service provider can assess its internal strengths and weaknesses, evaluate comparative advantages of leading competitors, identify the best practices of industry functional leaders, and incorporate these findings into a strategic action plan geared toward gaining a position of superiority (Min and Min, 1996). The establishment of a service benchmark typically consists of the following steps (Camp, 1989; Min and Min, 1996):

1. Identify and prioritize customer service attributes that influence the customer's perception of service quality.
2. Develop service metrics as performance standards.
3. Identify the best-practice service provider as a benchmark and compare its service performance with that of the benchmark using service gap analysis.
4. Develop a strategic action plan for continuous service improvement.

Formulating a Winning Customer Service Strategy in a Supply Chain

Customer service is an important prerequisite to customer satisfaction that leads to repeat business from the customer and subsequently increases sales revenue and market share. Increased sales revenue and market share impact the company's bottom line—namely, profitability. Considering the growing role of customer service in the company's profitability, customer service should be at the forefront of the company's business strategy. Regardless of the specific type and packages of services to be rendered to customers, a winning customer service strategy should include the following:

- Customers' perspective with respect to their needs, preferences, and judgment of the importance of certain service attributes or service bundles

- Systematic and objective measurement and monitoring of service performance relative to competitors
- Cost/benefit analysis (or return on investment) associated with service improvement initiatives
- Assurance of efficient and timely flow of information within the company, between the company and its customers, and across the supply chain (or among the partnering companies)

For example, in addition to offering the intangible facets of services such as comfort, relaxation, and pampering, some luxury hotels have attempted to make the unnoticed parts of its service package more visible by folding down the bed, in personalized welcome/thank-you notes, and gift-wrapped chocolates. This kind of strategy grabs the attention of customers instantly, even though it may not cost much for hotel management. More importantly, the note itself represents a line of communication that hotel management is trying to open to enhance customer contact, which in turn increases information flow between the hotel manager and the customer.

Generally speaking, a winning customer service strategy can be formulated by taking into account the following service dimensions:

- **Customer selectivity**—Some customers are more costly to service than others due to their inherent differences in character, nature, and special needs. For example, a healthcare insurance provider may shun smokers, obese people, senior citizens, and individuals genetically susceptible to diabetes and cancer, while targeting potential clients (e.g., nonsmokers) with low health risks. Likewise, some insurance companies may deny auto insurance to reckless drivers with a frequent accident history or may charge hefty insurance premiums to those drivers. Some third-party logistics providers may be reluctant to serve potential customers with less than a million-dollar account. Instead, they may want to focus on key customers with a lot of individual attention. Similarly, many airliners treat frequent flyers differently by prioritizing their boarding and upgrading their seat assignments.
- **Service mix/portfolio**—As discussed earlier, service quality is shaped not only by service products but also by other ingredients such as physical products, service environments, and service employees. Therefore, a service provider needs to determine which service mix best reflects the nature of the services it wants to provide in the particular business sector. For example, a coin-operated laundry would rather make an investment in upgrading washing machines and dryers than increasing the onsite assistance due to limited direct contact with its customers. On the other hand, tax consulting services and mortgage lending services, which require a high level of direct contact with their customers, may need to focus more on offering professional advice for each customer's unique financial

situation and giving personalized attention to the customer than on upgrading facility amenities.
- **Mass customization**—With the growing market base resultant from the globalization of business activities, customization on a mass scale is nothing new to many companies that offer a variety of service bundles. Ever since automobile manufacturers such as General Motors (GM) started offering a wide variety of car configurations (e.g., engine size, color, and other optional packages) to its customers, mass customization strategy has been applied to other sectors, including the service sector. For example, a regional cable operator often offers a wide choice of service bundles, such as access to basic television channels, premier movie channels, pay-per-view options, broadband Internet, caller identification, and long-distance telephone services, that cater to the different needs of customers at different prices.

Chapter Summary

The following summarizes key takeaways from this chapter:
- Customer service is a process for providing significant value-added benefits to the supply chain in a cost-effective way (La Londe et al., 1988). As such, customer service is a key to supply chain success. Nevertheless, many firms have taken customer service for granted and treated it as an after-thought that is needed to handle customer complaints. The lack of customer service awareness may stem from the difficulty in conceptualizing customer service. Indeed, the nature and scope of customer service may vary from one industry (or market) to another. Despite the different viewpoints, customer service can be shaped by the three stages of interaction between the company and its customer: pre-transaction, transaction, and post-transaction.
- La Londe et al. (1988) discovered that information has moved center stage in providing efficient and effective customer service. For example, many buyers would like to know order status information before the product is delivered. Likewise, many shippers are interested in knowing "real-time" shipment status during the delivery process. As customer demand for timely and accurate order status information increases, a growing number of companies have attempted to enhance their capability to communicate with their customers using electronic means such as email and the Internet. Thus, a growing number of companies began to view investment in information technology (IT) as a way to reach current and prospective customers and subsequently enhance their customer service. Regardless of IT infrastructure and expertise, the backing of supply chain partners for information sharing is essential for an uninterrupted information flow between the company and its customers.

- Without continuous customer feedback, a service provider will run into difficulty in identifying actual customer needs and preferences and therefore will fail to satisfy its customers. To solicit customer feedback, the service provider needs to develop a rapport with both existing and potential customers. That is why customer relationship management (CRM) has become an important part of customer service improvement.

Study Questions

The following are important thought starters for class discussions and exercises:
1. What are the differences between products and services?
2. How is customer service defined?
3. What constitutes customer service elements?
4. Illustrate a written customer service policy in the following service sectors:
   a. Supermarkets
   b. Post offices
   c. Banks
   d. Hospitals
   e. Fast-food restaurants
   f. Appliance stores
   g. Mass merchants
5. What is the role of customer service in creating and adding value for the supply chain? What are the expected payoffs from improving customer service?
6. Think of any service provider (e.g., McDonalds, Pizza Hut, Best Buy, U.S. Postal Service, Jiffy Lube) that you are familiar with and then graphically display their order cycle elements.
7. What is customer relationship management (CRM)? How can we measure the CRM success?
8. The firm's employee relationship often mirrors its customer relationship. How can we motivate employees to serve their customers better?
9. How would you identify and select "valued" customers? What are typical traits of valued customers?
10. What are segmenting variables for customer segmentation?
11. What are the main steps to take for service audits?

12. Think of a service bundle you have recently purchased from your local cable operator and determine whether you are satisfied with the given choices and whether such a choice is worth the money you have spent?
13. What are the key components of service quality? Can you assess their contribution to the improvement of service quality?
14. What are the main sources of service failures?
15. Explain the key differences among the four distinctive types of service gaps.
16. What are the main issues for measuring customer service?
17. How can you set service benchmarks? How can you identify your company's major competitors in the market? What are the greatest challenges for obtaining information regarding the service performance of your company's competitors?

Case: Shiny Glass, Inc.

This case was written based on a hypothetical scenario.

It was 7:00 a.m. on a Monday morning. George Einstein, a sales executive of Shiny Glass, Inc., sipped a latte in his roomy office overlooking Toledo's Maumee River and pondered what went wrong with his sales strategy. He was concerned about a growing problem with lowered sales expectations and a rapidly declining market share. Before reporting to his company's board about his sales performance, he was aware that something had to be done quickly to reverse recent sales trends. George acknowledged that sales were down due in part to a sluggish economy and increasing competition from cheap imported products from Asia. He was also bothered by a growing number of complaints from his major customers consisting of well-known retail chains such as Bed Bath & Beyond, Macy's, and J.C. Penney as well as leading food and beverage manufacturers, including Anheuser-Busch InBev, Brown-Forman, SABMiller, Bacardi, Tsingtao, PepsiCo, and Coca-Cola. Those complaints mostly stemmed from erratic delivery, frequent order fulfillment errors, and inconsistent post-sales services. These prompted George to revisit all of his company's order processing systems and customer service practices.

Logistics Network

Shiny Glass is a multinational firm (MNF) engaged principally in the manufacture, distribution, and sale of glassware, including wine glasses, pitchers, brandy snifters, and tumblers. Shiny Glass also started producing some medical devices such as catheters, glass hypodermic syringes, and optical glasses. The company operates 32 manufacturing plants and processing facilities across the globe, 15 of which are located in the United States (including one in Toledo, Ohio—the so called "Glass City"). Among three market segments—specialty markets, healthcare, and consumer products—that Shiny Glass is serving, its consumer products division represents 76% of company sales, with a whopping $1.2 billion in total sales for the past year.

Considering the important role of consumer products, Shiny Glass established its distribution center (DC) in Cincinnati, Ohio, where consolidation, packaging, labeling, and shipment of consumer products take place for worldwide distribution operations. Although Cincinnati looked to be an ideal location because of its proximity to Shiny Glass's headquarters in Toledo, it poses logistical challenges for a group of Shiny Glass's largest customers currently located in the Southeast U.S., Mexico, and Southern China, thereby adversely impacting its delivery services. The logistics problem is further compounded by the recent increase in the off-shore bottling of glass containers in China, Malaysia, Vietnam, and Australia. To ease the financial burden, both inbound and outbound shipments of the Cincinnati DC are performed mainly by either contract or common carriers rather than using the company's own private carriers. Shiny Glass is also experimenting with an optional plan to allow smaller and nearby customers to pick up their own orders at the Cincinnati DC. This experiment is gaining popularity for some customers looking for the deep discount that comes with this plan. This experiment, however, has frequently congested the company's loading docks, thereby further delaying outbound shipments. Seventy-two percent of all outbound shipments are made in volume-type (e.g., full-truckload) shipments, which takes more time to accumulate the orders and extends order cycle time.

The Antiquated Order Processing System

Shiny Glass's current order processing system handles most of the customer orders manually, although it was developed using Visual Basic for the front end and SQL (Structured Query Language) as the back end. A dozen customer service representatives (CSRs), who are not familiar with the computerized system and software, take 70% of customer orders through phone calls, faxes, or email from customers. Once orders are received, the order data is entered into a mainframe computer system, one by one. Approximately three hours after the customer order data is entered into the computer system, the data is checked for accuracy. If errors are found, a report is generated in such a way that it triggers reminder messages and "red" alerts for the sales department. Errors are commonly caused by human (e.g., CSR) error, data entry error, and miscommunication with customers. These errors can occur when the wrong information—such as an incorrect price, quantity, delivery schedule, or product code—appears on the order form without any follow-up or validation. Once an order is transmitted to the traffic department with the necessary packing and shipping instructions, it is shipped to customer locations immediately.

Even though not all orders are validated, the current order processing system is believed to work moderately well because the CSR thinks certain types of errors are unavoidable under any circumstances. However, George was somewhat surprised to discover that over the past three years, his company's record showed an order accuracy of 91.5%, which is below the industry average of 95%. Order accuracy represents the ratio of the number of error-free orders over the total orders (in dollar volume) shipped to customers in a given year. More often than not, order accuracy is checked by the sales department after the customer either complains about receiving a wrong product or threatens to file a claim. In other words, customer acknowledgement of receipt of orders without complaints is considered the evidence of error-free order fulfillment. The order fulfillment process described here is crucial for customer service, but is very time-consuming and labor intensive. As a cost center, order fulfillment and replenishment typically accounts for 50% to 65% of the sales department's personnel expenses (labor cost).

While carefully reviewing the current order processing system, George also suspects that Shiny Glass's adaptation of the new Omni-channel sales strategy may have something to do with order fulfillment errors. In an effort to recapture and expand customer bases, Shiny Glass began to sell some of its popular products such as beer mugs through its own discount outlet stores, catalog and mail-order companies, grocery chains, TV home shopping networks, and e-tailers such as Amazon and Overstock.com. Each of these sales channels requires a unique way of processing orders, distributing ordered products, and billing orders and thus creates unprecedented complexity associated with different types of customers and their service needs. To make matters worse, the aforementioned multichannel sales tend to result in the proliferation of many different products intended for different channels/segments, which causes another difficulty in forecasting demand. The impaired supply chain visibility resulting from forecasting difficulties contributed to frequent stockouts of popular items, which also takes a toll on customer service.

Discussion Questions

1. Do you think Shiny Glass would benefit from a new order processing system? If so, why? What is wrong with the current order processing system?
2. What is your assessment of Shiny Glass's customer service levels? If not positive, what may be the source of their current service problems? What kind of service gaps does Shiny Glass have?

3. What kind of recommendations would you make to improve Shiny Glass's supply chain operations that will eventually help improve its customer service?

Bibliography

Allen, N.J. and Meyer, J.P. (1990), "The Measurement and Antecedents of Effective, Continuance and Normative Commitment to the Organization," *Journal of Occupational Psychology*, 63, 1–18.

Bateson, J.E.G. (1992), *Managing Services Marketing: Text and Readings*, 2nd edition, Fort Worth, TX: The Dryden Press.

Bhattacharya, C.B. (1998), "When Customers Are Members: Customer Retention in Paid Membership Contexts," *Journal of the Academy of Marketing Science*, 26, 31–44.

Bowersox, D.J., Smykay, E.W. and La Londe, B.J. (1968), *Physical Distribution Management*, New York, NY: The McMillan Co.

Camp, R.C. (1989), *Benchmarking: The Search for Industry Best Practices That Lead to Superior Performance*, Milwaukee, WI: ASQC Quality Press.

Chakrapani, C. (1998), *How to Measure Service Quality and Customer Satisfaction: The Informal Field Guide for Tools and Techniques*, Chicago, IL: American Marketing Association.

Cooper, R. and Kaplan, R.S. (1991), "Profit Priorities from Activity-based Costing," *Harvard Business Review*, 74(3), 88–97.

Czarnecki, M.T. (1999), *Managing by Measuring: How to Improve Your Organization's Performance through Effective Benchmarking*, New York, NY: American Management Association.

Davis, H.W. (1971), "Four Reasons Why Customer Service Managers Can't Manage Customer Service," *Handling and Shipping*, November, 51–53.

Gosney, J.W. and Boehm, T.P (2000), *Customer Relationship Management Essentials*, Rockling, CA: Prima Publishing.

Gruen, T.W. (1995), "The Outcome Set of Relationship Marketing in Consumer Markets," *International Business Review*, 4, 447–469.

Gruen, T.W. (2000), "Membership Customers and Relationship Marketing," *Handbook of Relationship Marketing*, edited by J.N. Sheth and A. Parvatiyar, 355–380, Thousand Oaks, CA: Sage Publications, Inc.

Hakansson, H. and Snehota, I.J. (2000), "The IMP Perspective: Assets and Liabilities of Business Relationships," *Handbook of Relationship Marketing*, edited by J.N. Sheth and A. Parvatiyar, 69–93, Thousand Oaks, CA: Sage Publications, Inc.

Hopkins, D.S. and Bailey, E.L. (1970), *Customer Service: A Product Report*, New York, NY: The Conference Board.

La Londe, B.J., Cooper, M.C. and Noordewier, T.G. (1988), *Customer Service: Management Perspective*, Oak Brook, IL: The Council of Logistics Management.

La Londe, B.J. and Zinszer, P.H. (1976), *Customer Service: Meaning and Measurement*, Chicago, IL: National Council of Physical Distribution Management.

Lambert, D.M. and Lewis, C.M. (1980), "Meaning, Measurement and Implementation of Customer Service," *Proceedings of the Annual Conference of the National Council of Physical Distribution Management*, pp. 524–602.

Levy, M. (1981), "Toward an Optimal Customer Service Package," *Journal of Business Logistics*, 2(2), 87–110.

Lovelock, C. and Wright, L. (2002), *Principles of Service Marketing and Management*, 2nd edition, Upper Saddle River, NJ: Prentice Hall.

Min, H. (2006), "Developing the Profiles of Supermarket Customers Through Data Mining," *The Service Industries Journal*, 26(7), 1–17.

Min, H. and Min, H. (1996), "Competitive Benchmarking of Korean Luxury Hotels Using the Analytic Hierarchy Process and Competitive Gap Analysis," *Journal of Services Marketing*, 10(3), 58–72.

Myers, J.H. (1999), *Measuring Customer Satisfaction: Hot Buttons and Other Measurement Issues*, Chicago, IL: American Marketing Association.

Parasuraman, A., Zeithaml, V.A. and Berry, L.L. (1985), "A Conceptual Model of Service Quality for Future Research," *Journal of Marketing*, 49, 141–150.

Reichheld, F.F. and Sasser, Jr., W.E. (1990), "Zero Defections: Quality Comes to Services," *Harvard Business Review*, 69(1), 105–111.

Ronen, D. (1982), "Measures of Product Availability," *Journal of Business Logistics*, 3(1), 45–58.

Rust, R.T. and Oliver, R.L. (1994), "Insights and Managerial Implications from the Frontier," *Service Quality: New Directions in Theory and Practice,* edited by R.L. Rust and R.L. Oliver, Thousand Oaks, CA: Sage Publications.

Simon, L.S. (1965), "Measuring the Market Impact of Technical Services," *Journal of Marketing Research*, 2(1), 32–39.

Sisodia, R.S. and Wolfe, D.B. (2000), "Its Role in Building, Maintaining, and Enhancing Relationships," *Handbook of Relationship Marketing*, edited by J.N. Sheth and A. Parvatiyar, 525–563, Thousand Oaks, CA: Sage Publications, Inc.

Stock, J.R. and Lambert, D.M. (2001), *Strategic Logistics Management*, 4th edition, New York, NY: McGraw-Hill.

Storbacka, K. (1994), *The Nature of Customer Relationship Profitability: Analysis of Relationships and Customer Bases in Retail Banking*, Unpublished Research Report, Helsinki, Finland: Swedish School of Economics and Business Administration.

Storbacka, K. and Lehtinen, J.R. (2001), *Customer Relationship Management: Creating Competitive Advantage through Win-Win Relationship Strategies*, Singapore: McGraw-Hill Book Co.

Woodcock, N., Starkey, M. and Stone, M. (2003), "What is CMAT?" *The Customer Management Scorecard: Managing CRM for Profit*, edited by N. Woodcock, M. Stone and B. Foss, 11–20, London, United Kingdom: Kogan Page Limited.

Zikmund, W.G., McLeod, Jr., R. and Gilbert, F.W. (2003), *Customer Relationship Management: Integrating Marketing Strategy and Information Technology*, Hoboken, NJ: John Wiley & Sons, Inc.

4

Demand Planning and Forecasting

"If a man gives no thought about what is distant, he will find sorrow near at hand."
—Confucius

Learning Objectives

After reading this chapter, you should be able to:

- Understand various sources of demand and their impact on supply chain planning.
- Follow the steps involved in the demand management processes.
- Comprehend the differences among various types of forecasting methods
- Understand the underlying factors influencing the forecasting selection decision, and select the right forecasting method for a given organization and situation.
- Generate time series forecasts using exponential smoothing, moving average, and trend analyses.
- Recognize the connection between sales and operational planning and demand planning.
- Relate the collaborative commerce to supply chain management and understand its role in demand management.
- Understand the ways collaborative planning, forecasting, and replenishment work and evolve.
- Identify potential sources of the bullwhip effect and prevent the bullwhip effect.

Demand Management

Demand reflects what and how much customers want. Therefore, demand drives supply and subsequently production plans, which in turn dictate the company's financial, logistics, and marketing plans. Inaccurate demand information is the main source of business failures because it leads to insufficient supply, which makes customers unhappy, or excessive supply, which incurs the waste of valuable resources. However, accurate demand information is difficult to come by due to the volatile and uncertain nature of demand. A typical effort to obtain demand information involves predicting future demand ahead of time based on the past patterns of demand. Although demand forecasting is an important part of managing demand, its reliability often varies depending on a particular choice of forecasting methods, the length of time horizons, and the nature of demand. More importantly, if accurate demand information is not communicated back to manufacturers/service providers and their suppliers, it is meaningless to forecast. Also, the speed of communicating demand information among supply chain partners can impact the way demand should be managed. With this in mind, demand management may take the following four steps, as shown in Figure 4.1 (Crum and Palmatier, 2003, p. 11):

1. **Planning demand**—Involves more than just forecasting
2. **Communicating demand**—includes communicating the demand plan to supply chain partners across the entire supply chain
3. **Influencing demand**—Includes marketing and promotion plans, product positioning, and pricing
4. **Prioritizing demand**—Includes customer order management and customer profiling

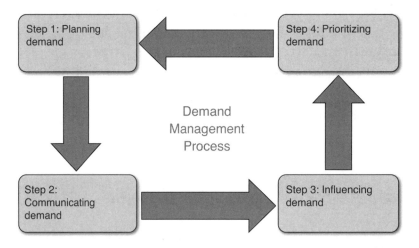

Figure 4.1. *Demand management processes*

These steps are discussed in detail in the following subsections.

Demand Planning

Demand planning refers to preparations for future demand that determine how much production and financial capacity are needed to meet such demand. Demand planning includes demand forecasting throughout the supply chain. Because demand planning requires up-to-date information about market dynamics, trends, and customer behavior, it relies heavily on marketing and sales data. For instance, a growing number of companies such as Walmart and Target use point-of-sales (POS) data to keep track of sales history and then customer demand patterns. Once marketing, sales, and customer data is fed into a forecasting system, the forecasting system should estimate demand in the future horizon and then become the basis of demand planning. Prior to finalizing a demand plan, the company needs to check whether such a plan is consistent with its long-term business strategy and make sure the plan aligns with its primary missions, as shown in Figure 4.2 (Crum and Palmatier, 2003, p. 29). For example, a company whose core competency is the production and distribution of high-end, niche-oriented cosmetics targeting rich female professionals may not want to enter a new market targeting budget-conscious customers simply based on demand projections.

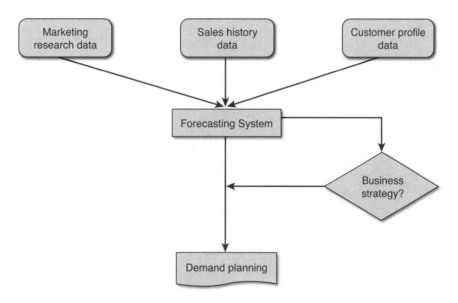

Figure 4.2. *Demand planning interfaces*

Communicating Demand

Accurate demand information is not worth a hill of beans if it is not communicated to the affected party in a timely manner. Even the unexpected cancellation of an order should be communicated to production planners and their suppliers. Likewise, the unexpected delay of a shipment should be communicated to the customer. Considering the random nature of demand, frequent changes in demand are not uncommon. Therefore, those changes need to be incorporated into the demand planning process. Nevertheless, the communication breakdown can occur for a number of reasons:

- A fear of delivering bad news to supply chain partners
- A fear of disseminating propriety information due to a lack of trust among supply chain partners
- An absence of managers who are responsible for managing demand information
- Multiple channels of communication (or multiple points of contacts) that lengthen and complicate the process of transmitting demand information to the affected supply chain partners
- Less efficient or poor means of communication (such as phone calls and faxes) that are less reliable and more time consuming than electronic transmission such as email and electronic data interchange (EDI)

To avoid potential communication breakdowns resultant from the aforementioned causes, the company may consider the following principles:

- A structured communication process should be developed to ensure that changes in demand are communicated in a timely manner and with sufficient details.
- A feedback mechanism that can acknowledge receipt of demand information should be established prior to opening a communication channel among supply chain partners.
- The demand information should be updated on a periodic basis, and the updated information should be communicated continuously to the effected parties.
- Individuals responsible for managing demand information and providing or receiving feedback should be designated as points of contacts. Each of these individuals should represent a single point of contact in his or her unit.
- Faster and more reliable means of communication such as email and EDI should be available to all supply chain partners with uniform standards.
- The primary means of communication should be backed up by a secondary means of communication in case of communication difficulties resultant from an unexpected system failure such as power failure, hacking, or computer downtime.

Influencing Demand

With the plethora of products available in the market today, customers have a wide variety of choices for their purchases. This means that, regardless of their appeal, many products and services will go unsold. In particular, new products introduced in the market are unlikely to receive the attention they need, unless the company influences customers to buy those products. Such influence includes incentives and extra rewards for customers who buy the company's products. Examples include promotional discounts for a limited period of time, coupons, introductory rewards (e.g., additional mileage for those customers who open a new credit card), and complimentary products (e.g., free toothpaste that comes with the purchase of a new bottle of mouthwash). Also, aggressive sales and marketing campaigns through various advertising media will help to promote unknown products and consequently influence demand.

The main purpose of influencing demand is to convince customers to buy enough products and services in such a way that the company can successfully achieve its sales and revenue goals. The process of influencing takes the four basic steps summarized in Table 4.1 (Crum and Palmatier, 2003, p. 63).

Table 4.1. Demand Influencing Process

Plan	Do	Check	Act
• Develop marketing, branding, and sales strategies and tactics. • Plan expected outcomes of strategies and tactics to create or add demand. • Input expected outcomes into the demand plan. • Review the financial feasibility of the demand plan. • Synchronize and integrate the demand plan across the supply chain. • Reach consensus and get approval for the demand plan from multiple stakeholders to execute the marketing, branding, and sales strategies and tactics.	• Execute marketing and branding activities for customer value and demand creation. • Execute the sales activities for retaining existing customers and attracting new customers.	• Measure the performance results of the marketing efforts, branding, and sales activities with respect to increases in market share, sales volume, revenue, and profits. • Develop customer feedback mechanisms and document customer responses to marketing, branding, and sales activities.	• Modify and adjust the marketing, branding, and sales plans based on performance results and customer feedback.

Prioritizing Demand

Prioritizing demand is needed when customers order more than anticipated, the delivery schedule of ordered products is tight, or the current production capacity of the company is too limited to meet the unexpected demand surge. There are a number of ways to cope with excessive demand that the company cannot accommodate given the limited production capacity and time frame. These may include the following (Crum and Palmatier, 2003):

- Give the customer the option of receiving the unplanned order later than originally requested.
- Accept the unplanned order and delay other customer orders, thereby increasing lead time or backlog.
- Encourage customers to buy alternative products with functions, features, and prices comparable to the products originally ordered.
- Decline the additional demand, especially if that demand represents non-core products or products with low profit margin.

Regardless of these options, prioritizing demand is a risky practice because it leads to deteriorating customer service and potential loss of sales/customers. Given its impact on customer services, prioritizing demand requires prioritizing customers. Customer priority can be set based on the customer's contribution to profitability, patronage, future business opportunities, partnership strengths, and the extent of opportunity loss.

Demand Forecasting

Demand forecasting is an essential tool for any kind of business activities. Without it, it is impossible to make a demand plan, which in turn influences a series of business plans, including marketing, sales, production, logistics, and financial plans. Therefore, the accuracy and timeliness of demand forecasting will dictate the efficiency and effectiveness of business plans and the subsequent business success. For example, the digital video recorder (DVR) pioneer TiVo, which was once a dominant market leader with nearly 4.4 million customers, stumbled and lost 145,000 subscribers after it failed to predict the explosive growth of the high-definition television (HDTV) market. Because

HDTV makes the traditional DVR without an HD function obsolete, TiVo struggled to increase its subscriptions. As a matter of fact, in 2007, TiVo ended up losing 145,000 subscribers and took an $11.2 million write-down for its leftover DVRs without HD features. This forecasting failure contributed to a 4.2% decline in its stock price and a loss of $17.7 million in revenue (Lieberman, 2007). When TiVo attempted to rebound from this mistake by introducing its own HD DVR to the market, it was too late.

Despite the significance of demand forecasting to business success, demand forecasting merely represents a "scientific estimate/conjecture" of future events and provides a means of gauging the "probable" direction of business activities. The detailed steps of demand forecasting are described in Figure 4.3 (DeLurgio, 1998, p. 20).

Regardless of scientific rigor, demand forecasting cannot be 100% accurate because it is often predicated on the premise that past trends will continue in the future. By analogy, you will get into an accident or become lost if you drive a car simply based on what you can see in the rearview mirror. One way to mitigate the potential risk of using demand forecasting is to select the right forecasting method under the right circumstances, or you might consider using multiple forecasting methods that complement one another. With that in mind, the following subsections discuss the pros and cons of some popular forecasting methods as well as the various factors (such as time horizon, decision hierarchy, sophistication, and the degree of reliance on historical data) that influence the accuracy of such forecasting methods.

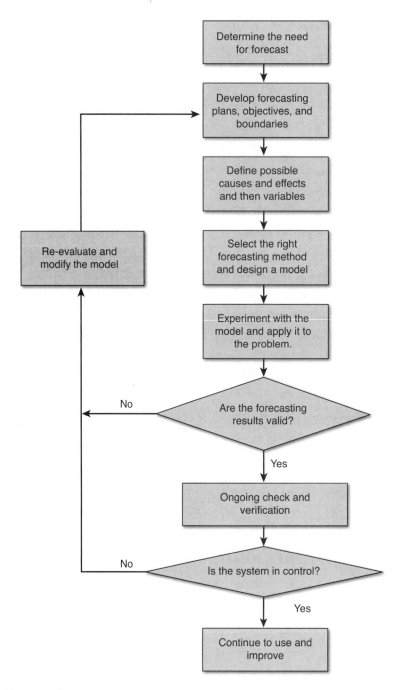

Figure 4.3. *Demand forecasting processes*

Exponential Smoothing

Exponential smoothing is a popular forecasting technique that uses a weighted average of past data to make a prediction of future events. Under the premise that the most recent past is more indicative of the future than what happened in the distant past, exponential smoothing gives greater weight to more recent data than the data in the more distant past. Exponential smoothing is intended for a short-term forecast with the presence of random patterns but no seasonal fluctuation (Shim, 2000). It is widely used to predict inventory levels or future sales in the retail sector. Its advantages include ease of use, the limited use of historical data, and relative accuracy for short-term forecasts. However, it tends to lag during a positive trend while overshooting the actual values during a negative trend. Also, it cannot take into account exogenous factors such as market conditions, price changes, and competitors' reactions when forecasting (Shim, 2000; Collier and Evans, 2007).

The basic exponential smoothing method is mathematically expressed as

$$F_{t+1} = \alpha A_t + (1-\alpha)F_t = F_t + \alpha(A_t - F_t) \tag{4.1}$$

where

F_{t+1} = the forecast for the next period, $t+1$

F_t = the weighted average of the forecast made for the prior period, t

A_t = the actual observation made in the prior period, t

α = smoothing constant ($0 \leq \alpha \leq 1$)

Here, a smoothing constant refers to a desired response rate that determines the level of smoothing and the speed of reaction to the differences between the forecast and the actual data (Chase et al., 2006). A smoothing constant is often adjusted by considering the nature of products and the degree of fluctuations in the past data series.

For example, let's suppose we want to predict the future sales of smartphones for a local retailer and have collected the monthly sales data contained in Table 4.2. Using the data summarized in the second column of the table, we generated the forecasting results in the third column. For simple illustrative purposes, we used $\alpha = .1$ initially and assumed that the average monthly sales volume for the previous year was 421. That is to say, the January forecast was 421. To begin with the initial series of forecasts, we calculated the February and March sales projections (F_2, F_3), respectively, using equation 4.1, as shown here:

February forecast = (0.1 × 429) + (0.9 × 421) = 422 items (rounded off)

March forecast = (0.1 × 510) + (0.9 × 422) = 431 (rounded up)

Likewise, we calculated the rest of the months' sales projections, where each month's forecast is used in the calculation of the next month's forecast. To see how the forecast results change, we experimented with α = .5. As Table 4.2 shows, forecasting results with α = .5 tended to produce better forecasts with smaller forecast errors than the results using α = .1 because the former better reflects growing sales patterns in the recent past with a heavier weight (emphasis) on the latest sales history. Regardless of the α value, however, exponential smoothing forecasting results tend to lag behind actual sales data. As such, exponential smoothing will lead to inaccurate forecasts unless actual sales volume remains constant and stable throughout the year. In other words, simple exponential smoothing is not suitable for forecasting product sales or inventory levels with seasonal variations or volatile demand fluctuations.

Table 4.2. The Results of Exponential Smoothing Forecasts for Smartphones

t = Period	A_t = Sales volume (in unit)	F_t = Sales projections when $\alpha = 0.1$	$A_t F_t$ = Forecast error	F_t = Sales projections when $\alpha = 0.5$	$A_t F_t$ = Forecast error
January	429	421	8	421	8
February	510	422	88	425	85
March	606	431	175	468	138
April	700	449	251	537	163
May	757	474	283	619	138
June	847	502	345	688	159
July	956	547	409	768	188
August	1,043	588	460	862	181
September	1,164	634	530	953	211
October	1,297	687	610	1,059	238
November	1,340	748	592	1,178	162
December	1,372	807	565	1,259	113

Moving Average

Moving average is a forecasting technique in which a consecutive number of the most recent observations are averaged and weighted either equally (simple moving average) or differently (weighted moving average). It is often suitable for a short-term forecast when past data exhibits a level pattern. In the moving average, the smaller the number of observations, the more responsive the forecast is to changes in past data. Also, moving average tends to respond highly to random changes and subsequently leads to huge errors (Reid and Sanders, 2005). Therefore, the forecasting result of the moving average may be sensitive to the number of observations chosen for the forecast.

The formula for a weighted moving average is mathematically expressed as

$$F_{t+1} = \sum_{t=1}^{n} w_t A_t = w_1 A_1 + w_2 A_2 + \ldots + w_n A_n \quad (4.2)$$

where

F_{t+1} = the forecast for next period, $t+1$

w_t = the weight placed on the actual observation in period t ($\sum_{t=1}^{n} w_t = 1$)

A_t = actual observation in the prior period, t

n = the total number of observations in the forecast

For the sake of simple illustration, let's suppose that the local small airport authority would like to predict the number of commuter airline passengers on a quarterly basis, with equal weight given to each quarter (i.e., $w_1 = 0.25$, $w_2 = 0.25$, $w_3 = 0.25$, $w_4 = 0.25$). Table 4.3 presents the actual passenger data for the last five years and summarizes the results of a quarterly (four-period) moving average. To begin with, the initial series of a four-quarter moving average (F_5) was calculated as follows:

(Four quarter moving total/4) = (878 + 1005 + 1173 + 883)/4 = 3939/4 = 985 passengers

The next quarter moving average was calculated by repeating the same process. As Table 4.3 shows, the forecast error varies from one quarter to another. It ranges from -2 to 376. In particular, a simple moving average with equation weight given to each previous period cannot capture dramatic variations of passenger numbers. Although heavier weight can be given to more recent data using a weighted moving average, this forecasting method is not effective in capturing fluctuating patterns of data such as a seasonal increase in the number of airline passengers due to weather and vacation schedules.

Table 4.3. The Forecasting Results of Airline Passengers Using Moving Average

Year	Quarter	A_t = Actual number of airline passengers	Four quarter moving total	F_t = Four quarter moving average	$A_t - F_t$ = Forecast error
2010	1	878			
2010	2	1,005			
2010	3	1,173			
2010	4	883	3,939		
2011	1	972	4,033	985	-13
2011	2	1,125	4,153	1,008	117
2011	3	1,136	4,116	1,038	-2
2011	4	988	4,221	1,029	-41
2012	1	1,020	4,269	1,055	-35

Year	Quarter	A_t = Actual number of airline passengers	Four quarter moving total	F_t = Four quarter moving average	$A_t - F_t$ = Forecast error
2012	2	1,146	4,290	1,067	79
2012	3	1,400	4,554	1,072	328
2012	4	1,006	4,572	1,139	-133
2013	1	1,108	4,660	1,143	-35
2013	2	1,288	4,802	1,165	123
2013	3	1,570	4,972	1,201	369
2013	4	1,174	5,140	1,243	-69
2014	1	1,227	5,259	1,285	-58
2014	2	1,468	5,439	1,315	153
2014	3	1,736	5,605	1,360	376
2014	4	1,283	5,714	1,401	-118

Trend Analysis

Trend analysis is a long-term time series technique that examines a long-term direction (trend) of a series of past data to project such a direction into the future. The direction can be linear (straight line) or nonlinear (curve) depending on the past pattern. In the case of a linear trend, its slope and intercept can be computed using the least squares regression method, as specified next. The equation that describes a liner trend is mathematically expressed as:

$$\bar{Y_t} = a + bt \qquad (4.3)$$

where

$\bar{Y_t}$ = the trend forecast for time period t

t = the number of time periods from 0

a = the estimate of the Y_t-axis intercept (i.e., the value of Y_t at $t = 0$)

b = the estimate of the slope of the trend line

The detailed steps for developing the linear trend line are as follows:

1. Determine slope b using the following equation:

$$b = \frac{\sum_{t=1}^{n} tY_t - n\bar{t}\bar{Y_t}}{\sum_{t=1}^{n} t^2 - n\bar{t}^2} \qquad (4.4)$$

where

$\bar{t}$ = average of t values
$\bar{Y}_t$ = average of Y_t
Y_t = the actual data (observed value) for time period t
n = the number of periods of data used in the least squares regression

2. Determine intercept a using the following equation:

$$a = \bar{Y}_t - b\bar{t} \qquad (4.5)$$

3. Generate the linear trend line:

$$\bar{Y}_t = a + bt \qquad (4.6)$$

4. Make a forecast for the appropriate time period (t).

To simplify the calculation of slope b, we can center the time value t by allocating 0 for a middle time period ($\frac{n+1}{2}$ th period) and then increase by one for each next period (e.g., 1, 2, 2) while decreasing one for each prior period (e.g., -1, -2, -3, …) if we have an odd number of time periods. If we have an even number of time periods, we can allocate -1 for $\frac{n}{2}$ th period and 1 for $\frac{n}{2}+1$ th period, respectively, and then increase by two for each next period (e.g., 3, 5, 7,…) while decreasing by one for each prior period (e.g., -3, -5, -7, …). Through this manipulation, Equations 4.4 and 4.5 will be converted into the following equations.

$$b = \frac{\sum_{t=1}^{n} tY_t}{\sum_{t=1}^{n} t^2} \qquad (4.7)$$

$$a = \bar{Y}_t \qquad (4.8)$$

Using these simplified equations, let's go back to the airline passenger data contained in Table 4.3 and then conduct a trend analysis. Table 4.4 illustrates the detailed calculations and then the forecasted values for each period. Using Equations 4.7 and 4.8, we obtain $b = 13.956$ and $a = 1179.3$, which produce a trend line of $\bar{Y}_t = 1179.3 + 13.956\ t$. By plugging t-value into this trend line, we can make a forecast of each period, as recapitulated in Table 4.4.

Table 4.4. The Forecasting Results of Airline Passengers Using Trend Analysis

Year	Quarter	t (translated time)	Actual number of airline passengers (Y_t)	tY_t	t^2	Forecasted value ($\hat{Y}_t$)	$\hat{Y}_t - Y_t =$ Forecast error
2010	1	-19	878	-16,682	361	914	-36
2010	2	-17	1,005	-17,085	289	942	63
2010	3	-15	1,173	-17,595	225	970	203
2010	4	-13	883	-11,479	169	998	-115
2011	1	-11	972	-10,692	121	1,026	-54
2011	2	-9	1,125	-10,125	81	1,054	71
2011	3	-7	1,136	-7,952	49	1,082	54
2011	4	-5	988	-4,940	25	1,110	-122
2012	1	-3	1,020	-3,060	9	1,137	-117
2012	2	-1	1,146	-1,146	1	1,165	-19
2012	3	1	1,400	1,400	1	1,193	207
2012	4	3	1,006	3,018	9	1,221	-215
2013	1	5	1,108	5,540	25	1,249	-141
2013	2	7	1,288	9,016	49	1,277	11
2013	3	9	1,570	14,130	81	1,305	265
2013	4	11	1,174	12,914	121	1,333	159
2014	1	13	1,227	15,951	169	1,361	-134
2014	2	15	1,468	22,020	225	1,389	79
2014	3	17	1,736	29,512	289	1,417	319
2014	4	19	1,283	24,377	361	1,444	-161
Total		0	23,586	37,122	2660		

Each of the forecasting methods introduced earlier has its pros and cons. Therefore, before choosing the appropriate forecasting method, one should understand their distinguished features, as summarized in Table 4.5 (Shim, 2000, p. 147).

Table 4.5. Summary of the Features of Common Time Series Methods

Features	Exponential Smoothing	Moving Average	Trend Analysis
Accuracy:			
Short-term (0 to 3 months)	Fair to very good	Poor to good	Very good
Medium-term (3 months to 2 years)	Poor to good	Poor	Good
Long-term (2 years or more)	Very poor	Very poor	Good
Identification of turning points or aberration	Poor	Poor	Poor

Typical applications	Production and inventory control, sales forecasts, exchange rate patterns, and stock price forecasts.	Inventory controls for low-volume items and sales forecasts with no data.	New product sales forecasts, economic trends, and product sales forecasts in the growth and maturity stages of the product life cycles.
Data required	A minimum of two years of history if seasonality is present; otherwise, less data is needed.	A minimum of two years of history if seasonality is present; otherwise, less data is needed.	A good rule of thumb is to use a minimum of five years' annual data to start with and, thereafter, complete history.
Key limitation	The forecast depends heavily on what happened most recently.	Erratic when data does not exhibit the level pattern.	Data should display a clear pattern over time, but this method tends to inflate forecasts when there is an upward trend.

Advanced Forecasting Techniques

Over the years, the basic forecasting methods introduced earlier have been refined and extended to capture the randomness and seasonality of some data patterns. Some of those advanced forecasting methods include double exponential smoothing, Winter's exponential smoothing, adaptive filtering, the Box-Jenkins method, state space forecasting, and agent-based forecasting. Among these, double exponential smoothing, Winter's exponential smoothing, and adaptive filtering share the same traits as the extension of the single exponential smoothing methods. Because the Box-Jenkins method is somewhat different from these three and has broad application potential, it will be discussed here.

The Box-Jenkins method is designed to account for repeated movements in the historical time series comprising the trend, seasonal, and cyclical patterns, thereby leaving a series made up of only random and irregular movements (Hurwood et al., 1978). To model the systematic patterns in the time series, it uses an iterative procedure that combines the autoregressive process with the moving average process and is then adjusted for seasonal and trend factors. After the appropriate weighting parameters are estimated, the model is repeatedly tested for its accuracy, and the process continues until a satisfactory level of accuracy is obtained. It is commonly referred to as the ARIMA (autoregressive integrated moving average) model because it relies on either past values obtained from the autoregressive process or past errors estimated by the moving average process. The Box-Jenkins method is widely used for short-term forecasts of daily stock prices, corporate earnings, energy prices, unemployment rates, and sales (Kim, 2000). The weaknesses of the Box-Jenkins method are its complexity, difficulty in updating the model, and inability to capture turning points in cyclical data series (Hurwood et al., 1978).

Selection of the Right Forecasting Method

Because there is no "one-size-fits-all" forecasting method, finding the right forecasting method can be an onerous task. Interested readers should refer to Chambers et al. (1971) for a detailed forecasting selection process. However, by closely examining the number of attributes associated with the forecasting methods, one can find a clue as to which forecasting method is suitable for which situation (decision environment) and which organization. According to Yokum and Armstrong (1995), here are some of these attributes, presented in order of importance:

1. Accuracy
2. Timeliness in providing forecasts
3. Cost savings resulting from improved decisions
4. Ease of interpretation
5. Flexibility
6. Ease in using available data
7. Ease of use
8. Ease of implementation
9. Incorporating judgmental input
10. Reliability of confidence intervals
11. Development cost resulting from computer usage and human efforts
12. Maintenance cost resulting from data storage and modifications

In addition, one can consider the market popularity of the forecasting method, but keep in mind what is best for some may not be best for everyone. Some of the industry surveys conducted by Dalrymple (1987) and Frank and McCollough (1992) have indicated that the most widely used forecasting method is the judgmental method, followed by trend analysis, moving averages, exponential smoothing, and Box-Jenkins. However, heavy reliance on market popularity in selecting a forecasting method is not wise given that different organizations have different forecasting needs. For example, although the judgmental method is most popular among practitioners, it is not known to be as accurate as quantitative forecasting methods such as trend analysis, moving averages, and exponential smoothing when enough data exists (Bailey and Gupta, 1999). Especially when there is little or no systematic change in data patterns, quantitative forecasting methods often tend to produce more accurate results than the judgmental method (Shim, 2000).

Sales and Operational Planning

Once demand projections are made using the forecasting techniques described in the prior section, they should be reviewed for accuracy, process accountability, resultant financial impacts, future risk assessments, capacity constraints, and resource allocation. This review process is often referred to as *sales and operational planning (S&OP)*. S&OP also can be used to create a consensus forecast when different functional units/departments (e.g., marketing vs. production) arrive at different or conflicting forecasting outcomes. Generally speaking, S&OP is an integrated decision-making process that periodically brings together all the functional plans (typically monthly) through collaboration and communication to produce agreed-upon sales forecasts and production plans. Simply put, S&OP aims to form the balance between the expected demand and the company's ability to ensure critical resources sufficient enough for this demand. By sustaining such a balance, S&OP enables the company to reduce wasteful production, reduce backorders, improve customer services, utilize resources, shorten customer response time, and maximize revenue.

Figure 4.4 shows how S&OP is typically structured and what its core inputs and outputs are. Core inputs include demand planning, whereas core outputs include optimal sales rates, production rates, and inventory levels. As shown in Figure 4.4, S&OP also links the company's business plans to its detailed processes, such as demand planning and master production schedules, and thus enhances teamwork and communication among managers representing different functional units. With enhanced communication, the company can make quick changes in its game plans (e.g., supply chain plans) before getting off the right track. Because S&OP periodically monitors the accuracy of data and performance indicators, continuous improvement of supply chain operations can be achieved. For example, the accuracy of a demand plan can be checked on a monthly basis, whereas the accuracy of master production schedules and material requirement planning can be measured on a weekly basis. Furthermore, the accuracy of inventory location, bill of materials, item master files, and shop floor control can be monitored on a daily basis, and quick adjustments can be made to improve supply chain performance. S&OP typically involves decision makers from the top and middle levels, such as the chief executive officer (CEO), chief financial officer (CFO), vice president of sales/marketing, vice president of operations, vice president of finance, production managers (one for each product family), and master scheduler (Sheldon, 2006).

According to Gartner's S&OP maturity model (Barrett and Uskert, 2010), it usually takes four stages to reach the perfect balance between demand and supply. To elaborate, the S&OP journey starts with stage one when the company simply reacts to demand without making a forecast, as summarized in Table 4.6. At stage two, the company attempts to match its supply to demand through structured forecasting, inventory control, and supply chain planning. At stage three, the company needs to go well beyond the traditional S&OP tools such as the forecasting and inventory optimization models that were used to balance supply and demand. In other words, the company needs to change not only its process and technology, but also address its organizational culture, ownership, and core belief through the integration of cross-functional processes. Once the S&OP process reaches the final stage (stage four), the company's collaboration extends its own enterprise and breaks organizational silos to achieve the creation or addition of end-to-end value. At this stage, S&OP should be fully incorporated into strategic planning and be managed and coordinated by a group of top-level executives representing each member of the supply chain.

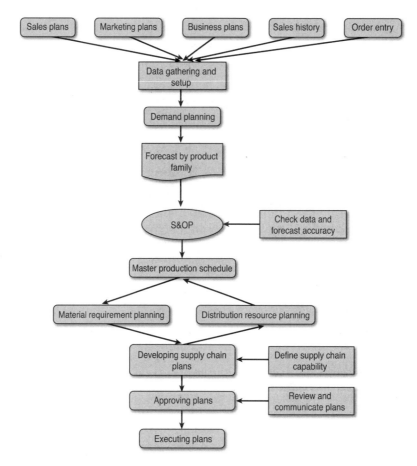

Figure 4.4. *S&OP processes*

Table 4.6. Four Stages of The S&OP Maturity Model

Strategic features	Stage 1: Reacting	Stage 2: Anticipating	Stage 3: Collaborating	Stage 4: Orchestrating
Balance level				
Key goals	Sales reviews and development of operational plans	Supply and demand matching	Profitability and what-if scenario planning	Demand sensing and cross-enterprise collaboration
Primary tools	Pre-S&OP meetings, Excel-based reports, and Enterprise Resource Planning (ERP)	Demand forecasting and inventory optimization models	Collaborative planning, forecasting, and replenishment (CPFR), point-of-sales information sharing	Risk analysis, value analysis, simulation
Focus	Sales data collection, addressing data accuracy, lead time analysis, and production capacity planning	Marketing intelligence gathering, sales forecast accuracy, assurance of data integrity, and linking S&OP to strategic execution	S&OP as the forum for strategic decision-making process, capturing demand signals, effective communication among multiple stakeholders, consensus build-up	Cultural shifts (e.g., move to openness, transparency, rewarding the right behavior), change management, performance metric development, strategic alliances

Collaborative Commerce

Because a superior supply chain is often created via the joint effort of a network of business partners, a growing number of companies have attempted to develop a set of collaborative processes that will integrate various business activities within and across the supply chain. One of those processes includes collaborative commerce, which aims to enhance interactions among supply chain partners by building many-to-many, technology-enabled business networks. As displayed by Figures 4.5a and 4.5b, collaborative commerce can stretch a traditional linear supply chain to a nonlinear supply chain with an interwoven network of supply chain partners who can create powerful synergy. These networks can be an important source of competitive advantage, because the collective power created by the networks is much harder to compete against than the individual power created by a single company. As a matter of fact, Porter (1980) indicates that superior firm performance is functionally related to a firm's membership (i.e., the extent and scope of partnerships). Considering the competitive advantages created by collaborative commerce, collaborative commerce can be a driving force behind strong business growth. In particular, collaborative commerce can be an effective tool for a firm with limited resources that has to rely on external resources, such as information technology infrastructure and expertise, to fully exploit e-commerce. For example, Dell and Volkswagen leveraged public electronic links to facilitate interactions with their suppliers and streamline/automate business transactions. Such links helped reduce operating expenses and compress cycle time. Deloitte's survey conducted in 2001 indicated that firms that linked their business processes with other companies enjoyed 70% higher profitability than those firms that did not integrate with trading partners via electronic networks (Deloitte Consulting, 2001).

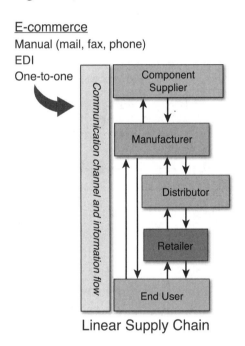

Figure 4.5a. *The linear supply chain linked by e-commerce*

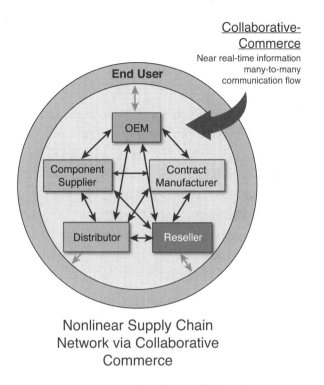

Figure 4.5b. *The nonlinear supply chain linked by collaborative commerce*

Although the successful formation of collaborative commerce can lead to improvement in revenue, profit, return on investment, brand recognition, and productivity, the following questions should be addressed prior to forming collaborative commerce (Dyer, 2000):

- What are the appropriate boundaries of the extended network? To what extent (e.g., partnerships versus arm's-length relationships) should the company integrate?
- What strategies should the company employ in managing partnerships?
- What are the processes (e.g., investment in information technology, knowledge sharing, technology transfer) involved in collaborating effectively with supply chain partners in order to achieve a competitive advantage?

With these questions in mind, the following subsections introduce efficient consumer response (ECR) and collaborative planning, forecasting, and replenishment (CPFR) as tools that facilitate collaborative commerce.

Efficient Consumer Response

Efficient consumer response (ECR) is an extension of quick response (QR) that aims to increase customer value, eliminate non-value-adding costs, reduce order cycle time, and more be responsive to customer needs by using technological tools such as electronic data interchange (EDI), bar coding, and point-of-sale (POS) systems. In a nutshell, ECR focuses on the reduction in order cycle time and the improvement of supply chain efficiency throughout the entire supply chain, thus resulting in increased financial leverage. Although ECR was initially introduced in the grocery retail sector, its usage has been widespread to other industry sectors due to its benefits. For example, goods that once took a couple of months or longer to be ordered and received are ordered and delivered on a weekly basis under ECR. Also, reduction in order cycle time will lead to a substantial decrease in on-hand inventory levels. According to Food Marketing Institute (http://www.fmi.org/supply/ecr), ECR is composed of four components:

- **Efficient store assortments**—To minimize inventory and maximize store space at the consumer interface, this component addresses how many items to carry in a category, what type of items and in what sizes/flavors/packages, and how much space to give to each item.
- **Efficient replenishment**—To minimize time and cost associated with the replenishment system, this component focuses on shortening the order cycle and eliminating costs in the supply chain by utilizing point-of-sale information, cross-docking, and electronic ordering systems.
- **Efficient promotion**—To maximize the total system efficiency of trade and consumer promotion, this component addresses inefficient promotional practices that tend to inflate inventory values, which may not be fully passed through to the end consumers to influence their purchasing decisions.
- **Efficient product introductions**—To maximize the effectiveness of new product development and introduction, this component improves the entire process of introducing new products and consequently preventing unnecessary development efforts and costs.

Regardless of the ECR structures just described, ECR cannot be successful without ensuring partnerships among supply chain members (especially manufactures and retailers) that allow for joint product development, introduction, promotion, and inventory replenishment.

Collaborative Planning, Forecasting, and Replenishment

Collaborative planning, forecasting, and replenishment (CPFR) is generally referred to as a nine-step joint demand planning process that aims to enhance supply chain visibility by improving order forecasts and fulfillment through continuous communications among multiple supply chain partners (Min, 2008). The process is composed of the following nine steps (Ackerman, 2000):

1. Develop a front-end agreement.
2. Create joint business plan.
3. Create sales forecasts.
4. Identify exceptions for sales forecasts.
5. Resolve/collaborate on exception items.
6. Create order forecasts.
7. Identify exceptions for order forecasts.
8. Resolve/collaborate on exception items.
9. Generate orders.

The detailed steps of CPFR are shown in Figure 4.6 (Ackerman, 2000).

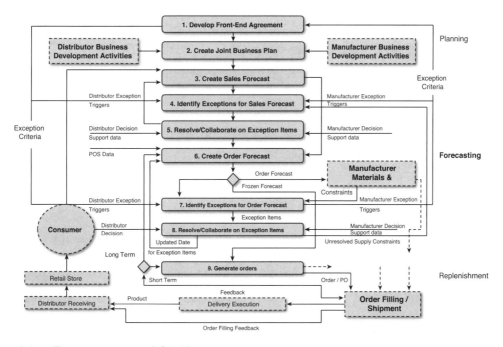

Figure 4.6. *The nine steps of CPFR*

Although CPFR evolved from traditional collaborative tools—such as electronic data interchange (EDI), vendor managed inventory (VMI), and efficient consumer response (ECR)—it differs from others in that it brings mutual benefits to all the supply chain partners involved by utilizing more interactive, broader communication processes throughout the supply chain rather than relying on limited transaction-level automation. Other benefits of CPFR include higher inventory turnover, lower stockout rate, improved order fill rate, improved cash flow, more accurate production scheduling, more amicable business relationships among supply chain partners, reduced cycle time, reduced order picking/receiving costs, reduced labor costs, and quicker response to customer needs (Sherman, 1998; Williams 1999; Barratt and Oliveira, 2001; McCarthy and Golicic, 2002; Andraski and Haedicke, 2003; Min, 2008).

For example, the pilot implementation of CPFR by Walmart and its supplier Warner-Lambert increased average in-stock rates from 87% to 98%, reduced lead time from 21 to 11 days, and increased sales volume by $8.5 million in 1995 (Fahrenwald et al, 2001). A German office supplies manufacturer named Herlitz AG, which shared information with its retailer Metro through CPFR, reduced its inventory by 15%, curtailed its stockouts by 50%, and increased its annual sales by 3% (Andraski, 2000). Similarly, Johnson & Johnson reported an increase in its in-stock rate, from 91.5% to 93.8%, after adopting CPFR (Inventory Management Report, 2002). After CPFR was implemented for Sears and Michelin in 2001, they reported that in-stock levels at the Sears stores were improved by 4.3%, and Sears' distribution centers-to-store fill rate was increased by 10.7%, while the combined Sears and Michelin inventory levels were reduced by 25% (Steermann, 2003). In addition, Smaros (2003) observed that CPFR made it possible to detect discrepancy between actual and forecasted sales at the early stage of the supply chain process through exception alerts. For example, sharing of early sales information in the apparel industry improved the forecasting accuracy from 45% to 92% (Fisher et al., 2000).

Despite the aforementioned promises, CPFR is not without its obstacles. These obstacles may include cultural and technical incompatibility among supply chain partners, lack of trust among supply chain partners, lack of scalability, lack of internal alignment, inadequate software and technology support, disintegrated demand and supply management processes, substantial startup investment for building a communication infrastructure, antitrust laws, legacy systems, and difficulty in real-time coordination of information exchange (Mentzer et al., 2000; Barratt and Oliveira, 2001; McCarthy and Golicic, 2002; Seifert, 2003; Crum and Palmatier, 2004). To overcome such obstacles, the company needs to carefully decide which type of collaboration it would like to engage in. According to Campbell and Goold (1999), there are six different types of collaboration to choose from:

- **Shared know-how**—This form of collaboration involves sharing of best-in-class business practices in certain business processes, or leveraging expertise in functional areas, or pooling knowledge about how to succeed in specific geographical regions.

- **Shared tangible resources**—This form of collaboration uses a common manufacturing facility, equipment, and storage space to eliminate the duplication of physical assets and then create economies of scale for supply chain partners.
- **Pooled negotiation power**—This form of collaboration intends to increase the company's bargaining power and negotiation leverage to obtain the best possible deal.
- **Coordinated strategies**—This form of collaboration involves aligning the strategies of two or more supply chain partners to coordinate reactions against common competitors.
- **Vertical integration**—This form of collaboration involves coordinating the flow of products and services from one channel to another in order to reduce pipeline inventories, utilize capacity, and improve market access.
- **Combined new business creation**—This form of collaboration can occur when supply chain partners create new businesses by combining know-how from different companies, and by extracting resources from different companies to invest them into new businesses.

The Bullwhip Effect

Poor demand planning and forecasting in a supply chain often leads to an imbalance between supply and demand that, in turn, results in either product shortages or overstocked inventory. Figure 4.7a illustrates an amplified demand forecasting error further downstream in the supply chain, when each supply chain member is blindsided by the lack of accurate demand information. For example, when the actual customer demand is 1,000 units, with 10% set aside for a safety stock, the supplier furthest upstream of the supply chain ends up absorbing 1,464 units—that is, 364 units more than what is actually needed, as shown in Figure 4.7b. This phenomenon is called the *bullwhip effect* (Austin et al., 1997). The bullwhip effect (or *whiplash* or *whipsaw phenomenon*) is generally referred to as an inverse ripple effect of forecasting errors throughout the supply chain that often leads to amplified supply and demand misalignment, where orders (perceived demand) to the upstream supply chain member (e.g., the supplier) tend to exaggerate the true patterns of end-customer demand because each chain member's view of true demand can be blocked by its immediate downstream supply chain member.

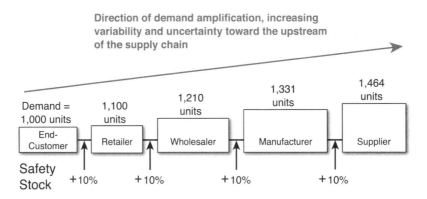

Figure 4.7a. *Demand planning error caused by the supply chain invisibility*

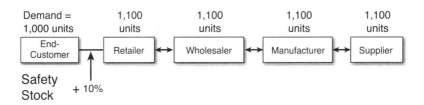

Figure 4.7b. *Reduced demand error resulting from the enhanced supply chain visibility*

The supply/demand misalignment caused by the bullwhip effect is graphically displayed in Figure 4.8. The common symptoms of the bullwhip effect include delayed new product development, constant shortages and backorders, frequent order cancellations and returns, excessive pipeline inventory, erratic production scheduling, and chronic overcapacity problems (Min, 2000). For example, in the $300 billion grocery industry, the bullwhip effect may account for $75 to $100 billion worth of unproductive pipeline inventory, because 5% variation in end-customer consumption can result in 300% to 400% variation in downstream suppliers' production (Artzt, 1993). In general, the bullwhip effect can increase total business operating costs by 12.5% to 25% (Lee et al., 1997). Metters (1997) also estimates that the elimination of the bullwhip effect could increase company profits by an average of 15% to 30%.

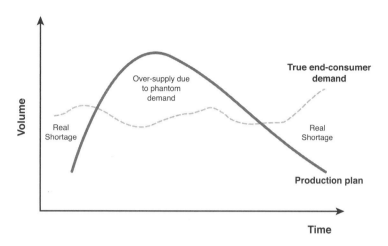

Figure 4.8. *Supply-demand misalignment caused by the bullwhip effect*

Potential Causes of the Bullwhip Effect

The bullwhip effect can be caused by a number of management failures and mistakes, as summarized next (Min, 2000):

- **Information failure**—Product proliferation, diversity, and demand uncertainty make it increasingly difficult for manufacturers, suppliers, and retailers to predict and plan for orders and production volume. In particular, during the period of continuous product shortages, overzealous downstream suppliers and retailers tend to over-order and stockpile their shelves with excessive inventory to catch up with previous demand, thus creating "phantom demand," which in turn triggers over-production in the supply chain. Therefore, a lack of accurate demand information is one of the primary causes of the bullwhip effect.

- **Chain complexity**—In a typical chain structure, middle men such as distributors are considered sales agents of manufacturers and often represent multiple manufacturers with diverse products. As such, channel intermediaries such as distributors by nature cannot match the quantities and characteristics of products desired by the ultimate end customers to those of products provided by manufacturers. For this reason, the presence of channel intermediaries complicates the supply chain and subsequently increases order cycle time throughout the entire supply chain. Therefore, the chain complexity that further blurs supply chain visibility can add up to forecasting difficulty and the subsequent bullwhip effect.

- **Product proliferation**—As many mass-merchants such as Walmart, Sears, and Target are demanding more diversified product lines and tailored services to meet the specific needs of different segments of customers, there has been an explosive proliferation of products offered to customers in terms of color, design, and

function. Such product proliferation may significantly increase the chain complexity and demand volatility, which in turn causes the bullwhip effect.
- **Sales promotion**—Today's savvy customers tend to shop only when a sales promotion in the form of coupons, rebates, and seasonal discounts is available. This shopping pattern often results in peaks and valleys of customer orders and correlates with frequent price changes. Although a sales promotion is intended to benefit end customers, many downstream distributors also tend to stockpile huge amounts of promotional inventory with an expectation of a future price increase. Therefore, a sales promotion is likely to create phantom demand throughout the supply chain, contributing to over-order, over-production, or temporary product shortages.
- **Economies of scale**—To exploit cost-saving opportunities resulting from economies of scale, many firms prefer to order on a periodic basis so that they can accumulate orders large enough for volume purchasing or freight consolidation. Such periodic ordering is likely to cause a so-called "hockey stick" phenomenon (order surges at the end of month, quarter, or year), which not only amplifies order variability but also increases order cycle time. For instance, with the widening spread between truckload (TL) and less-than-truckload (LTL) rates in the wake of transportation deregulation, a growing number of firms are prompted to aggregate small orders into larger shipments by delaying shipments until a sufficient amount of volume is created. Such freight consolidation can substantially increase order cycle time, thus exacerbating the bullwhip effect.
- **Speculative investment behavior**—In years of economic upturns, many firms tend to perceive product shortages as the loss of market share or revenue, thereby triggering a period of over-investment and production. For instance, encouraged by various forms of government incentives such as increased local content rules, preferential loans, and government bailouts, many automakers in Asia substantially increased their production capacity in the early 1990s and were left with a huge amount of inventory as the growth rate of auto sales declined. Such a speculative investment behavior may have contributed to the serious financial trouble and subsequent major restructuring of the Korean auto giant Kia and the Indonesian national auto-maker, Bimantara in the 90s. Because speculative investment behavior often impedes a firm's ability to match supply and demand, it aggravates the bullwhip effect.

Chapter Summary

The following summarizes key takeaways from this chapter:
- Demand planning enables the company to develop production plans that, in turn, impact capacity, marketing, logistics, financial, and workforce plans. Therefore,

poor demand planning is the main source of business failures. Poor demand planning is often the byproduct of inaccurate demand forecasting, misunderstanding of demand priorities, lack of information sharing among trading partners, and bullwhip effects.

- Demand forecasting is necessary to project future demand and adjust the current production capacity to meet future demand. It also determines whether future demand is sufficient enough to warrant constant production and employment of workers.

- Forecasting techniques are classified into qualitative and quantitative methods. Qualitative methods are primarily based on human judgment, opinions, and are subjective and therefore can be biased and relatively inaccurate. Examples of qualitative methods include the Delphi method, expert opinion, consumer panels, market research, and historical analogy. Qualitative methods are often used for long-range forecasts and forecasts for new products.

  On the other hand, quantitative methods are based on mathematical formulas and statistical analyses, which can incorporate a large amount of data at one time into the forecasting process and can automate the forecasting process using computer programs. When quantifiable information is available, quantitative methods are more accurate than qualitative methods because the former can eliminate inadvertent human errors. One of the well-known forms of the quantitative methods is the time-series method, which includes exponential smoothing, moving average, and trend analysis. Another form of quantitative method is a causal method that makes a forecast based on the functional relationship between the forecasted event and the histories of external factors. Examples of the causal method are regression analysis and econometric methods.

- None of the forecasting techniques is perfect. In other words, there is no "one-size-fits-all" forecasting technique. Therefore, the selection of the right forecasting method for a given situation is crucial for successful demand forecasting. The common forecasting selection criteria include forecasting accuracy, time horizons, ease of use, flexibility, and data requirements.

- Collaborative commerce is a wave of the future for communicating demand to supply chain partners. It facilitates information sharing among supply chain partners and enhances demand forecasting accuracy throughout the supply chain.

- With the advent of e-commerce—especially electronic data interchange—CPFR (collaborative planning, forecasting, and replenishment) was introduced as a way to share demand information among supply chain partners and make a joint forecast for future orders. CPFR is known to be useful for reducing order fulfillment errors, stockout situations and cycle time while improving inventory turnover and cash flow.

- The bullwhip effect is the leading cause of supply-demand misalignment and excessive inventory pileups, which incur high inventory carrying costs. The elimination of the bullwhip effect requires the integration and coordination of the supply chain to allow for timely flow of demand information throughout the supply chain.

Study Questions

The following are important thought starters for class discussions and exercises:

1. Explain what steps are required to manage demand.
2. What comprises the demand influencing process?
3. Why is demand forecasting important for business decisions?
4. What are the differences among the time-series methods?
5. Compare and contrast three popular time-series methods (i.e., exponential smoothing, moving average, and trend analysis).
6. What are the main criteria for selecting the right forecasting method?
7. Illustrate forecasting needs in the following business sectors:
   - Automobile manufacturers
   - Tire makers
   - Stock markets
   - Trucking firms
   - Music CDs
   - Computer software
   - Apparel stores
   - Supermarkets
   - Fast-food restaurants
   - Ice cream vendors
   - Hot dog vendors at a baseball game
   - Appliance stores
8. Imagine that you need to introduce a new gadget such as the Apple Watch. Determine which forecasting method is appropriate for projecting the future demand of such this gadget.
9. Why do we need collaborative commerce? How is it related to e-commerce?
10. What are the key differences between traditional forecasting methods and CPFR?

11. What are the primary causes of the bullwhip effect?
12. How would you prevent or mitigate the bullwhip effect?

Case: Seven Star Electronics: Demand Planning

As a leading global manufacturer, Seven Star Electronics makes and sells a variety of consumer electronics, including different types of televisions (e.g., LCD, plasma, LED, OLED, and UHD TVs), DVD players, smartphones, stereo equipment, digital cameras, refrigerators, washers, and dryers. Its sales and marketing department is hungry for new ideas that will reverse the recent trend of Seven Star's declining sales and market share. However, the market for consumer electronics is commoditized and becomes extremely competitive because of the short lifespan of products and the growing number of competing manufacturers in low-cost sourcing countries such as China, India, Malaysia, Thailand, and Indonesia. In particular, Chinese firms such as Haier are selling their electronic products abroad in ever-greater numbers, and a battle is set to be fought that may reshape the global marketplace. Adding to concerns, more consumers are demanding new gadgets with fancy design, making some of the old but once popular products such as flat-screen analog TVs and 60-inch projection TVs obsolete and thus reducing demand for such products. These changes in customer preferences pressure Seven Star to lower its product prices to an unprofitable level while sustaining high-quality and speedy market responses.

Fierce Competition

Consumer electronics represent a mature industry in the United States, Western Europe, Japan, and the "Four Dragons" (Korea, Hong Kong, Singapore, and Taiwan) in Asia. Although emerging economies such as China, India, Brazil, and Russia have begun to pick up demand for some electronic devices such as smartphones, tablets, and digital cameras, most of the consumers in the emerging markets still do not have sufficient purchasing power to acquire expensive electronic appliances and high-definition TVs. Regardless, Seven Star management believes that the best way to attract new customers and retain old customers is to provide superior customer value with innovative product design and technological breakthrough. This strategy puts pressure on the short-manned, scattered manufacturing plants across Far East Asia, Southeast Asia, and the North America. Due to varying workmanship (worker skills), production schedules, logistics infrastructure, and labor productivity in different locations of manufacturing plants, it was increasingly difficult for Seven Star to meet stricter quality standards in the so-called OECD (Organization for Economic Co-operation and Development) countries, including the United States. As a result, rejects for substandard items have been increasing at alarming rates over the last several years and a growing number of unsold products were written off. To make matters worse, Seven Star has been receiving a number of calls and emails from long-time U.S. customers such as Best Buy, hhgregg, Sears, Target, and Walmart, complaining about late shipments and lack of post-sale repair services. These mass-merchants often advertise promotions for the sale of consumer electronics during the holiday seasons, such as the Independence Day, Thanksgiving, and Christmas holidays. Therefore, their promotional success depends heavily on on-time delivery and error-free order fulfillment. Seven Star management knows that losing big customers such as Best Buy, hhgregg, Sears, Target, and Walmart would be disastrous. Given that Seven Star invested a huge amount of money to establish its brand recognition in the North American market, the loss of those high-profile customers would spell serious trouble. To deal with the current crisis, Seven Star has decided to focus on the curved ultra-high-definition (UHD) 4K (4,000-pixels-wide) TV as a case in point because it is a high-margin new product and potential cash cow with mesmerizing technical features, including connection to online interactive media, wireless audio speakers, optional 3D views, and instant access to movies with Netflix, videos with YouTube, music with Pandora, photos with Flickr and Picasa, and social apps such as Facebook.

Tricky Demand Planning

Demand planning starts with the marketing division, which determines the forecast for flat-panel HDTVs (both LCD and plasma) by month for the next year. Then they pass it along to the production division in the North American manufacturing plant in Mexico. Because the marketing division's forecasts are usually off target, the production managers want to increase the manufacture of products by 10% or more as safety stock. However, given the very short shelf life of new products, any inventories of new products exceeding the six months' supply are considered "obsolete." Fearful of rapid obsolescence, Seven Star's production managers always wonder how conservative they should be in projecting future demand. From time to time, conservative projection has been a major source of late delivery problems and the subsequent loss of sales. In particular, product shortages resulting from delayed shipments increased missed sales opportunities for Seven Star's major customers (especially) during the end-of-the year holiday season.

Because the marketing division provides crucial information related to future customer demand and sales projections, a team of production managers, including Mr. John Little, decided to ask for demand-related information from the marketing division on a periodic basis. However, Shirley Simmons, the company's chief marketing officer, indicated to John that her forecasts were often based on the limited sales history of the old analog TVs and she never knew what future demand might be for relatively new products such as curved UHD 4K TVs. She seemed to presume that things do not change much from year to year and from one TV model to another, despite the fact that the sale of big-ticket items such as UHD 4K TV could be seasonal due to its high price tag, ranging from $1,300 to nearly $40,000. For example, in the United States, many end customers tended to defer their purchase of HDTVs until they receive their tax refunds sometime during late spring or early summer. Otherwise, they tended to buy those products near or during the holiday seasons as special gifts to their families, friends, relatives, and acquaintances. Although worldwide shipments of UHD 4K TVs have soared around 20-fold during last year, their sales momentum can get stalled due to rising cable service fees and the unexpected delay of UHD broadcasting and 4K streaming. Considering the aforementioned complicated forecasting dynamics, Shirley's simplistic thought was not very helpful for John in addressing the ongoing demand forecasting problem. Therefore, John was somewhat frustrated.

Instant Lesson

While attending the recent supply chain management conference in San Diego, John learned from one of the renowned keynote speakers that Walmart and Target successfully implemented the so-called "collaborative planning, forecasting, and replenishment" (CPFR) to improve demand forecasts for their upstream supply chain partners (i.e., suppliers) such as Warner Lambert. After he heard that story, John felt as if he saw light at the end of the tunnel. Because both Walmart and Target are Seven Star's customers, John was wondering if his company could embrace CPFR. With limited knowledge of CPFR, John does not know what it takes to adopt CPFR, to whom he needs to talk, whether its benefits are warranted, and how costly it will be to implement CPFR. John calls a series of meetings with his marketing counterpart, Shirley, and purchasing executives from both Walmart and Target. In the meantime, he needs to develop an immediate forecast for a newly designed 55-inch curved UHD 4K TV, which is considered to be one of the largest screen HDTVs on the U.S. market. At this juncture, the only information available to him from the marketing division is the past four-year sales history of the flat-panel 42-inch LCD TV. This information is summarized in Exhibit 4.1.

Exhibit 4.1. Four-Year Sales History for the 42-inch LCD TV in the U.S. Market

	Past Sales Patterns			
Month	Year 1	Year 2	Year 3	Year 4 (Latest)
1	55,212	39,873	42,167	67,555
2	57,315	47,566	54,329	65,674
3	23,457	34,675	25,340	47,897
4	28,363	43,056	50,092	71,235
5	34,679	37,891	41,034	55,543
6	47,890	44,123	39,098	49,875
7	51,237	48,145	41,790	58,761
8	27,894	21,367	30,989	45,670
9	34,525	29,876	27,896	35,987
10	44,790	35,784	36,098	44,982
11	67,899	57,980	65,541	69,845
12	73,598	74,675	69,312	98,310

Note: The sales figures shown in the table are the number of units actually sold each month. Some products returned in the following month may have been included in preceding monthly figures.

Discussion Questions

1. Can you suggest any changes in the forecasting practices that will improve the company's forecast and customer services?
2. How would you formulate a long-term global supply chain strategy that will help the company retain valued customers such as Best-Buy, hhgregg, Sears, Walmart, and Target?
3. Which factors will impact the forecasting accuracy of Seven Star's curved UHD 4K TV?
4. Which forecasting method is most appropriate for projecting the demand of a new product like Seven Star's curved UHD 4K TV?
5. How would you make your own forecast for the newly designed 55-inch curved UHD 4K TV in the U.S. market for each month for the next year? Justify your forecast and explain why it makes sense.
6. How would you develop a supply chain strategy that can help Seven Star increase the market share in the U.S. market? This strategy may include reconfiguration of the current distribution network, relocation/consolidation of current manufacturing plants, contract manufacturing, logistics outsourcing, and strategic alliances with business partners.

Bibliography

Ackerman, K.B. (2000), "CPFR—How it Could Change the Warehouse," *Warehousing Forum*, 15(10), 1–2.

Andraski, J.C. and Haedicke, J. (2003), "CPFR: Time for the Breakthrough?" *Supply Chain Management Review*, 7(3), 54–60.

Artzt, E.L. (1993), "Customers Want Performance, Price, and Value," *Transportation and Distribution*, 32–34.

Austin, T.A., Lee, H.L., and Kopczak, L. (1997), *Unlocking Hidden Value in the Personal Computer Supply Chain*, Unpublished White Paper, San Francisco, CA: Andersen Consulting Creative Services.

Bailey, C.D. and Gupta, S. (1999), "Judgment in Learning-curve Forecasting: A Laboratory Study," *Journal of Forecasting*, 18, 39–57.

Barratt, M., and Oliveira, A. (2001), "Exploring the Experiences of Collaborative Planning Initiatives," *International Journal of Physical Distribution & Logistics Management*, 31(4), 266–289.

Barrett, J. and Uskert, M. (2010), *Sales and Operations Planning Maturity: What Does It Take and How to Get There?*, Unpublished Report, Gartner RAS Core Research Note G00207249, Stamford, CT: Gartner Group.

Campbell, A. and Goold, M. (1999), *The Collaborative Enterprise: Why Links between Business Units Often Fail-and How to Make Them Work*, Reading, MA: Perseus Books Group.

Chambers, J.C., Mullick, S.K., and Smith, D.D. (1971), "How to Choose the Right Forecasting Techniques," *Harvard Business Review*, 49(4), 45–74.

Chase, R.B., Jacobs, F.R., and Aquilano, N.J. (2006), *Operations Management for Competitive Advantage*, 11th edition, New York, NY: McGraw-Hill/Irwin.

Collier, D.A. and Evans, J.R. (2007), *Operations Management: Goods, Services and Value Chains*, 2nd edition, Mason, OH: Thomson South-Western.

Crum, C. and Palmatier, G.E. (2003), *Demand Management Best Practices: Process, Principles and Collaboration*, Boca Raton, FL: J. Ross Publishing, Inc.

Dalrymple, D.J. (1987), "Sales Forecasting Practices: Results from a United States Survey," *International Journal of Forecasting*, 3, 379–391.

Deloitte Consulting (2001), *Collaborative Commerce: Going Private to Get Results*, New York, NY: Deloitte Consulting and Deloitte & Touche.

DeLurgio, S.A. (1998), *Forecasting Principles and Applications*, Boston, MA: Irwin McGraw-Hill.

Dyer, J.H. (2000), *Collaborative Advantage: Winning through Extended Enterprise Supplier Networks*, New York, NY: Oxford University Press, Inc.

Frank, H.A. and McCollough, J. (1992), "Municipal Forecasting Practice: 'Demand' and 'Supply' Side Perspectives," *International Journal of Public Administration*, 15, 1,669–1,696.

Food Marketing Institute (2006), *ECR: Efficient Consumer Response*, http://www.fmi.org/supply/ecr, Chrystal City, Virginia: Food Marketing Institute.

Hurwood, D.L., Grossman, E.S., and Bailey, E.L. (1978), *Sales Forecasting*, New York, NY: The Conference Board, Inc.

Inventory Management Report (2002), "CPFR—No Longer a Question of If But When," *IOMA Inventory Management Report: Improving Logistics and Supply Chain Management*, December, 1–14.

Ireland, R. and Bruce, R. (2000), "CPFR: Only the Beginning of Collaboration," *Supply Chain Management Review*, 4(4), 80–88.

Lee, H.L., Padmanabhan, V., and Whang, S. (1997), Information Distortion in a Supply Chain: The Bullwhip Effect, *Management Science* 43, 546–558.

Lieberman, D. (2007), TiVo Takes Hit from HDTV Growth, *USA Today*, August 30, 1B.

McCarthy, T.M. and Golicic, S.L. (2002), "Implementing Collaborative Forecasting to Improve Supply Chain Performance," *International Journal of Physical Distribution & Logistics Management*, 32(6), 431–454.

Mentzer, J.T., Foggin, J.H., and Golicic, S.L. (2000), "Collaboration: The Enablers, Impediments, and Benefits," *Supply Chain Management Review*, 4, 52–58.

Metters, R. (1997), Quantifying the Bullwhip Effect in Supply Chains, *Journal of Operations Management* 15, 89–100.

Min, H. (2000), "The Bullwhip Effect in Supply Chain Management," *Encyclopedia of Production and Manufacturing Management* edited by P. Swamidass, Norwell, MA: Kluwer Academic Publishers, 66–70.

Min, H. and Yu, V. (2008), "Collaborative Planning, Forecasting and Replenishment: Demand Planning in Supply Chain Management," *International Journal of Information Technology and Management*, 9(1), 71–78.

Porter, M.E. (1980), *Competitive Strategy*, New York, NY: The Free Press.

Reid, R.D. and Sanders, N.R. (2005), *Operations Management: An Integrated Approach*, 2nd edition, Hoboken, NJ: John Wiley & Sons.

Seifert, D. (2003), *Collaborative Planning, Forecasting, and Replenishment: How to Create a Supply Chain Advantage*, New York, NY: AMACOM.

Sheldon, D.H. (2006), *World Class Sales & Operations Planning*, Ft. Lauderdale, FL: J. Ross Publishing.

Sherman, R. J. (1998), "Collaborative Planning, Forecasting & Replenishment (CPFR): Realizing the Promise of Efficient Consumer Response Through Collaborative Technology," *Journal of Marketing Theory and Practice*, 6(4), 6–9.

Shim, J.K. (2000), *Strategic Business Forecasting: The Complete Guide to Forecasting Real World Company Performance*, revised edition, Boca Raton, FL: St. Lucie Press.

Steermann, H. (2003), "A Practical Look at CPFR: The Sears-Michelin Experience," *Supply Chain Management Review*, 7(4), 46–53.

Williams, S.H. (1999), "Collaborative Planning, Forecasting, and Replenishment," *Hospital Material Management Quarterly*, 21(2), 44–51.

Yokum, T. and Armstrong, J.S. (1995), "Beyond Accuracy: Comparison of Criteria Used to Select Forecasting Methods," *International Journal of Forecasting*, 11, 591–597.

5

Inventory Control and Planning

"There are two ways to extend a business. Take inventory of what you are good at and extend out from your skills. Or determine what your customers need and work backward, even if it requires learning new skills."
—Jeff Bezos, CEO of Amazon

Learning Objectives

After reading this chapter, you should be able to:

- Understand the basic elements of inventory management.
- Evaluate the impact of inventory management on the level of customer services.
- Comprehend the different types and functions of inventory.
- Identify the components of inventory carrying costs and ordering costs and then estimate costs associated with inventory.
- Classify inventory into ABC categories and learn to use an appropriate inventory review system relevant to ABC categories.
- Develop the tools to measure inventory performances such as inventory turns and accuracy.
- Determine optimal order quantity and reorder quantity under different business environments.
- Find ways to improve inventory turns.
- Understand the differences among EOQ, MRP, DRP, and JIT and their suitability for various kinds of inventory planning and control.
- Understand new ways of controlling and planning inventory, including vendor-managed inventory.

The Principles of Inventory Management

Inventory is an idle asset, in various forms, held for current and future use (sale or distribution). Because an idle asset represents an underutilized resource, inventory can end up being a financial burden for the company. Despite the risks associated with inventory, many companies want to keep inventory to meet anticipated customer demand and not lose sales opportunities. Without inventory, it is difficult to make products available to customers when they are needed. Indeed, inventory is regarded as a viable alternative for future production and procurement. Therefore, inventory is often perceived to be a *necessary evil* for meeting customer service requirements. In particular, inventory can be problematic when it exceeds actual customer demand or it becomes obsolete. As a general rule of thumb, inventory is considered to be "excess" when the amount of inventory at hand exceeds the 12-month projected customer demand. Inventory is considered to be "obsolete" when it has no forecasted usage and little or no usage in the past six months. Either excess or obsolete inventory will be wasted, because many companies will sell such inventory at a discounted price or write it off to free up storage space and ease their tax burden. Considering these two contrasting faces of inventory, the careful policies, planning, and control are necessary to keep the right amount of inventory at the right place essential for customer service and to keep the cost of maintaining inventory at minimum. Therefore, inventory management is more than inventory control. In general, an inventory management principle is composed of a three-step process:

1. Developing inventory policies
2. Planning inventory
3. Controlling inventory

These are further subdivided into the various components described in Figure 5.1 (Bernard, 1999).

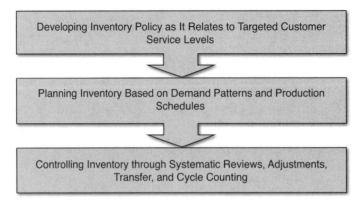

Figure 5.1. *Key components of the inventory management principle*

Functions of Inventory

Despite the pitfalls of holding inventory, there are a number of reasons why many organizations keep inventory. These reasons are directly tied to the functions of inventory described next:

- **Buffer inventory**—This is an extra cushion that is kept over an extended period to prepare for unexpected demand surge and subsequently prevent potential stockouts. This kind of inventory is often needed to reduce the risk of shutting down the production line and distribution disruptions. The buffer inventory is also called *safety stock*.

- **Lot-size inventory**—This is inventory accumulated as a result of over-purchasing and over-production to take advantage of economies of scale. To elaborate, many organizations purchase some products in volume to exploit quantity (volume) discount or freight discount opportunities. Also, a company can reduce per-unit setup costs by mass-producing goods at once. This kind of inventory is also called *cycle stock*.

- **Anticipation inventory**—This is inventory built up in anticipation of future estimated demand, price increases, promotional campaigns, seasonal fluctuations, labor strikes, or plant shutdowns. This kind of inventory helps maintain level production or distribution that prevents overtime or idle time in the future. The seasonal inventory that is kept for the upcoming Thanksgiving holiday or Christmas season is a good example of anticipation inventory. Another example includes speculative inventory that is kept to protect the company from future price inflations or product shortages.

- **Pipeline (or transit) inventory**—This is inventory that exists due to transportation time lag. In other words, products in transit can be considered to be inventory because products that were pre-ordered but have not reached warehouses, distribution centers, or retail stores are still not available for customers. This kind of inventory can be reduced by using a faster mode of transportation.

- **Decoupling inventory**—This is inventory that makes every stage of the supply chain independent of any unexpected supply chain disruptions. For instance, machine breakdowns during the manufacturing stage will shut down the production process, which can subsequently create either shortages of work-in-process inventories at the next production stage or product shortages at the following distribution stage. Therefore, decoupling inventory kept between various stages of the supply chain will allow paced production and distribution without interruptions.

Types of Inventory

In terms of product transformation processes, inventory can take a variety of forms. These forms are generally classified into raw materials, parts/components, work in process, finished goods, and supplies. Here are detailed descriptions of these forms:

- **Raw materials**—Raw materials are virgin physical resources or items that are purchased from external organizations or extracted from natural sources such as mines to be used for making parts/components or products. Examples include steel (or iron ore) for the nuts and bolts, plastic for the automobile dashboard, glass for windows, wheat for bread, aluminum for a bucket, and rubber for tires.
- **Parts/components**—Parts and components are major ingredients that will make up a finished product and make it function properly. Examples include a battery and engine for an automobile, a hard drive for a personal computer, a dust bag for a vacuum cleaner, and a knob for a desk drawer.
- **Work in process**—Work in process (WIP) refers to items that are still in the manufacturing process and are waiting to be processed within the supply chain. WIP often plays the role of decoupling inventory.
- **Finished goods**—Finished goods are completed items that are ready for sale or distribution. In other words, finished goods are the ones that do not require any more of the transformation process.
- **Supplies**—Supplies are items (including finished goods) needed to get engaged in the production process, but do not comprise the finished product. Examples include maintenance, repair, and operating (MRO) supplies; pencils, pens, and paper needed for drawing the blueprint of a newly designed automobile; and light bulbs that allow the worker to get engaged in the process of producing an automobile engine inside a dark manufacturing plant.

Inventory Classification

With the growing need for customization of products, many manufacturers have diversified their product lines. For instance, milk products can be categorized into several different stock-keeping units (SKUs) such as whole milk, 2% low-fat milk, zero-fat milk, lactose-free milk, chocolate flavor milk, and so forth. Such product proliferation can overwhelm the task of managing all SKUs with equal attention. Therefore, priorities should be set to handle inventories. One of the systematic ways of classifying inventories with respect to their priorities includes a classic Pareto principle, where approximately 10% to 20% of the inventory accounts for approximately 60% to 80% of the total inventory value, or 80% of the sales volume is typically accounted for by 20% of the products (customers). This principle originated from the Pareto principle, named after a nineteenth-century Italian economist Villefredo Pareto, who discovered that a "vital few" (about 20% of the population) controlled a vast majority (80%) of the wealth in Milan, Italy. The Pareto principle, based on the observation that many situations were dominated by a very few elements, was broadened to classify inventory into several different categories with respect to their priorities. The most popular inventory classification schemes include critical value analysis and ABC inventory analysis, which will be discussed in detail in the following subsections.

Critical Value Analysis

Critical value analysis is frequently used by military organizations, which "subjectively" classify inventory items by assigning point values for three to five categories. The manner in which inventories are classified by critical value analysis is summarized in Table 5.1. Examples of the top-priority items might include ammunition and rifles, whose shortages can jeopardize solders' lives. Examples of high-priority items might include military vehicles such as Humvees and armored vehicles, whose shortages can undermine the soldiers' fighting capability. Examples of medium-priority items might include military tents and shovels whose shortages would not create life-threatening situations for soldiers but are needed for their comfort. Examples of low-priority items might include military uniforms, shirts, and underwear, which are not essential for combat readiness.

Table 5.1. An Inventory Classification Made by Critical Value Analysis

Category	Item Features
1. Top priority	No stockout—critical items
2. High priority	Essential, but limited stockouts permitted
3. Medium priority	Necessary, but occasional stockouts permitted
4. Low priority	Desirable, but stockouts allowed

ABC Inventory Analysis

ABC inventory analysis classifies inventory items or SKUs into three categories—"A" items, "B" items, and "C" items—according to the total annual dollar usage. ABC inventory analysis was initially developed in 1951 by H. Ford Dicky of General Electric to categorize inventory items into three classes according to relative sales volume, cash flows, lead time, or stockout costs. Potential criteria for the ABC classification scheme include the following:

- Annual dollar volume of the transactions for an item
- Unit cost (price)

- Scarcity of materials used in producing an item
- Availability of resources (e.g., manpower, facilities) to produce an item
- Lead time
- Storage requirements (e.g., refrigeration) for an item
- Pilferage/shrinkage risk, shelf life
- Cost of running out of stock
- Engineering design volatility

A rule of thumb for illustrating the ABC classification scheme is shown in Table 5.2.

Table 5.2. An Illustration of the ABC Classification Scheme

Question	Classification
1. Annual usage > $50,000 2. Unit cost (price) > $500 3. Lead time > 6 months 4. Shelf life < 3 months	A
1. $10,000 ≤ annual usage ≤ $50,000 2. $100 ≤ unit cost ≤ $500 3. 3 months ≤ lead time ≤ 6 months 4. 3 months ≤ shelf life ≤ 6 months	B
1. Annual usage < $10,000 2. Unit cost < $100 3. Lead time < 3 months 4. Shelf life > 6 months	C

Figure 5.2 graphically depicts the results of ABC classification scheme. Table 5.3a summarizes the degree of control and planning associated with the ABC inventory classification scheme. Table 5.3b provides specific guidelines for managing different types of inventories (*A Comprehensive Guide on Materials and Supply Chain Management*, 2007).

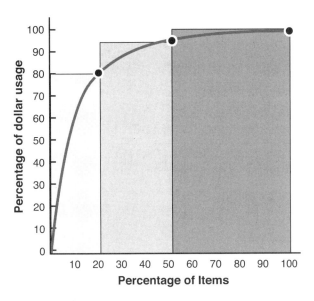

Figure 5.2. *The graphical display of the ABC classification scheme*

Table 5.3a. ABC Inventory Control and Planning

Item	Degree of Control	Type of Records	Lot Size	Frequency of Review	Size of Safety Stocks
A	Tight	Accurate and complete	Small	Continuous (frequent)	Small
B	Moderate	Good	Medium	Occasional	Moderate
C	Loose	Simple	Large	Infrequent	Large

Table 5.3b. ABC Inventory Management Guidelines

Item	Management Responsibility	Purchasing Strategy	Potential Sources of Supply	Frequency of Deliveries	Follow-ups
A	Senior manager	Centralized purchasing	Multiple	Weekly	Frequent
B	Middle level manager	Either centralized or decentralized	Double or more	Every three months	Periodic
C	Delegated to a floor-level employee or inventory clerk	Decentralized purchasing	Single or double	Every six months	Occasional

Chapter 5: Inventory Control and Planning

Because a shortage or unavailability of higher-dollar usage or fast-moving items can have a greater adverse impact on the customer service level, different customer service strategies should be developed in accordance with the ABC classification, as illustrated in Table 5.4.

Table 5.4. Customer Service Strategies in the ABC Classification Scheme

Item	Percent of Sales (1)	Customer Service Level (2)	Weighted Customer Service Level (3) = (1) × (2)	Management Strategy
A	70%	99%	69.3%	• Frequent evaluation of forecasts • Frequent (at least once a month) cycle counting • Daily updates of records • Frequent review of order quantities • Small orders • Frequent follow-up
B	20%	95%	19%	• Quarterly cycle counting • Medium size orders
C	10%	90%	9% Total: 97.3%	• Periodic review or visual replenishment (two bin system) • Large order quantities • Count items infrequently

Independent Demand Inventory Control and Planning

Because the inventory characteristics tied to demand patterns require a different inventory control mechanism, inventory is often classified as either independent demand inventory or dependent demand inventory. Independent demand inventory refers to an item whose demand is unrelated to other items and is therefore not affected by the demand patterns of the other items. Examples include finished goods representing the outbound flow of the supply chain, such as automobiles, furniture, apparel, appliances, televisions, and personal computers. It also includes supplies such as pens, pencils, and paper. On the other hand, dependent demand inventory refers to an item whose demand is a direct result of a need for some other item and is therefore not subject to random demands of end customers. For example, the need for Halloween candies will drive a demand for sugar. Similarly, the need for sofas will increase a demand for fabrics or leather. Examples of dependent demand inventory are raw materials (e.g., oil, chemicals, corns, steel), parts/components (e.g., batteries, tires, windows, computer chips), and subassemblies. Dependent demand inventory typically represents the inbound flow of the supply chain and often requires bulk shipments. In the following subsections, several inventory models intended for controlling independent demand inventory are introduced.

Economic Order Quantity Model

In determining the level of inventory replenishment and the order timing, one of two types of multiple-period inventory systems can be used: a perpetual system (fixed order quantity or continuous inventory or Q-model) or a periodic system (fixed order interval or P-model). As summarized in Table 5.5, in a perpetual system, a new order is triggered whenever the inventory level reaches a specified (predetermined) level, called a *reorder point*. Because the inventory level is constantly monitored under this system, the reaction time to stockouts will be quicker and consequently the risk of stockouts will be reduced. Therefore, a perpetual system is appropriate for controlling more expensive and critical items such as replacement parts and supplies, although it would be time-consuming to manage. Here are the pros and cons of the perpetual system:

- Advantages
  - The order size (e.g., economic order quantity) is optimal and meaningful.
  - It is relatively insensitive to forecasting results and parameter changes.
  - Safety stock is needed only for the lead time period.
  - Easier to uncover shrinkage and theft.
  - Stock level at any given point in time is updated on continuous or real-time basis.
- Disadvantages
  - The order quantity may not be changed for years.
  - Clerical errors and mistakes in record keeping make the system impotent.
  - Numerous independent orders may result in high purchasing and transportation costs.
  - Opportunities for placing large aggregated orders, which can result in discounts, may be reduced.

On the other hand, in a periodic system, a new order is initiated on a periodic basis (e.g., every Tuesday, at the end of every month) to bring the inventory level back up to a desired (predetermined) target level. Depending on demand fluctuations, order size will vary from one order to the other. Also, due to less-direct control and little inventory recordkeeping, this system may result in either larger inventory levels or inventory shortages. This system is often used in grocery stores, drug stores, and mom-and-pop retail stores. Between these two contrasting systems, an economic order quantity (EOQ) makes more sense for a perpetual system under which the same order amount that minimizes total inventory costs can be placed throughout the year.

Table 5.5. The Differences between the Perpetual System and the Periodic System

	Perpetual	**Periodic**
Order Quantity	Fixed	Variable
Order Period	Variable	Fixed
Predetermined Value	Reorder point	Target inventory
Safety Stock	Less	Large
Stockout Possibility	During lead time (less stockout possibility)	During lead time and cycle time
Time to Maintain	Lengthy due to perpetual recordkeeping	Shorter
Type of Items	High-priced, critical, important items	Low-priced, less important, and multiple type goods

An economic order quantity (EOQ) is referred to as an optimal order quantity that minimizes total annual inventory cost and answers the questions of how much and when to place a replenishment order. Its concept has been around since the early 1900s and is generally credited to Ford Harris of Westinghouse. EOQ is derived under the following assumptions:

- Demand for products is known with certainty and is constant over time.
- Lead time (time between the placement of an order and its receipt) is constant.
- The unit price of a product is constant over time. Also, the unit price of the product remains the same regardless of order quantity; that is to say, there are no economies of scale.
- Ordering and setup costs are fixed and constant.
- All demands for products will be met; therefore, no shortage or backorder will be allowed.
- Products are ordered independently of each other.

Under the preceding assumptions, EOQ can be determined by minimizing the total annual inventory cost, which is composed of annual inventory carrying (holding) cost and annual ordering cost. The annual inventory carrying cost is further subdivided into four cost components: cost of capital, inventory service, storage space, and inventory risk, as shown in Figure 5.3, which also illustrates the detailed cost components of the inventory carrying cost. Among these, cost of capital is usually the largest component of the inventory carrying cost. The inventory carrying cost can be calculated by multiplying the average inventory level by the annual carrying cost per unit. The annual carrying cost per unit can be estimated by multiplying the annual inventory carrying charge (in percentage) by the unit price of the product.

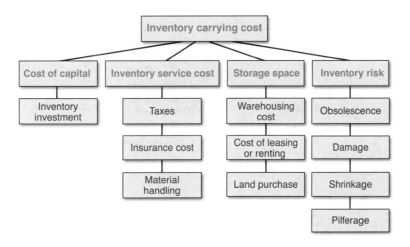

Figure 5.3. *Key components of annual inventory carrying cost*

As opposed to inventory carrying cost, which includes both fixed and variable cost components, ordering cost is a variable cost incurred from the point of order placement to the point of order receipt. Therefore, the ordering cost is typically composed of order placement and order receipt costs (see Bernard, 1999). These costs can be further subdivided into a variety of cost components, as shown in Figure 5.4.

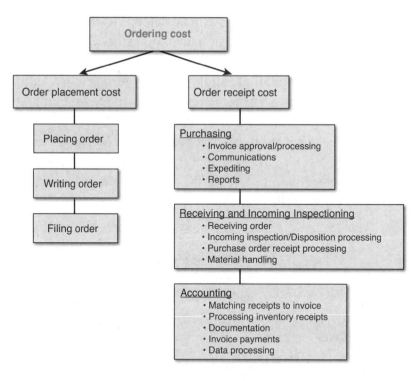

Figure 5.4. *Key components of annual ordering cost*

By summing both inventory carrying cost and ordering cost, the company can calculate total annual inventory cost (TC). It is mathematically expressed as follows:

TC = (inventory carrying cost) + (ordering cost)

= (average inventory level) × (unit price × inventory carrying charge in percentage) + (order frequency) × (ordering cost per order)

$$= \left(\frac{Q}{2} P \cdot i\right) + \left(\frac{D}{Q} S\right) \tag{5.1}$$

where Q = order quantity to be determined

P = unit price of product

i = inventory carrying charge in percentage

D = annual demand

S = ordering cost per order

By taking derivatives with respect to Q, the optimal value of Q can be determined as follows:

$$\frac{\partial TC}{\partial Q} = \frac{P \cdot i}{2} - \frac{DS}{Q^2} \tag{5.2}$$

This equation can be rewritten as:

$$0 = \frac{P \cdot i}{2} - \frac{DS}{Q^2} \quad (5.3)$$

$$\frac{P \cdot i}{2} = \frac{DS}{Q^2} \quad (5.4)$$

$$EOQ \text{ (optimal order quantity } Q) = \sqrt{\frac{2DS}{P \cdot i}} = \sqrt{\frac{2DS}{H}} \quad (5.5)$$

where H = unit inventory carrying cost

Once EOQ is determined, order cycle time can be determined like so:

Order cycle time = (annual working days)/(order frequency)

$$= \frac{W}{D/EOQ} = \frac{W \cdot EOQ}{D} \quad (5.6)$$

where W = annual working days

The reorder point at which a new order should be placed to replenish inventory can be calculated as follows:

$$R = \text{(expected daily demand)} \times \text{(lead time in days)} + \text{safety stock}$$
$$= d \cdot LT + \text{safety stock} \quad (5.7)$$

where

d = expected daily demand

LT = lead time in days

To illustrate how the EOQ model works, let's suppose that an auto part store in Chicago would like to replenish its battery products for the upcoming winter season, during which many automobiles are expected to replace their dead batteries. Based on the last several years of sales history, the store's estimated annual demand for batteries is 1,800 units. The unit price of a new battery is $100, and its inventory carrying charge is estimated to be 25%. Its ordering cost per order is $25. Table 5.6 summarizes all the parameter values needed for determining the optimal order quantity.

Table 5.6. Parameters for Determining EOQ

Parameter	Value
Annual demand	1,800 units
Unit price (P)	$100
Inventory carrying charge (i)	25% (or .25) annually
Ordering cost per order (S)	$25
Annual working days (W)	300 days
Lead time (LT)	5 days

Using Equation 5.5 and the data contained in Table 5.6, EOQ equals $\sqrt{\frac{2(1,800)(25)}{(100)(0.25)}}$, or 60 units.

If this store orders 60 units of new batteries, it can minimize its total inventory cost. By plugging EOQ into Equation 5.1, we can calculate the minimum total inventory cost.

First, the annual inventory carrying cost would amount to $750, or $\frac{(60)(100)(.25)}{2}$. Second, annual ordering cost would amount to $750, or $\frac{(1800)(25)}{60}$. Summing up these two costs, we see that the minimum total inventory cost will be $1,500. Notice that annual inventory carrying cost is tantamount to annual ordering cost, when the optimal order quantity of 60 units of new batteries is ordered. This result indicates that our EOQ calculation is correct, as displayed by Figure 5.5. Using Equation 5.6, order cycle time turns out to be $\frac{(300)(60)}{1800}$, which is ten days. This means that every ten days, 60 units of batteries should be ordered to replenish inventory. Also, using Equation 5.7, we see that the reorder point turns out to be $\frac{(1800)}{300} \times 5$, or 30 units. This means that whenever the inventory level reaches 30 units with no need for a safety stock, a new order should be triggered to avoid a stockout situation.

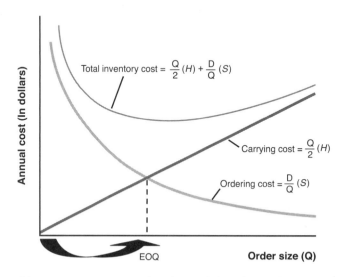

Figure 5.5. *The total inventory cost curve for determining the optimal order quantity*

Quantity Discount Model

The EOQ model introduced earlier is based on six assumptions. However, some of those assumptions might not be representative of actual situations in practice. For example, if you buy two pairs of shoes, it is not uncommon that some shoe stores will offer you a 50% discount for second pair during the sales period. Likewise, many suppliers offer volume discounts on products if a certain volume of product (so-called price-break quantity) is purchased at once. In this situation, because the order quantity affects the unit price of order product, the annual purchasing cost should be added to the total inventory cost. Therefore, the total inventory cost equation can be expressed as follows:

TC = inventory carrying cost + ordering cost + purchasing cost

$$= (\frac{Q_k}{2} P_k \cdot i) + (\frac{D}{Q} S) + P_k \cdot D \qquad (5.8)$$

where

Q_k = order quantity k^{th} discount interval

P_k = unit price of product at price break quantity k

D = annual demand (daily demand × annual working days)

From this equation, a straight EOQ formula cannot be derived due to the changing unit price of the product according to order quantity. Therefore, as displayed in Figure 5.6, the total inventory cost curve with multiple price breaks would be discrete, as opposed to the continuous curve associated with the basic EOQ model shown earlier in Figure 5.4. That is to say, a simple application of the EOQ formula to a quantity discount situation will not produce an optimal order quantity. The multiple step procedure summarized in Figure 5.7, however, can guide us to find the optimal order quantity under various price breaks. To illustrate how this procedure works, let's suppose a local moving company buys boxes. Its box supplier quotes a price of $1.00 per box for quantities of 10,000 boxes or more, $1.10 per box for orders of 5,000 to 9,999 boxes, and $1.20 per box for lesser quantities. This moving company operates 200 days per year. Daily demand for boxes is 250 units, and ordering cost per order is $50. If annual carrying charges are 20% of the unit purchasing price, what will be the optimal order size? To answer this question, we first calculate the EOQ for the lowest price break point by following the steps described in Figure 5.6. Initial EOQ (P_1 = 1.00) is 5,000 (units), which equals $\sqrt{\frac{2(50)(50000)}{(1.00)(.20)}}$. However, if the company orders a quantity of 5,000 boxes, it cannot take advantage of a 20-cent discount from the regular price. In other words, this EOQ is not on the lowest price curve and is therefore considered "infeasible." Therefore, this EOQ cannot be an optimal order quantity, so we need to take the next step and compute another EOQ with the next lowest unit price ($1.10). EOQ ($P_2$ = 1.10) is 4,767 (units), which equals $\sqrt{\frac{2(50)(50000)}{(1.10)(.20)}}$. This EOQ is still infeasible, because 4,767 is lower than the next lowest price break quantity of 5,000. Therefore, we keep calculating EOQ for the highest unit price which is a regular price of $1.20. This EOQ turns out to be 4,564, or $\sqrt{\frac{2(50)(50000)}{(1.20)(.20)}}$. Because the EOQs for two lower-price break quantities are infeasible, we have to calculate and then compare the total inventory costs using Equation 5.8. The results of this calculation are as follows:

TC ($P_3 = \$1.20$, $Q_3 = 4{,}564$) = $\frac{4564}{2} \times (1.20)(.20) + \frac{50000}{4564} \times 50 + (1.20)(50{,}000)$
= $61,096

TC ($P_2 = \$1.10$, $Q_2 = 5{,}000$) = $\frac{5000}{2} \times (1.10)(.20) + \frac{50000}{5000} \times 50 + (1.10)(50{,}000)$
= $56,050

TC ($P_1 = \$1.00$, $Q_1 = 10{,}000$) = $\frac{5000}{2} \times (1.00)(.20) + \frac{50000}{5000} \times 50 + (1.00)(50{,}000)$
= $51,000

From these results, it is apparent that the order quantity of 10,000 units will minimize the total inventory cost because it yields the lowest total cost among the three options. As such, the optimal order quantity will be 10,000 units. This example illustrates that quantity discounts may lead to a larger purchase quantity to exploit purchase cost savings. However, it should be noted that the lowest total inventory cost will not always result from the largest order quantity associated with the lowest price break.

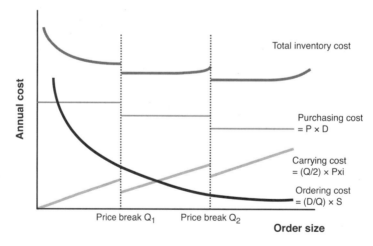

Figure 5.6. *The total inventory cost curve when quantity discounts are available*

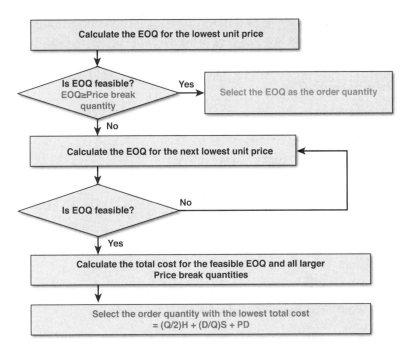

Figure 5.7. *The steps for finding the optimal order quantity under price breaks*

Quantity and Freight Discount Model

The increasing commitments of companies to Just-In-Time (JIT) principles forces them to reassess the basic EOQ model and its variations that fail to solve the dilemma of small orders incurring high freight costs versus large orders incurring high inventory carrying costs. The dilemma has become more serious than ever before because recent transportation deregulatory laws broaden cost differentials between small and large orders. Such a dilemma can be solved by developing an improved inventory model that not only maintains a simple principle of the EOQ model, but also identifies an optimal order quantity when both volume and shipping discounts are available.

Prior to presenting the model, we make the following underlying assumptions:

- Demand is known and continuous at a constant rate.
- The discount price or freight-rate applies to "all units" purchased. That is, once the ordered quantity exceeds the specified minimum amount for the discount or freight rate, the same unit price or freight rate is applied to every ordered item.
- The buyer is responsible for paying the freight cost. In other words, the buyer incurs all transportation costs, including in-transit carrying costs, because he or she bears the risk of damage or loss in transit.

- Transit (lead) time is known and deterministic. More importantly, in-transit inventory carrying costs are directly proportional to average transit time.

Within the framework of the inventory theoretic model proposed by Baumol and Vinod (1970), the total annual inventory cost is expressed as follows:

Total Annual Inventory Cost

= total annual direct shipping cost

+ total annual in-transit inventory carrying cost

+ ordering cost + recipient's inventory carrying cost   (5.9)

Adding purchasing cost, we see that Equation 5.9 becomes

Total Annual Inventory Cost

= total annual direct shipping cost

+ total annual in-transit inventory carrying cost

+ ordering cost + recipient's inventory carrying cost

+ purchasing cost

or, more specifically,

= (unit freight rate) × (order size)

+ (cost per unit of time) × (average transit time)

× (order size) + (cost per order) × (number of orders)

+ (average inventory level) × (per unit per year carrying cost)

+ (unit purchasing cost) × (order size)   (5.10)

In the presence of volume and shipping discounts, Equation 5.10 can be mathematically stated using the notations defined in Table 5.7.

Table 5.7. Summary of Notations

Symbol	Description
Q_i	Order quantity in units in the ith discount interval
D	Annual demand requirement
C_t	Carrying cost of in-transit inventory per unit per year
L_i	Average transit time (lead time) in the ith discount interval
H_p	Inventory carrying charge (fraction) per dollar of inventory per year
F_i	Unit freight rate in i^{th} discount interval
P_i	Unit purchasing cost at the vendor
S	Ordering cost per order
B_i	The ith price-break quantity
t_i	The ith freight-break quantity

$$TC(Q_i) = F_i D + C_t L_i D + S\frac{D}{Q_i} + \frac{Q_i}{2}(P_i + F_i)H_p + P_i D \quad (5.11)$$

By grouping terms in Equation 5.11, we obtain this:

$$TC(Q_i) = (F_i + C_t L_i + P_i)D + S\frac{D}{Q_i} + \frac{Q_i}{2}(P_i + F_i)H_p \quad (5.12)$$

By taking the derivative of the annual total cost with respect to the order size (Qi) and setting it equal to zero, we get the following optimal order size (Qi*):

$$Q_i^* = \sqrt{\frac{2DS}{(P_i + F_i)H_p}} \quad (5.13)$$

With all-units quantity and freight-rate breaks, the optimal lot size Qi* is valid (feasible) only if it falls within the discount interval corresponding to its unit purchase and shipping cost. Consequently, the "local" optimum order size Q_l^* corresponding to the ith discount interval is as follows:

$$Q_l^* = \begin{cases} b_{i-1} & \text{if } Q_i^* < b_{i-1} \text{ and } b_{i-1} > t_{i-1} \\ t_{i-1} & \text{if } Q_i^* < b_{i-1} \text{ and } b_{i-1} < t_{i-1} \\ Q_i^* & \text{if } b_{i-1} \leq Q_i^* \leq b_i \text{ and } t_{i-1} \leq Q_i^* \leq t_i \\ b_i & \text{if } Q_i^* > b_i \text{ and } b_i < t_i \\ t_i & \text{if } Q_i^* > b_i \text{ and } b_i > t_i. \end{cases} \quad (5.14)$$

If Q_i^* is not valid, the global optimum order quantity, Q_g^*, is the local optimum order quantity, which minimizes the annual total cost function (Equation 5.14) at the corresponding Q_l^*. The following sample problem illustrated in Table 5.8 shows the aforementioned logical procedure for determining the global optimum order size.

Table 5.8. Sample Data for the Example

Annual demand requirement = 36,000 units				
Ordering cost per order = $100				
Carrying cost fraction per unit per year = .10				
In-transit carrying cost per day = $2				
Order Size	Unit Purchasing Cost	Unit Freight Rate	Average Transit Time	Transportation Mode
Less than 500	$50	$20	15 days	LTL
500–999	$48	$20	15 days	LTL
1,000–1,999	$45	$20	15 days	LTL
2,000–3,999	$45	$12	12 days	TL
4,000 or above	$45	$5	16 days	CL

The sample data represents both quantity and freight rate discount schedules with three price breaks and freight rate breaks. The purchased items can be shipped via truck at an LTL rate ($20 per unit) or a TL rate (12 per unit). Herein, a 40% discount is applied to the TL rate to reflect actual situations. In practice, TL shipment is known to be faster than LTL shipment (Ballou, 1985). Accordingly, the average transit time of the LTL shipment is estimated to be 15 days, whereas that of the TL shipment is estimated to be 12 days. In addition, we consider the case where the purchased items can be shipped via rail at a CL rate ($5 per unit). Because rail is the slowest mode, average transit time of the CL shipment is estimated to be 16 days.

To evaluate these alternative discount schedules, we first calculate the EOQ for the lowest price and freight rate. By solving the EOQ formula given in Equation 5.13, we obtain the following:

$$EOQ = \sqrt{2(36000)(100)/50(0.10)} = 1200 \text{ units}.$$

This EOQ is invalid because it is less than the lowest freight rate break quantity (i.e., 4,000 units). Because this EOQ value is still less than 2,000 units, the EOQ for the second lowest freight rate will be invalid. The largest feasible EOQ occurs at an order quantity of 1052 units because it is within the range of 1,000 and 1,999 units. Based on equation (5.14), the local optimum quantities to consider are the largest feasible EOQ and two freight rate break quantities larger than that amount. To choose the global optimum order quantity among these, we compute the following total costs using Equation 5.12:

TC (1,052) = [20 + 2(15) + 45] (36,000) + 100(36,000)/1,052 + 1,052(65) (0.10)/2
= $3,426,841

TC (2,000) = [12 + 2(12) + 45] (36,000) + 100(36,000)/2,000 + 2,000(57) (0.10)/2
= $2,923,500

TC (4,000) = [5 + 2(16) + 45] (36,000) + 100(36,000)/4,000 + 4,000(50) (0.10)/2
= $2,962,900

Because the order quantity of 2,000 units yields the lowest total annual cost of $2,923,500, the global optimum order quantity is 2,000 units.

Risk Pooling Model

Risk pooling refers to a strategy of aggregating demand across different stocking locations by centralizing stocking locations (e.g., warehouses) to reduce demand variability and subsequently decrease safety stocks. In a risk pooling model, there are two measures of demand variability: standard deviation, which measures the absolute variability of customer demand, and coefficient of variation, which measures the variability of customer demand relative to the average demand. Coefficient of variation is calculated as follows (Simchi-Levi et al., 2008):

Coefficient of variation = standard deviation/average demand

The higher the coefficient of variation, the greater the benefit gained from risk pooling. Also, pooling lead time risk by splitting orders simultaneously may enable companies to reduce inventories without sacrificing service or significantly increasing order-processing costs. Another form of risk pooling is to share inventory among different retail outlets.

Dependent Demand Inventory Control and Planning

As summarized in Table 5.9, the demand characteristics of dependent demand inventory are quite different from those of independent demand inventory. Therefore, the inventory control and planning for dependent demand require different approaches. These include material requirements planning (MRP), which is referred to as a time-phased, priority-dependent inventory control and planning system, which calculates the material requirements and schedules orders to meet changing demand while minimizing unnecessary inventories. MRP consists of a set of logically related procedures, decision rules, and records designed to translate a master production schedule into time-phased net requirements, and the planned coverage of such requirements, for each dependent demand inventory item needed to implement this schedule (Orlicky, 1975).

Table 5.9. Independent Demand versus Dependent Demand Inventory

Characteristics	Independent Demand	Dependent Demand
Demand pattern	Is influenced by market conditions; originates from outside the company.	Is related to the demand for another item; originates within the company.
Order pattern	Is often made up of numerous small orders (continuous and uniform).	Is often lumpy (discrete).
Estimation	Might occur at uncertain times; should be forecasted.	Can be better controlled by the company when it occurs; should be calculated (derived).
Example	Finished goods, supplies, spare parts, etc.	Work in process, raw materials

The main objectives of MRP are to ensure the availability of materials, components, parts, and subassemblies for planned production by planning manufacturing activities, delivery schedules, and ordering processes ahead of the actual needs. The principal functions of MRP are as follows:

- Order the right material—i.e., what to order
- Order in the right quantity—i.e., how much to order
- Order at the right time—i.e., when to order

The basic needs of MRP include

- **Master production schedule**—A forecast of what products need to be made in the next few months
- **Bill of material**—A recipe that tells you which materials are used, in what quantities, to build each product
- **Lead time**—The time required to obtain or manufacture all products and materials
- **Batch size**—The maximum amount that can be processed at any one time
- **Inventory balance**—The on-hand stock balance of all your products and materials

Among these needs, as shown in Figure 5.8, the three primary inputs of MRP are a master production schedule (MPS), bill of materials (BOM), and the inventory record/status file (or item master file). To elaborate, MPS is a statement summarizing the exact quantity and timing of producing finished products that will meet the anticipated demand. MPS basically drives the MRP process. BOM is often dubbed as a product structure tree that hierarchically lists all the raw materials, components/part, and subassemblies required to produce a finished product. The detailed functions of BOM are to

- Provide tree views of needed parts, components, and materials (e.g., manufacturer part numbers, defined alternate approved materials, and sequencing for materials).
- Automate the planning and purchasing of materials with start and finish dates.
- Allow users to create work orders by final end quantity or on full or partial batch multipliers.
- Allow users to easily add, change, or delete the components required for any work order where changes are frequent.
- Support calculation of costs that include a variety of labor and overhead calculations

The inventory record file represents a database that contains detailed information about the amount of inventory on hand, the amount and timing of scheduled order to receive, and the amount and timing of future orders that will meet the demand requirements.

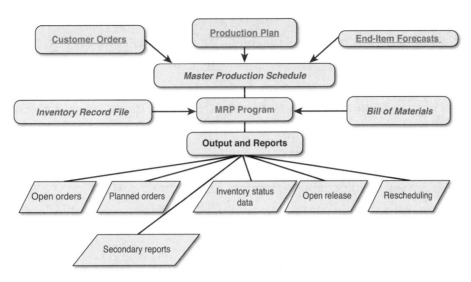

Figure 5.8. *MRP inputs and outputs*

As recapitulated in Table 5.10, MRP is radically different from EOQ in many respects. Therefore, MRP requires different processes.

Table 5.10. A Comparison of EOQ with MRP

EOQ	MRP
Independent demand	Dependent demand
Continuous/uniform demand	Discrete/lumpy demand
Continuous lead time demand	No lead-time demand
Reorder point ordering signal	Time-phased ordering signal
Historical demand base	Future production base
Forecast all items	Forecast end items only
Quantity-based system	Quantity and time-based system
Safety stock for all items	Safety stocks for end items only
Items for sale	Items for internal usage
Reactive	Proactive

The MRP Calculation Process

Per Russell and Taylor (2006), MRP is designed to address the following questions by taking three calculation steps: *netting* (a process of subtracting on-hand inventory and

Chapter 5: Inventory Control and Planning   149

scheduled receipts from gross requirements to produce net requirements), *lot sizing* (a process of determining the quantities in which items are usually made or purchased) and *time phasing* (a process of subtracting an item's lead time from its due date to know when to order an item):

- Which parts, components, and materials are needed to produce the end item?
- Do we have any of those parts, components, and materials already built or secured? How much?
- Which parts, components, and materials called for by the BOM are in stock and which ones need to be bought or made, and when do they have to be here?
- How can we schedule production or purchase orders to meet future needs/demands?

To answer these questions, each entry in the MRP matrix should be either predetermined or calculated. Entries in the MRP matrix include *gross requirements,* which are the total amount required for a particular item that can be derived from either the MPS or planned order release of the parent for parts, components, and materials. Therefore, for dependent demand inventory, gross requirements are tantamount to the amount of planned order release of the parent. *Scheduled receipts* represent orders such as work-in-process or in-transit items that have been already ordered and that are scheduled to arrive at future periods. Changing these receipt quantities or dates will incur real cost. *On-hand inventory* is the amount of inventory that is currently on hand or projected to be on hand and available for use at the end of each period. On-hand inventory can be calculated as follows:

On-hand in the previous period

- gross requirements in the previous period
+ scheduled receipts in the previous period
+ planned order receipts in the previous period
- safety stock
= on-hand inventory in the current period

Therefore, if on-hand inventory turns out to be negative, the negative amount of on-hand inventory represents an inventory shortage that requires additional production or order. *Net requirements* are the actual amount of a particular item that needs to be produced or ordered after taking into account on-hand inventory and scheduled receipts. Therefore, net requirements can be calculated as below:

Net requirements = gross requirements - scheduled receipts - on-hand inventory

(Only the positive figure is considered; if it is negative, it should be left out of the calculation.)

It should be noted that if the result of net requirements is a negative figure, the net requirements are regarded as zero. *Planned order receipts* are the amount of an order that needs to be placed to meet net requirements. It should be noted that this amount can differ depending on the particular lot-sizing rule that is used. This includes either a static or dynamic rule. A static rule can be subdivided as follows:

- An EOQ rule that orders minimum quantity
- A periodic order quantity rule that orders maximum quantity

On the other hand, under a dynamic lot-for-lot (L4L) rule, the exact quantity will be ordered. With L4L ordering, planned order receipts are tantamount to net requirements. Finally, *planned order releases* are planned order receipts offset by the lead time. Planned order releases are so-called "pretend" orders whose changes will not incur any cost.

Example of MRP

To better understand how an MRP system works, let's consider the case of a furniture manufacturer that assembles a folding chair. Figure 5.9 shows its product structure tree or BOM (diagrammatic recipe). This also can be summarized in a non-graphical fashion, as illustrated in Table 5.11. The BOM shows that the gross requirements for one finished chair include one seat, one seatback (back support), and four legs. Let's suppose that 100 units of folding chairs should be delivered to this manufacturer's customer in eight weeks from now, with one week of lead time. The lead time needed to buy seats and backseats is two weeks, whereas the legs need only one week of lead time to acquire from the supplier. The lead time required to procure all other parts, such as tapping screws, river nuts, washers, crossbeams, and angel stops, is one week. After checking inventories for chair components, this manufacturer discovered that there are 20 units of seat inventory on hand, 30 units of seatback inventory, 150 units of tapping screws, 40 units of front legs, and 30 units of back legs on hand. There is no inventory on hand for the other components and parts. To figure out what components, how much of them, and when they should be ordered to meet the production schedule, this manufacturer creates the inventory record file illustrated in Figures 5.10, 5.11, and 5.12 using the MRP calculation process explained earlier.

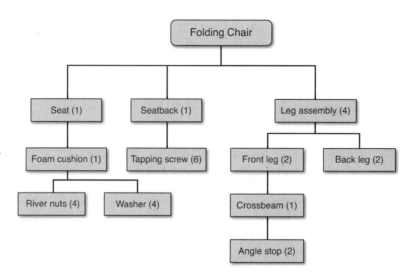

Figure 5.9. *A product structure tree for a folding chair*

Table 5.11. Bill of Materials for a Folding Chair (in a Non-graphical Format)

Item Number	Description	Material Type	Total Weight in Product (grams)	Purchased Item?
1	Folding chair			No
2	Seat	Steel	1,366	Yes
3	Seatback	Plastic	279	Yes
4	Legs	Steel		Yes
5	Front leg	Steel	1,360	Yes
6	Back leg	Steel	394	Yes
7	Crossbeam	Steel	312	Yes
8	Seat cushion	Polyurethane	56	Yes
9	Tapping screw	Stainless steel	3	Yes
10	River nut	Steel	4	Yes
11	Washer	Steel	2	Yes
12	Angle stop	Rubber	5	Yes

Week	1	2	3	4	5	6	7	8
Quantity								100

[Inventory record file for seat, LT = 2 weeks]

Week	1	2	3	4	5	6	7	8
Gross requirements							100	
Scheduled receipts								
On hand from prior period	20	20	20	20	20	20	(-80)	
Net requirements							80	
Planned order receipt							80	
Planned order release					*80*			

Figure 5.10. *Inventory record file for the seat component*

Week	1	2	3	4	5	6	7	8
Quantity								100

[Inventory record file for seatback, LT = 2 weeks]

Week	1	2	3	4	5	6	7	8
Gross requirements							100	
Scheduled receipts								
On hand from prior period	30	30	30	30	30	30	(-70)	
Net requirements							70	
Planned order receipt							70	
Planned order release					*70*			

Figure 5.11. *Inventory record file for the seatback component*

From Figures 5.10 and 5.11, we learn that the furniture manufacturer should place an order of 80 units of seats and 70 units of seatbacks in week 5 to meet the production schedule. Figure 5.12 also indicates that 270 additional units of tapping screws should be ordered in the third week to avoid any shortages of tapping screws for scheduled production. Using the same logics and procedures, the company can develop inventory record files for the remaining components and parts for the folding chairs. As illustrated, MRP helps the company determine what components and parts are needed to produce the finished product and which components and parts called for by the BOM are in stock and which ones the company has to buy, and when they have to be available for production.

Week	1	2	3	4	5	6	7	8
Gross requirements							100	
Scheduled receipts								
On hand from prior period	30	30	30	30	30	30	(-70)	
Net requirements							70	
Planned order receipt							70	
Planned order release					70			

[Inventory record file for tapping screws, LT = 1 week]

Week	1	2	3	4	5	6	7	8
Gross requirements				420				
Scheduled receipts								
On hand from prior period	150	150	150	(-270)				
Net requirements				270				
Planned order receipt				270				
Planned order release			270					

Figure 5.12. *Inventory record file for the tapping screws*

Pros and Cons of MRP

MRP is known to bring the following benefits:

- It maintains a reasonable level of safety stock.
- It improves customer service (quicker reaction to market demand).
- It increases sales.
- It minimizes or eliminates inventories.
- It reduces idle time.
- It reduces setup costs.
- It identifies process problems.
- It schedules production based on actual demand.
- It aids capacity planning.
- It coordinates materials ordering.
- It is most suitable for batch or intermittent production schedules.

Despite numerous merits, MRP may suffer from the following drawbacks:

- It's computer intensive.
- Its planning "brain" is predicated on a "push" system rather than the more popular "pull" system of manufacturing.
- It's known to be ineffective at the micro level
- It's difficult to make changes once operating.
- It cannot capture rising ordering and transportation costs.
- It's usually insensitive to short-term fluctuations in demand.
- It frequently becomes quite complex.
- It may not work exactly as intended.

Distribution Resource Planning

Distribution resource planning (DRP) is referred to as the application of MRP scheduling and time-phasing logic to the management of distribution inventories, transportation, and material flows across the distribution pipelines. The emphasis of DRP is on scheduling and people rather than ordering and techniques. The main objectives of DRP are as follows:

- To determine the needs of inventory stocking locations and ensure that supply sources will meet the demand

- To establish or improve the integration between a firm's distribution function and its manufacturing source of supply
- To give users visibility up and down the complete distribution network

The inputs of DRP are composed of the following (Martin, 1995):

- Sales forecasts by stocking-keeping unit (SKU) at the inventory stocking location
- Customer orders for current and future delivery
- Available inventory for sale by SKU at the inventory-stocking location
- Outstanding purchase orders and/or manufacturing orders by product
- Logistics, manufacturing, and purchasing lead time
- Modes of transportation used as well as deployment frequencies
- Safety stock policies by SKU at the inventory-stocking location
- Normal minimum quantity of product to be purchased, manufactured, and distributed

By processing these inputs, DRP can answer the following questions:

- Which product is needed, how much, and where and when is it needed?
- How much of the transportation capacity, warehousing space, manpower, and equipment are needed to fulfill customer orders?
- What actions must be taken to expedite or delay purchases and/or production to synchronize supply and demand?
- How to integrate manufacturing, purchasing, and logistics?

In contrast with MRP, DRP starts with customer demand and works backwards toward establishing a time-phased plan for securing the necessary finished products. The detailed comparisons of MRP and DRP are summarized in Table 5.12.

Table 5.12. MRP versus DRP

	MRP	DRP
Focus	Manufacturing oriented.	Distribution oriented.
Demand type	Dependent demand inventory plan.	Independent demand inventory plan.
System	Push system (maker driven).	Pull system (customer driven).
Structure	Bill of materials (from overviews to details).	Bill of distribution (from details to overviews).
MPS's Role	MPS is its input.	Its output is input to MPS.
Approach	Order-based schedule.	Shipping-schedule based.

Despite their differences, MRP and DRP are linked to each other to synchronize the flows of products from the point of production to the point of consumption. MRP is responsible for an inbound flow of materials, parts, and components, whereas DRP is responsible for an outbound flow of finished products.

In a nutshell, DRP aims to predict future shortages and recommend preemptive actions to avoid any imbalances between supply and demand.

One of the key inputs for DRP is a bill of distribution (BOD) that provides some visibility across the distribution pipeline by inducing a customer and supplier linkage. BOD is a counterpart of BOM in the MRP system, as illustrated in Figure 5.13.

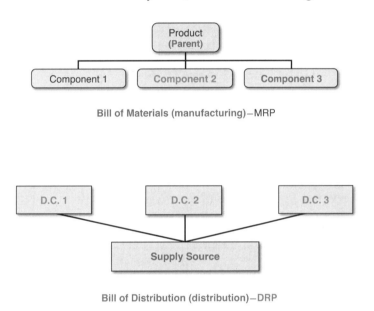

Figure 5.13. *BOM versus BOD*

Example of DRP

Although the way that DRP works is very similar to MRP in that both are time-phased inventory control systems, DRP relies on the bill of distribution instead of BOM. To illustrate the differences between DRP and MRP, let's consider the situation where a consumer product manufacturer would like to develop its shipping schedule of bottles of mouthwash. This manufacturer sells and distributes its mouthwash products to multiple retail franchise stores across the U.S. through two distribution hubs, based in Las Vegas and Louisville. As shown in Figures 5.14 and 5.15, in the beginning of the first week of a schedule, there are 600 units of on-hand inventory available at the Las Vegas distribution center (DC) and 450 units available at the Louisville DC. In each week, the projected on hand was calculated by subtracting the forecasted sales volume from the previous projected on hand, whereas products in transit would be added to the projected on hand balance in the week during which they are due to arrive. The same logistics will ripple through the shipping schedule. Every time the projected on-hand balance drops below the safety stock level, an additional shipment of 300 units should arrive to replenish the safety stock and thus prevent any potential product shortages. For example, in the third week, the projected on-hand balance at the Las Vegas DC dips below the safety stock level of 300 units and thus additional 300 bottles of mouthwash should be received by the end of the third week. As shown in Figures 5.14 and 5.15, it is noted that the planned shipment ship date in the last row of the DRP table will be fed into the master production schedule, which triggers the MRP process.

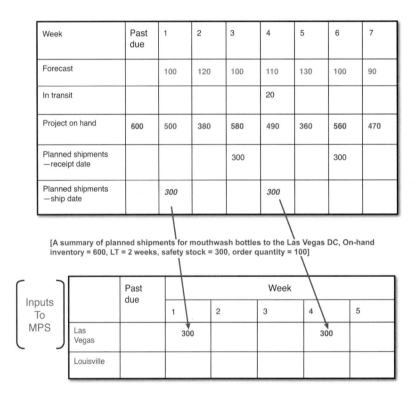

Figure 5.14. *DRP table for mouthwash products at the Las Vegas DC*

158 Chapter 5: Inventory Control and Planning

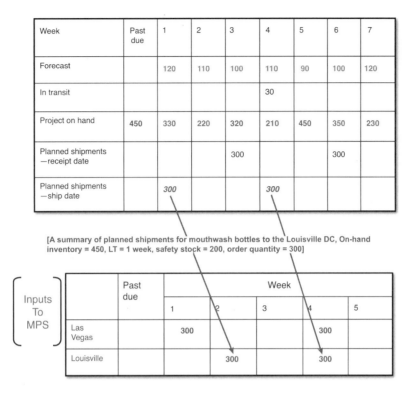

Figure 5.15. *DRP table for mouthwash products at the Louisville DC*

DRP Functionality and Benefits

By feeding the input information into the DRP system, we can perform the following functions (Smith, 1991):

- Identify, cancel, reduce, and redistribute excess inventory.
- Project resource needs in value, weight, cube, hours, etc., and provide manufacturers and suppliers with advance notice about future resource needs, including orders.
- Use exception reporting.
- Create action messages.
- Handle multiple plant, multiple warehouse inventory management.
- Calculate future needs of the distribution centers/warehouses from the plant.
- Update the master production schedule.

The potential benefits of DRP include the following:

- Reduced freight costs due to fewer rush or premium shipments and better carrier management and planning

- Reduced inventories
- Reduced warehouse space due to lower inventory
- Improved obsolescence controls
- Reduced backorders
- Enhanced coordination between manufacturing and distribution
- Improved budgeting tools

Just-In-Time Inventory Principles

Just-In-Time (JIT) is a management philosophy that involves purchasing or producing exactly what is needed at the precise time (i.e., just when it is needed and not long before or long after) to eliminate the waste of expensive resources. Such waste represents anything that adds cost and does not add value, in any workplace setting, to the purchasing and/or production processes. Waste typically includes the following:

- **Defects**—Defects undermine quality, requiring work to be redone and consequently causing costs overrun.
- **Delays**—Delays make customers wait and irritate them.
- **Non-value-adding processes**—Such processes prolong work completion time.
- **Idle motion**—This lowers labor productivity.
- **Overproduction**—This creates oversupply and makes extra products useless.
- **Long and redundant transportation**—This slows services flow and increases lead time.
- **Unused inventory**—Unused inventory occupies space and delays the discovery of defects and other complications.
- **Underutilized talent**—This undermines employee morale and productivity.

To make JIT a viable strategy, four principles should be followed (Kinsey, 1992):

- The empowerment of people is a main source of improvement.
- Perfection of product quality helps achieve zero defects.
- Flexibility is needed for economical small-lot production or small-order purchasing.
- Responsive production and order schedules are needed for exact customer demand.

To elaborate, JIT requires the following prerequisites:

- Small lot size (e.g., a lot size that ranges from 5% to 20% of daily usage, but seldom exceeds 10% of daily usage)

- Short lead time
  - Geographic concentration of suppliers and their close proximity to JIT manufacturers (e.g., inbound transit time less than one day)
  - Short, reliable delivery and frequent shipment (e.g., several times a day)
- Short setup time
  - Quick tool change (e.g., automated press line)
  - Multi-skilled workers with cross-functional training and job security
  - No labor strikes (e.g., nonunionized labor)
- Dependable quality (e.g., zero-defect principle)
  - Quality at the source (*Jidoka*) with a smaller number of suppliers with long-term contracts and early supplier involvement in new product development
  - Total quality management and consensus management through employee involvement.

As discussed, the basic tenets of JIT are quite different from those of the traditional inventory management control and planning system under the push system. Such differences are summarized in Table 5.13. Thus, JIT, which represents the pull system, differs from MRP, which represents the push system in various respects, as summarized in Table 5.14. Because the pull system is based on signals indicating need (or demand), JIT utilizes *kanban*, which means "signboard" in Japanese and authorizes production or shipment from the downward stream only if there is a need for production or shipment. A *kanban* can be a card, a piece of paper, a light, a flag, or a verbal signal that is often attached to the fixed-size container. It contains all the necessary information for an order, such as part number, description, production quantity, delivery time, and so forth.

Table 5.13. Differences between the Traditional Push System and JIT-based Pull System

"Push" (Just-In-Case)	"Pull" (Just-In-Time)
Large lots are efficient.	Ideal lot size is one unit.
Discrete production is encouraged because "economies of scale" is a priority.	Uniform, balanced production is necessary because "producing more than necessary" is nothing more than waste.
Job shop manufacturing environments are appropriate.	Flow shop or cellular manufacturing environments are appropriate.
Inventory provides safety.	Safety stock is a waste.
Inventory smoothes production.	Inventory hides production or distribution problems. Problems present opportunities for improvement.
Responds to the anticipation of demands.	Responds to actual demands.

"Push" (Just-In-Case)	"Pull" (Just-In-Time)
Individual efforts.	Team efforts are crucial ("boundary spanning" communication or coordination is common).
Freight consolidation saves transportation cost.	Freight consolidation increases lead time.
Suppliers are regarded as adversaries. Multiple sourcing is the norm.	Suppliers are viewed as coworkers. Supplier base reduction is a norm.
Tolerates some scraps.	Zero defects.
Machine maintenance is *not* critical. Requires little effort to maintain. Focused on corrective maintenance.	Machine breakdown should be kept minimal. Requires great effort to maintain. Focuses on preventive maintenance.
Slow changeover.	Quick setups and mixed-model runs. A wide variety of parts can be made more frequently.

Table 5.14. JIT versus MRP

MRP	JIT
Push system (maker driven) • "Pushes out" replenishment from the maker before materials are needed. • Large lot sizes may be automatically delivered without the need for a request from the user.	Pull system (user driven) • Only authorizes the replenishment of materials from production if there is a need for them. • When the user needs additional materials, he posts *kanban*, thereby triggering a "pull" for more supply.
Inventory = "asset"	Inventory = "liability"
Computer-based system	Manual method (trial-and-error)
Lead time should be offset due to time-phased planning.	No lead time is offset because lead time is so short.
Long planning horizon • MPS typically covers 6 months of time periods. • Aggregate plan covers 12–16 months of time periods.	Short planning horizon • MPS typically covers 2–3 months of time periods. • Aggregate plan covers less than 12 months of time periods.
Recordkeeping (future production base)	No recordkeeping (current demand base)

If JIT is used properly for a right setting, it can bring a number of benefits:

- Improved problem-solving capability.
    - JIT exposes problems that industry elsewhere tends to bury in excess materials. Exposed problems become solved problems in JIT environments.
- Reduced lead time resultant from smaller order/lot sizes.

- Increased equipment utilization as a result of producing smaller lot sizes without making unnecessary extras.
- Increased product quality.
  - In a JIT setting, quality is "worker centered," which is far more effective than quality inspection made by a delegated quality control department.
- Reduced paperwork and simple planning systems.
  - Product mix or volume can be changed by adjusting the number of cards (*kanbans*) in the JIT system.
- Increased inventory turnover ratio.

As a matter of fact, Toyota was able to shorten the time needed to produce its car from 15 day to one day after utilizing JIT. Also, Apple computer is reported to save $20 million of inventory costs after using JIT. Similarly, General Motors (GM) increased its production by 100%, but its inventory increased by only 6% after implementing JIT. As illustrated by the successful implementation of JIT by Norfolk Southern and Ryder Integrated Logistics, JIT can also be applied to a specific logistics setting. JIT logistics is generally referred to as a pinpoint time-based logistics philosophy that delivers materials/finished goods to points along the production/distribution/consumption pipeline in the exact amount and at the precise time that such materials/finished goods are required. Its detailed inner-workings are displayed in Figure 5.16.

Although there are numerous JIT success stories, JIT is by no means a cure-all. Its major drawbacks include the following:

- It requires an atmosphere of close cooperation and mutual trust between the workplace and management.
  - Non-union labor is preferred.
- It requires a large number of production setups and frequent shipments of purchased items from suppliers.
- It is not well-suited for irregularly used parts or specially ordered products.
  - JIT cannot respond quickly to changes in the schedules because there is little extra inventory or capacity to absorb changes.

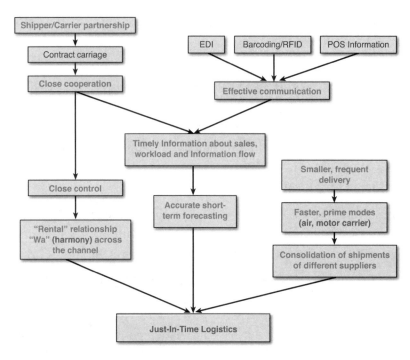

Figure 5.16. *Inner-workings of JIT logistics*

Basics of Cycle Counting

Missing or unaccounted inventory can be costly because it increases time looking for items, it delays or fails to fulfill customer orders, it lengthens production time, it raises expediting (order and transportation) costs, and it increases inventory carrying costs. To prevent missing or unaccounted inventory, many companies practice *cycle counting,* which generally refers to counting inventory physically on a cyclical schedule to check the inventory accuracy. Thus, cycle counting involves comparing physical quantity on hand with the quantity on record and identifying causes that contribute to count variance. Such causes may include multiple inventory locations, off-site inventory locations, unrecorded inventory transactions, poor inventory classifications, picking errors, misplacement, off-hour (e.g., midnight) requisitions, untrained/careless counters, missing documents, too many expedited shipments, and unclear tolerance limits.

Also, the two primary methods for tracking count variances are on-hand tracking and transaction tracking. *On-hand tracking* is the most commonly used method and gives you a snapshot of what your inventory accuracy is at that specific point in time (Piasecki, 2003). It is helpful in projecting the impact your variances might have on subsequent operations. However, the on-hand method cannot pinpoint process problems because it relies heavily on current on-hand balances. For example, if on a given day you count a certain part and then discover a shortage of 30 units out of an on-hand balance of 100, the inventory accuracy rate is supposed to be 70%. However, if you did not count this item until the following day and an additional 2,900 units were received prior to the next cycle count, your accuracy rate would be 99%, even though you are still missing those 30 units. *Transactional tracking* compares the variance amount to the amount consumed during the count period and is therefore far more useful in determining process problems than on-hand tracking. Unfortunately, this method can be more difficult to implement than on-hand tracking (Piasecki, 2003).

Cycle counting can be more efficient when counting a certain small percentage of the total number of items daily with prescribed frequency for each item and assigning responsibility for reporting inventory transactions to specific employees. If accurate inventory records can be maintained over several consecutive years, the company may be able to eliminate periodic physical inventory counts. The key issues of cycle counting include the following:

- What to count?
  - A-items or items with high pilferage risk
- How often to count it?
  - Weekly count is generally desired. Otherwise, at least once every 12–13 weeks is appropriate.
- When to count?
  - Nonshipping days or less heavy picking days or when the inventory is at the lowest level.
- How to count?
- Who should count it?
  - A cycle counter, warehouse operator, or inventory control manager can handle it.
- Which record to look at?

To make cycle counting successful, the company should ensure the following:

- All stock receipts have been placed in their proper bin/shelf location.
- All printed sales orders and transfers for stock material have been filled.
- Computer records for these receipts, sales orders, and transfers have been updated and kept.
- All product returns have been processed and restocked.
- No customer orders are filled and no inventory is relocated or moved until all inventory counts are completed.

Here are some typical procedures that may enhance cycle counting efficiency:

- Set dates early so that the counter can have sufficient preparation time.
- Determine who will count and train that person properly.
- Create a map of warehouses/stock rooms.
- Recount and audit initial counts.
- Use barcoding and/or radio frequency identification (RFID).
- Perform the count during off-hours.
- Set the "reasonable" count tolerance (e.g., 1% tolerance level for "A" items, a 2%–5% tolerance level for "B" items, and a 5%–10% tolerance level for "C" items).

Figure 5.17 illustrates the step-by-step cycle counting procedures.

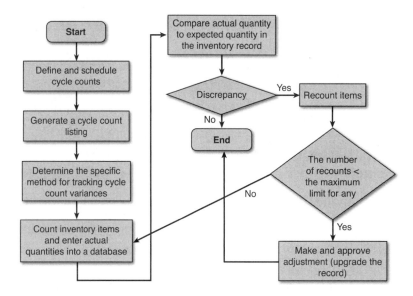

Figure 5.17. *Typical cycle counting procedures*

In particular, if recounting is necessary during the cycle counting procedure, the following actions can be taken:

- Develop good audit procedures.
- Audit while people are counting.
- Select an experienced and trained auditor.
- Recount with the team (e.g., an original counter plus at least one other).
- Have the auditor review and sign off on the results of the recount.
- Check whether count errors are discovered in particular areas or bins (e.g., inventory errors are likely to occur for counting the fastest moving items).

Vendor Managed Inventory

Vendor managed inventory (VMI) is an inventory management principle whereby the supplier not only provides its customers (manufacturers, distributors, or retailers) with products but also decides which products (items) to replenish, in what quantities, and when, based on the agreed ground rules. Per Roberts (2006), the common ground rules of VMI are as follows:

- The primary responsibility for maintaining target inventory levels lies with the suppliers of those parts.
- The classical discrete purchase order processes are replaced by a dynamic replenishment process.
- The time of the title transfer of inventory items from the supplier to the customer is delayed until near the time the part is used in manufacturing.
- Finished goods inventories at predefined levels for a particular customer are reserved until they are needed, usually by placing those inventories in a location closer to the manufacturing site.

In VMI, the supplier owns inventory and takes responsibility for the management of the customer's inventory. This means that the supplier may take the financial burden of maintaining inventory. In exchange for the burden of maintaining inventory, the supplier is often rewarded with increased or more stable business opportunities. The customer that takes advantage of VMI reaps various benefits. These benefits include increased inventory turnover, increased sales, increased order fill rates, improved customer service, shorter lead time, higher cash flows, decreased liabilities, and improved inventory planning based on actual customer sales. Indeed, a survey conducted by Datalliance (2007) indicated that the distributors utilizing VMI experienced an average increase in sales of 47%, an average increase in inventory turnover of 38%, and an average reduction in stockouts by 45% over a two-year span.

Despite the potential benefits of VMI, its success often hinges on several factors:

- Information flow
    - Compatible information system (e.g. Electronic Data Interchange) standards facilitate the smooth and reliable transmission of information between the customer and the supplier.
    - Accurate demand forecasts minimize unnecessary safety stocks for suppliers.
- Material flow
    - Sufficient process capability shortens lead time.

- Strong partnership (mutual trust) between the customer and the supplier
    - The customer (e.g., retailer) usually initiates VMI and provides its supplier with crucial sales data.
    - The supplier should be committed to avoid overstocks and out-of-stocks, while ensuring just-in-time delivery of parts and supplies.
    - The clear service-level agreement between the customer and the supplier should be made on the treatment of overstocks or order errors.
    - The supplier should be assured of receiving timely payment.

Also, it should be noted that VMI may fall apart if increased administrative cost and the lack of special pricing opportunities such as volume discounts eventually overburden the supplier without sufficient benefits that offset the cost increases. On the other hand, VMI can increase the risk of the loss of control and flexibility on the part of the customer.

Chapter Summary

The following are key lessons learned from this chapter:
- Inventory can be a valuable asset that enables a company to serve its customers in a timely manner by playing the role of buffer against unpredicted demand surges, quality failures, and supply chain disruptions. On the other hand, inventory can be a liability that wastes limited company resources by incurring inventory carrying cost, tying up company personnel in managing inventory, and hiding the sources of quality/logistics problems.
- Considering these two conflicting sides of inventory, the company must decide what the right level (size) of inventory is. In determining such a level, the EOQ model is often called for. The EOQ model, however, is not a "cure-all" inventory control technique. It primarily works for controlling the so-called independent demand inventory. Therefore, MRP was introduced to handle dependent demand inventory items.
- As U.S. firms witnessed the success of the Toyota Production System in controlling inventory, the Japanese lean principle called "JIT" gained popularity among U.S. firms in the 1980s. As a viable alternative to EOQ and MRP, JIT can be leveraged to generate more accurate demand signals, reduce volume flexibility, minimize lot sizes, and consequently absorb any supply chain shock better with improved supply chain visibility. However, JIT is not a panacea for solving the inventory problems and is not intended for certain specific business environments. Therefore, JIT adoption should be preceded by careful review of the potential user's business environments and corporate culture.

- The efficiency and effectiveness of any inventory control and planning processes often hinge on the accuracy of the inventory record. To maintain the accurate inventory record, cycle counting is necessary.
- VMI has proliferated over the last few years as a way of improving a company's financial performance. In a typical VMI program, a supplier generates orders for its customer based on demand information, such as point-of-sale data transmitted by the customer. In other words, the customer shifts responsibility for planning and replenishing inventory to its supplier. Thus, VMI can be viewed as a collaborative partnership based on mutual trust, benefit sharing, and continuous communication between the customer and its supplier.

Study Questions

The following are important thought starters for class discussions and exercises:

1. Explain the role of inventory in various organizational settings (e.g., retail stores such as Walmart, grocery stores such as Kroger, bookstores such as Books-A-Million, automobile manufacturers such as Ford, and electronics manufacturers such as Phillips).
2. State the purposes and roles of the different types of inventories positioned at the different stages of a supply chain.
3. How does inventory affect the company's bottom line (e.g., financial performance)? How would you estimate total inventory costs (e.g., inventory carrying costs)? What constitute inventory carrying costs?
4. How would you use ABC analysis to classify the inventory? What are the appropriate criteria for categorizing inventory as A, B, or C items? Under the premise that your company's inventory classification is accurate, what kind of inventory planning and control strategies would you use for a different category of inventory?
5. Explain in what instances the EOQ model can be utilized to determine the optimal order quantity. What are the necessary assumptions of the EOQ model?
6. How sensitive can the EOQ model be to changes in its parameters?
7. How does the quantity (volume) discount affect the inventory decision? How do you determine the optimal order quantity when the quantity and/or freight discounts are available? Discuss the potential impact of price breaks on the company's supply chain strategy.
8. Explain the differences between the independent demand inventory and the dependent demand inventory. Also, list the examples of the dependent demand inventory in various organizational settings.

9. Compare and contrast between EOQ and MRP. Explain in what instances MRP can be used.
10. What constitute the primary inputs of MRP?
11. Construct a simple bill of materials (BOM) for a typical desktop personal computer and then discuss how MRP explosion works.
12. Discuss whether or not MRP can be applied to service organizations.
13. What are the similarities and differences between MRP and DRP? How can MRP be integrated with DRP? Explain what kind of roles both MRP and DRP play in making ERP work.
14. What are the main prerequisites to JIT success? How does JIT differ from MRP? Can JIT be used concurrently with MRP? What are the greatest challenges for implementing JIT?
15. Why is cycle counting important for inventory control?
16. How can vendor managed inventory (VMI) improve the supply chain efficiency? What are the key ingredients for the successful implementation of VMI?

Case: Sandusky Winery

Sandusky Winery has been at the heart of Ohio wineries for over 75 years as a family-controlled business. It produces, distributes, and markets a variety of wine products, including sparkling wine, champagne, and wine accessories such as wine refrigerators/chillers, wine racks, glassware, and corkscrews. Riding on a rapid growth in wine sales in the U.S., which reached a new record of 360 million 9-liter cases with an estimated retail value of $34.6 billion, Sandusky Winery's marketing department wants to expand its product lines and then aggressively promote its new products. Sandusky's marketing department is eager and able to move large volumes of wine products during its upcoming promotional campaigns. Also, Sandusky's logistics group prides itself on its on-time delivery record and a high order fulfillment rate of 98.7%.

Out of Space

After cranking up its production volume and piling up inventory too much, Tony Jackson, the production manager, discovered that his production facility has run out of space and patience. His stock (storage) room is filled to the brim from floor to ceiling with a variety of old wine products and promotional items, including ones commemorating the 1996 Summer Olympic event in Atlanta. With each production run, he must help his inventory manager find storage space for leftover products from previous promotional runs. Some of these leftover products, which should be kept within temperature controlled storage rooms (ideally at 62°F to 68°F) temporarily ended up with the empty boxcars outside its manufacturing plants in Ohio. These products range from less than 10 dollars each to several hundred dollars for high-end vintage wine products. Although Sandusky has continually purged some of the less popular products to reduce the inventory and storage burden, it does not want to lose potential sales opportunities by running out of stock. In the past few years, Sandusky managed inventory by using a so-called "earn and turn" concept, which computes the earnings margin ratio [(sales price - cost of goods sold)/sales price] multiplied by inventory turns. So far, however, the "earn and turn" concept has not helped Sandusky identify popular (or fast-moving) products and control the excessive supply of unpopular products.

Peter Newman, the company's vice president of finance and head controller, is also frustrated. At the urging of the company's newly hired consultants, Peter conducted detailed cost analyses of this practice prior to the approval of any new product configuration. Peter figured out various cost components including production setup costs, quality assurance, sales/promotion costs, transportation costs, packing cost, labeling cost, and warehousing costs. Despite these prudent precautions, Sandusky's profits and cost savings never materialized at the projected levels.

An Overwhelming Sense of Déjà Vu

Juan Carlos, the warehousing manager, got frustrated with the continued overflow and the subsequent mess in his warehouse and the inability to accommodate additional production output. He has been sick and tired of working in overcrowded and dangerous conditions day in and day out. He recently discovered that a list of items kept in inventory had grown to over 100 as each product run became a separate stock-keeping unit. Unless Sandusky dramatically changes its production schedule and adjusts its production capacity, it is absolutely necessary to either build or rent additional warehousing space to store further production output. Given a lack of empty lots near Sandusky's plant and existing warehousing facility in Ohio, it does not seem to be sensible for the company to build a new additional warehouse nearby. The option of changing the current layout of its manufacturing plant to better utilize its space does not seem to be attractive because it will disrupt the current product operations and will not create much needed space. With their hands tied, Sandusky considered the option of renting public warehouses. Unfortunately, the nearest temperature controlled public warehouses are located in Columbus, Ohio and Louisville, Kentucky.

Before making any strategic decisions regarding additional warehousing space, a group of consultants recently hired by Sandusky conducted a quick analysis of its inventories. Their analysis reveals that inventories are recorded separately for every single item in the inventory record files. Financial records, on the other hand, lump together all inventories relating to the same products. This practice makes it virtually impossible to determine the fair value of meanwhile-obsolete inventories. The consultants came to the conclusion that it would be fairly simple to resolve the warehouse space problems by asking Sandusky's marketing/sales department for help. All salvageable items can be sorted out and then some promotional wine products can be relabeled at the cost of three dollars per unit and sold at discounted prices, while all remaining obsolete inventories can be discarded and written off. Considering Sandusky's deteriorating financial ratios with growing debts, this will be a painful option. Adding more headache and heartburn, even the aforementioned options will not address the current inventory and warehousing problems. Given its expertise in the matter, Sandusky's purchasing department has long played a major role in determining total inventory costs. Purchasing is also responsible for establishing the inventory budget. Peter accordingly asks Jackie Carson, the senior buyer for wine materials, to research all possible components of inventory costs. In particular, he asks her to search for hidden costs that were not included in past cost analyses. In the meantime, both Tony and Juan are assigned the task of finding better inventory control and planning strategies.

Discussion Questions

1. What could be the main sources of the inventory mess created by Sandusky?
2. Which inventory cost components are missing in the calculation of total inventory cost?
3. What can be the consequences of writing off excessive inventory? Is such inventory salvageable? Is inventory write-off justified? If so, why or why not?
4. How would you change the current inventory management practices (including inventory classification schemes) to ease storage space problems?

Bibliography

Ballou, R. (1985), *Business Logistics Management: Planning and Control*, Englewood Cliffs, New Jersey: Prentice-Hall.

Baumol, W.J. and Vinod, H.D. (1970), "An Inventory Theoretic Model of Freight Transport Demand," *Management Science*, 16(7), 413–421.

Datalliance (2007), *Vendor Managed Inventory: Two-year Cumulative Study*, Unpublished White Paper, Cincinnati, Ohio: Datalliance.

Kinsey, J.W. (1992), *The Just-In-Time World*, Troy, Michigan: Constantine Associates.

Lewis, C.D. (1997), *Demand Forecasting and Inventory Control: A Computer Aided Learning Approach*, New York, NY: John Wiley & Sons, Inc.

Martin, Andre J. (1995), *Distribution Resource Planning: The Gateway to True Quick Response and Continual Replenishment*, 2nd edition, New York, NY: John Wiley & Sons.

Orlicky, J. (1975), *Material Requirements Planning: The New Way of Life in Production and Inventory Management*, New York, NY: McGraw-Hill.

Piasecki, D.J. (2003), *Inventory Accuracy: People, Processes, & Technology*, Ops Publishing.

Roberts, C. (2006), "The Rise of VMI," *Asia Pacific Development*, 99–101.

Russell, R. and Taylor III, B.W. (2006), *Operations Management: Quality and Competitiveness in a Global Environment*, Danvers, MA: John Wiley & Sons, Inc.

Simchi-Levi, D., Kaminsky, P., and Simchi-Levi, E. (2008), *Designing and Managing the Supply Chain: Concepts, Strategies and Case Studies*, 3rd edition, Boston, MA: McGraw-Hill/Irwin.

Smith, B.T. (1991), *Focus Forecasting and DRP: Logistics Tools of the Twenty-first Century*, New York, NY: Vantage Press.

6

Warehousing

"The process of laying out a warehouse is a lot like putting a puzzle together. Like a puzzle, it is difficult to complete until all the pieces have been defined and assembled."
—Edward H. Frazelle, *World-Class Warehousing and Material Handling (2002)*

Learning Objectives

After reading this chapter, you should be able to:

- Understand the basic principles of warehouse management.
- Comprehend the different types and roles of warehouses.
- Grasp the emerging warehousing trends and their impact on the new roles of warehouses.
- Identify the components of warehousing costs.
- Design an optimal warehouse network within the supply chain.
- Design an optimal warehouse layout in such a way that it minimizes costs and maximizes space efficiency.
- Develop relevant warehousing performance metrics such as order fill rate, picking accuracy, shipping accuracy, putaway accuracy, and labor productivity.
- Understand the various technological tools and equipment to automate warehousing processes.
- Cope with the challenges of recruiting, hiring, and retaining productive warehousing employees.
- Leverage the warehouse management system to enhance warehousing productivity.
- Understand ways of controlling hidden warehousing costs associated with returned products.

Warehouses in Transition

A warehouse is an important part of a firm's logistics system that is used to store products (e.g., raw materials, parts/components, goods-in-process, finished goods) at and between the point of origin and the point of consumption. It provides information to management on the status, condition, and disposition of items being stored to create time and place utility for the firm's customers. As illustrated in Figure 6.1, the traditional functions of a warehouse include the following:

- Receiving
  - The orderly receipt of all materials coming into the warehouse
  - Providing the assurance that the quantity and quality of such materials as ordered
  - Disbursing materials to storage or to other organizational functions requiring them
- Prepackaging
  - Packaging bulky products in merchandisable quantities or in combination with other parts to form kits or assortments
- Putaway
  - The act of moving material from the point of receipt (e.g., receiving dock or quality inspection unit) to a storage area (e.g., material handling, item data and location verification, or product placement)
- Cross Docking
  - Involves receiving bulk shipments and then breaking these shipments into smaller orders, packing them, and shipping them immediately without them ever being stored in the warehouse
- Unitizing
  - Consolidating a number of individual items onto one shipping unit for easy handling
  - Packaging merchandise in appropriate shipping containers
  - Accumulating orders by outbound carriers

These traditional warehouse functions, however, have been evolving more rapidly over the past two decades than in any previous decade due to changes in the business environment and management philosophy. Such changes are sparked by several trends that have shaped the U.S. industry: the increasing popularity of Just-In-Time principles, where inventories are kept at a minimum, meaning the role of warehouses as inventory storage facilities has been fading; customer-centric business practices that require increased value-added services; and the advent of e-commerce, which demands speedy and error-free fulfillment of online orders via increased automation and paperless transactions.

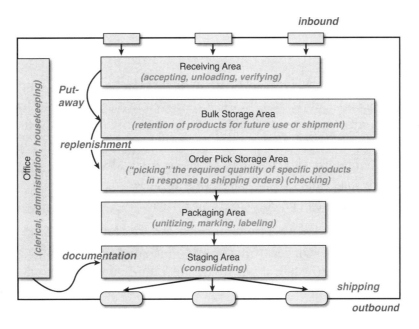

Figure 6.1. *Traditional functions of the warehouse*

Reflecting these trends, a growing number of warehouses in the U.S. are engaged in "value-added services" that go beyond the traditional warehousing functions. These value-added services encompass a variety of product customization activities, including packaging, labeling, kitting (i.e., light assembly of component parts), marking, pricing, repair, shrink wrap bundle packing, sequencing (i.e., organizing the items in a load), and return processing. In other words, today's warehouses have become "flow-through" staging points where a majority of products move through warehouses without being ever stored and go through some transformation process for increased customization. To elaborate, the following characteristics represent today's warehouses (Frazelle, 2002):

- They provide more customization.
- They offer more value-added services.
- They process more returned products.
- They execute more, smaller transactions due to increased B2C e-fulfillment.
- They handle and store more diverse products with international origins and destinations due to offshoring activities.
- There is less time to process orders.

- There is less margin for error ("do it right the first time").
- There are more less-skilled, nonnative workers (e.g., new Hispanic immigrants).

Types of Warehouses

One of the important warehousing decisions involves selecting a particular type of warehouses that impacts cost and flexibility. Basically, warehouses are classified into one of three categories: private warehouses, public warehouses, or contract (third-party) warehouses. A *private warehouse* can be either leased or owned by the firm whose products are stored in it. A *public warehouse* is owned and operated by a firm engaged in the business of offering space and diverse warehousing services on a "for-hire, as needed" basis. A public warehouse can take a variety of forms:

- **General merchandise warehouse** —Intended for storing any kind of products.
- **Refrigerated (cold storage) warehouse**—Intended for preserving perishable items such as grocery store items in a temperature-controlled storage environment.
- **Household goods warehouse (e.g., mini-storage facility)**—Intended for storage of personal property.
- **Special commodity warehouse**—Designed for some agricultural products such as grains, salt, wool, and cotton.
- **Bulk storage warehouse**—Designed for dry products such as coal, sand, and chemicals.
- **Bonded warehouse**—Storage for imported goods. The company doesn't have to pay customs duties and excise taxes until those goods are released and sold to the U.S. domestic market.

A contract warehouse provides a fixed amount of space and customized supplemental services exclusively for a customer for an extended period of time (e.g., six months or longer).

Pros and Cons of Private Warehousing

Because a private warehouse is considered the property of the owning firm, the owning firm of the private warehouse has the following advantages:

- Greater direct control over warehousing employees, operations, and scheduling in filling customer orders

- Greater degree of freedom to meet the specific needs (e.g., temperature- or humidity-controlled storage, hazardous materials storage and handling, layout changes to e-fulfillment operations) of the owning firm
- Greater economies of scale in large-volume operations and the subsequent cost savings over the long term
- A potential financial gain from the long-term appreciation in property value
- Some tax benefits resultant from depreciation allowances on the building and equipment as well as local real-estate tax savings granted to the owning firm investing in the local community and creating jobs

On the other hand, a private warehouse can be a financial burden during a time of downsizing and strategic changes. Its drawbacks include the following:

- Inability to adapt to changes in the owning firm's product mix, strategic locations, and layout design
- Increased difficulty in managing warehousing labor problems and maintaining the warehouse equipment and building
- Huge startup investment tied up with the purchase of warehousing property

Pros and Cons of Public Warehousing

In public warehousing, the cost per unit of goods handled and stored in a warehouse is proportional to the volume of business transactions (or size of rented space) made by the user; therefore, large fixed costs are not incurred. As a result, public warehousing presents the following advantages:

- Conservation of financial capital and the subsequent improvement of cash flow
- A greater flexibility to increase warehouse space to cover peak demands or decrease warehouse space with relative ease to discontinue unprofitable operations
- A reduced risk of dealing with labor disputes, employee turnover, and natural disasters (e.g., fire, earthquake, tornado, flooding, hurricane)
- More predictable cost of storing and handing products due to periodic billing based on the amount of rented space
- More professional management by the commercial warehousing firm that specializes in warehousing operations
- No responsibility for real-estate property taxes and insurances

Despite these advantages, public warehousing is not without its shortcomings:

- Lack of control over warehousing operations and increases in public warehousing charges

- Diseconomies of scale due to limited rented space
- Potential risk of a communication breakdown between the user and the warehouse owner, which may lead to slower customer response time and order fulfillment error
- Limited location choices and lack of attention to a particular user

Types of Warehouse Leases

As a compromise between private and public warehousing, leasing all or part of a warehouse space can be a good business move. The leasing option gives a company greater flexibility because it frees up capital and reduces sunk cost. Despite numerous potential benefits, entering a commercial lease agreement for warehousing space can be complex due to the customized nature of leasing contracts and various compliance requirements such as zoning, permits, and facility access. For example, depending on the lease terms, a lessee (lease user, depositor, or tenant) may be responsible for covering some expenses such as insurance, maintenance, and utility costs. As such, the lessee should carefully assess various types of leases and select the most ideal one among the following options (see Bolten, 1997):

- **Gross lease**—This is a type of commercial lease where the tenant pays one lump-sum amount to the landlord (lessor), who is responsible for expenses related to the ownership of the property, such as real-estate taxes, building maintenance, and property insurance costs. Utilities, however, may or may not be included within a gross lease.
- **Net lease**—Generally speaking, a net lease allows the tenant to absorb fewer fixed and variables expenses (e.g., a portion or all of the taxes, fees, and maintenance costs for the property in addition to rent). There are three categories of net leases: single (net), double net (net-net), and triple net (net-net-net). In a popular case of the triple net lease, the tenant pays a set rental amount (usually reduced) to the landlord, but pays a share of the landlord's real estate taxes, property insurance, maintenance expenses, and utilities. In this lease term, the tenant assumes the risk of property tax and insurance increases, while being liable for most personal injuries that occur on the leased property.
- **Flat rental lease**—In this lease term, a tenant pays a fixed and equal amount (e.g., a set rental fee) to the landlord at specified rental periods (e.g., monthly) throughout the duration of the contract.
- **Adjustable (graduated or graded)**—In this lease term, the tenant pays an agreed rental fee that, after an initial period at a fixed rate, can be escalated or deescalated periodically for successive lease periods to reflect changes in the

leased property's appraised value or changes in a certain economic indicator such as consumer price index (CPI).
- **Percentage lease**—In this lease term, the tenant pays an agreed minimum rental fee (e.g., annual fixed rent) plus a percentage (negotiated between the landlord and the tenant) of any sales revenue earned by the tenant or throughputs produced by the tenant on the leased property. For instance, in the event the tenant's gross sales exceed a certain amount (so-called "breakpoint"), the landlord will receive a certain percentage of the gross sales over the breakpoint as the additional rent.

Warehousing Costs

A strategic decision as to whether a company's warehouse should be built, leased, or rented hinges heavily on warehousing costs. In general, warehousing costs are composed of real-estate costs, equipment costs, labor costs, administrative expenses, and hidden miscellaneous costs. Real-estate costs are the costs associated with the ownership of a real estate property. These costs include the following:

- Land purchase cost, leasing cost, or rental fee
- Property taxes
- Property insurance
- Utilities
- Interior maintenance and renovation/modification
- Exterior (e.g., ground) maintenance associated with mowing, snow plowing, cleaning, painting, and parking lots
- Depreciation

Equipment costs are incurred from the purchase and use of storage and handling equipment such as fork-lift trucks, conveyor systems, pallets, and tote boxes. These costs can be broken down as follows:

- Maintenance (including fuel charges, lubrication, service part replacement) and repair
- Depreciation
- Detention/demurrage
- Leasing or rental cost

Given the labor-intensive nature of warehousing operations, warehouse labor costs are classified into direct labor costs, compensation and fringe benefits, payroll taxes, and other expenses. These cost elements are further subdivided as follows:

- Direct labor costs
- Wages
- Bonuses
- Overtime pay
- Compensation and fringe benefits
- Health and worker compensation insurance
- Pension
- Life and accidental death insurance
- Long-term disability benefit
- Vacation
- Holidays
- Sick leave
- Personal leave (including funeral and maternity leaves)
- Jury duty
- Statutory payroll taxes
- Worker's compensation
- Federal Insurance Contributions Act (FICA) tax, such as social security and Medicare tax
- Federal unemployment tax
- State unemployment tax
- Other taxes
- Training and education
- Employee turnover costs, including recruitment, interviews, and referral bonuses

Administrative expenses are tied to payroll set aside for warehousing executives, administrative staff, and office maintenance. To elaborate, these expenses are as follows:

- Executive salaries
- Supporting staff (e.g., secretaries) salaries
- Office maintenance (e.g., dedicated office space, office equipment maintenance)
- Janitorial services
- Securities
- Office supplies
- Data processing fees
- Postage, telephone, copying, printing, courier services, and office utilities

Hidden costs represent expenses that are often overlooked when operating a warehousing facility. These include the following:

- Cost of vacancy or underutilization
- Damage, loss, and pilferage
- Trash hauling
- Pest control
- Legal expenses
- Loss of insurance deductibles
- Executive travel expenses
- Security system installation and maintenance
- Relocation and moving expenses.

Warehouse Network Design

Traditionally, warehouses have played three primary roles in supporting logistics functions (e.g., Maltz, 1998; Min and Melachrinoudis, 2001):

- **Storage role**—Makes long production runs more economical and bridges temporal gaps between demand and supply by making products available for customers on a timely basis
- **Consolidation role**—Reduces total transportation cost by aggregating small orders into large shipments and then taking advantages of economies of scale
- **Customization (postponement) role**—Reduces inventory carrying cost and enhances customer services by delaying the final stage of fabrication, assembly, and distribution until customers actually order specific types of products

With the increasing adaptation of Just-In-Time principles, which minimize storage needs, and the globalization of business activities, which diversifies customer bases, the traditional roles of warehousing are changing. Such changes in warehousing roles often necessitate the reassessment of the existing warehouse network. In addition, a strategy of searching for new warehouses should be altered based on changes in global supplier and customer bases, corporate mergers/acquisitions, downsizing, economic conditions, regional labor markets, and government legislation. In the following subsections, two contrasting strategies of designing a warehouse network (i.e., centralization and decentralization) are introduced, along with their pros and cons.

Warehouse Centralization Strategy

One of the most noticeable trends in today's warehousing operations is increased velocity. Increased velocity is credited to increasing adaptation of more effective inventory management practices such as Just-In-Time (JIT) inventory systems, improved forecasts, more accurate inventory-tracking systems through barcoding and radio frequency identification (RFID), and warehouse management systems. Indeed, a survey conducted by Speh (1999) showed the average of inventory turns by 30% from 1995 to 1998. Similarly, the wholesale trade and distribution industry, including the warehousing sector, improved its average inventory turns from 69.4 to 72.8 between 2009 and 2010 (Gilmore, 2011). Also, order-to-delivery time tended to decrease over time and consequently increase inventory velocity.

For example, order-to-delivery cycle time decreased 31%—from 5.2 days to 3.6 days—during 2003 and 2004 (Harrington, 2005). Given the increased inventory velocity, many firms no longer require a large number of stocking points, which consequently allows them to consolidate existing stocking points into fewer locations. In general, warehouse centralization strategy involves consolidation of regional warehouses into a small number of master stocking points and the subsequent phase-out of redundant regional warehouses. Along with better capacity utilization, aggregated inventory, and higher throughput of centralized warehouses, such a strategy often brings a substantial amount of savings in warehousing and inventory carrying costs (Min and Melachrinoudis, 2001). For instance, the square-root rule of inventory consolidation allows a firm to estimate the amount of savings in inventory investment as a result of centralized warehousing operations. Under the premise that the firm relies on economic order quantity (EOQ) rules for inventory control and planning and that all stocking points of the firm carry the same amount of inventory, the simplest form of the square-root rule can be mathematically expressed as follows (Ballou, 1999, pp. 352–354):

$$I_T = I_i \sqrt{n} \quad (6.1)$$

where

I_T = the optimal amount of inventory to stock, if consolidated into one location in dollars, pounds, cases, or other units of measurement

I_i = the amount of inventory in each of n locations in the same units of measurement as I_T

n = the number of stocking locations before consolidation

Another benefit includes reduced material-handling costs resultant from bulk storage and picking at the centralized locations. In addition, central administrative costs can be reduced through less effort being spent in managing fewer warehouses. Similarly, cross-hauling transportation expenses between different regional warehouses can be reduced due to the reduction in a number of regional warehouses and the subsequent decrease in multiple transfers among them. On the other hand, warehouse centralization strategy may lengthen response time to customers and increase direct shipment costs due to a fewer number of regional warehouses in the proximity of customer locations (Min and Melachrinoudis, 2001)

Warehouse Decentralization Strategy

New ventures that would like to expand their businesses in new territories need to disperse their warehouses in wider geographical areas. Such dispersion of warehouses would allow them to get closer to regional customer bases and respond to customer demand promptly. In general, warehouse decentralization strategy aims to shorten customer response time and improve order-fill rates by positioning inventory locations at the lowest downstream of the supply chain. Such a strategy is often used by e-tailers whose online direct sales require direct-to-home or store delivery services on a quick response basis. Despite its merits, the warehouse decentralization strategy leads to increased inventory carrying costs and warehousing costs due to a large number of stocking points. It can also increases transportation costs due to smaller, more frequent shipping requirements (Min and Melachrinoudis, 2001). Furthermore, multiple, small-sized regional warehouses would be harder to manage and automate the processes due to diseconomies of scale.

Profile Analysis for a Warehouse Network Design

To select a particular warehouse network design between the centralization and decentralization strategies, a profile analysis should be performed. The profile analysis is intended to provide the management team with a tool to identify the supply chain needs and the degree of fit between the company's network design criteria such as level of investment, time horizons, market positioning, government regulations, labor market conditions, taxes and subsidies, community attitude, and access to logistics infrastructure (Min and Melachrinoudis, 2001). The profile analysis is composed of four steps (Hill, 1994):

1. Select the appropriate aspects of products (e.g., types of product, product range, life cycles), markets (e.g., sales/promotional tools, delivery service requirements), logistics (e.g., volume shipments via consolidation), financial investment (e.g., start-up investment, level of inventory), and corporate strategy (e.g., corporate missions).
2. Display the tradeoffs between centralization and decentralization.

3. Develop the profiles of products and targeted market segments to see alignment between those profiles and strategic choices between centralization and decentralization.
4. Illustrate the degree of consistency between the characteristics of products/markets and the strategic choice. The straighter the profile, the more consistent the chosen strategy is with characteristics of products/markets.

Warehouse Layout

The generation of layout alternatives is a critical step in the warehouse planning process, because the chosen layout will serve to establish the physical relationship between warehousing activities (Tompkins et al., 1996). As such, warehouse layout ("inside the box") impacts the efficiency of space planning, the flow of material handling, warehouse throughput, equipment utilization, and labor productivity. Put simply, the primary objective of a warehouse layout is either to maximize resource utilization while satisfying customer requirements or to maximize customer services subject to given resource constraints (Tompkins et al., 1996). The key warehouse resources include space, equipment, and personnel. Therefore, a warehouse layout begins with space planning that aims to estimate space requirements and to optimally allocate space according to a product's storage and handling requirements. Herein, warehouse spaces are classified into four different types (Ackerman, 1999):

- **Open spaces**—Include aisles and empty spaces that are subject to "honeycombing"—wasted (empty) space created by blocked or underutilized storage areas.
- **Rack spaces**—Usually assigned according to the size of a fully loaded pallet. Some rack spaces can be dedicated to specific stock-keeping units (SKUs), whereas others can be used randomly for any SKUs.
- **Bin spaces**—Generally used for nonpalletized material.
- **Order pick spaces**—Used to accommodate only a small quantity of the items necessary to fill an order. These spaces are usually established when there are constant order-fill demands for the items.

Given the aforementioned warehouse spaces, a warehouse layout decision is concerned with determining the optimal stack heights, the number of aisles, the size of staging areas, order picking lines, and reserve storage spaces necessary to perform warehousing tasks. Other important factors for the warehouse layout decision include the usage/demand rates (e.g., fast moving, medium moving, and slow moving) of stored products, similarity of products, size (e.g., load dimension and weight, stacking height), characteristics, space constraints, and inventory profiles (e.g., unit loads per stock keeping unit, cubic velocity, special handling requirements). To elaborate, fast-moving items should be stored such that the order picker's travel distance to those items is minimized; therefore, they are stored in deep racks (e.g., double or triple deep racks). Also, it is common that items received together tend to be stored in the same area to minimize the order picker's travel distance. Likewise, items that are destined to the same store are often stored together. To ease handling, it is normal to store large-sized, bulky, heavy items close to the point of use or in the areas having the lowest stacking heights. Product characteristics such as shelf life, shape, and potential danger need to be taken into consideration for the layout because perishable, fragile, flammable, and dangerous products should be segregated to the designated storage area. Otherwise, some warehouse employees may be exposed to serious dangers and contaminations. Finally, to minimize vertical and horizontal honeycombing, the layout decision should consider space constraints imposed by the ceiling heights, posts and columns, floor loads, trusses, and sprinkler system (Tompkins et al., 1996). The warehouse layout can take four different forms, depending on the patterns of warehousing activities. These forms are u-shape, straight-thru, modular-spine, and multistory flows (Frazelle, 2002).

U-Shaped Flow

In a U-shaped layout, the floor space is allocated in such a way that receiving areas are located closer to unloading/loading docks in the front of the warehouse, storage and retrieval areas are reserved in the back of the warehouse, and unitizing/shipping areas are located near the loading/unloading docks in the same side of the warehouse, as illustrated in Figure 6.2. In this layout, the wide central aisle is typically arranged in a form of a U, which consists of two vertical aisles connected by a horizontal cross aisle. This layout can facilitate cross-docking operations because receiving and shipping docks are adjacent to one another and receiving and shipping processes can share the same dock doors. Other advantages of this layout include better utilization of lift trucks because putaway and retrieval trips are easily combined, and improved security because a single side of the building is used for entry and exit (Frazelle, 2002). This layout is popular in the retail distribution sector.

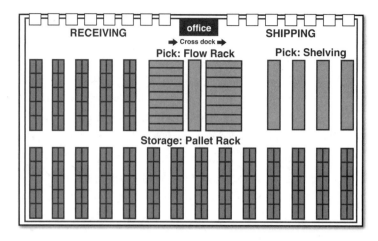

Figure 6.2. *An illustrated example of a U-shaped layout*

Straight-Thru Design

In the straight-thru layout, the warehouse is considered a "flow-through" terminal where receiving docks are located on one side of the warehouse, whereas shipping docks are located on the opposite side of the warehouse, as shown in Figure 6.3. In between the receiving and shipment staging areas, this layout positions sorting, assembly, and temporary inventory holding areas. This layout may be ideal for Just-In-Time operations where the storage of the materials is brief, but it can nullify a storage design based on ABC inventory classification and tends to underutilize docks due to two separate dock locations on opposite sides. This layout is more common in the manufacturing sector.

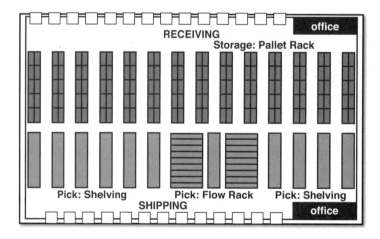

Figure 6.3. *An illustrated example of a straight-through layout*

188   Chapter 6: Warehousing

Modular-Spine Design

The modular-spine design is suited for large-scale warehousing operations where customized activities need to be added to the warehousing process with great flexibility. This layout is desirable especially when standalone and uniquely designed buildings are needed for high-speed sortation, as well as for customizing operations such as monogramming, pricing and marking, and unit-load automatic storage and retrieval (AS/RS) systems (Frazelle, 2002). The modular-spine design works better for ABC-based storage schemes.

Multistory Design

The multistory layout is common in Asian and European countries where real estate is limited and expensive. However, the multistory layout is not conducive to smooth material handling due to multiple bottlenecks and a lack of automation opportunities created by unconnected flow spaces.

Warehouse Asset Management

The key resources in the warehouse include various assets such as the building, equipment (e.g., lift trucks), fixtures (e.g., racks), pallets, bins, and inventory. In particular, inventory in the warehouse represents goods in transit and therefore its ownership may change during the movement and the storage. In other words, goods stored in the warehouse need to be tracked for the transfer of responsibility for potential loss and damage.

According to Section 7-204 of the Uniform Commercial Code (UCC), a warehouse operator is liable for loss, or damage to the goods as a result of his or her failure to exercise proper care in regard to them, as a reasonably careful person would exercise under similar circumstances. However, unless otherwise agreed, he or she is not liable for damages such as natural disasters that could not have been avoided by the exercise of such reasonable care (Ackerman, 2000). Also, the UCC stipulates that the extent of liability for loss or damage may be limited depending on a contractual term in the warehouse receipt or storage agreement which can set forth a specific liability per article or item, or value per unit of weight. No increased valuation shall be permitted contrary to a lawful limitation of liability contained in the warehouse operator's tariff, if any. Reasonable provision as to the time and manner of presenting claims and instituting actions based on the bailment may be included in the warehouse receipt or tariff. Herein, examples of loss and damage to the depositors' goods may include: minor damage as the result of improper handling, packaging, or minor accident; incidental loss as a result of theft or pilferage; unavoidable loss as a result of fire, flood, roof collapse following an unusually heavy snowfall, or other natural catastrophes. Depositors, therefore, should measure the potential risk of such damages or losses before specifying contractual terms and entering storage agreements (Bolten, 1997).

To minimize loss and damage, a variety of automated solutions can be utilized. These solutions include barcodes, active and passive radio frequency identification (RFID), cell, global positioning systems (GPS), and automatic identification and data capture (AIDC) technologies, which can track and sense products stored in warehouses or other mobile resources such as returnable containers, equipment, racks, parts, and people. Thus, these automated solutions can enhance asset visibility and security and consequently improve asset utilization, capital deployment, resource tracking, and regulatory compliance.

Material Handling

Material handling refers to various activities of moving, storing, protecting (e.g., packaging, unitizing), sequencing, and picking the right material in the right condition at the right time and at the right place to create time and place utility in the warehouse. Material handling typically accounts for 25% of all employees' activities, 55% of all factory space, and 87% of production time (Frazelle, 1986). As such, material handling represents 15 to 70% of the total cost of a manufactured product and is the main cause of 3 to 5% of damaged products (Tompkins, 1996).

In general, the design of a material-handling system requires the following steps:

- **Step 1: Assess the current process for potential improvement**—This step includes the evaluation of the current warehousing design, existing layout, order profiles (e.g., order frequency/popularity, similarity, size, and characteristics), SKU cubic velocity, and current space.
- **Step 2: Define objectives and scope of the proposed material handing system**—This step analyzes the material handling needs (including equipment needs) and develops design plans based on space availability, throughput requirements, capacity and budget constraints, and project timelines. In particular, if the SKU cubic velocity and material-handling labor cost are high, the use of a conveyor system makes sense.
- **Step 3: Generate alternative designs**—This step develops alternative material-handling designs, which include material flow block diagrams, warehouse layouts, unit load configuration, information flows, and process maps. Also, the pros and cons of these alternative designs should be evaluated.
- **Step 4: Select the best design among the alternatives**—This step involves selecting the material-handling design that best meets the main objectives of the

material-handling system with respect to the characteristics (e.g., size, weight, shape, flammability, fragility) of the materials to be handled, material flow patterns, types of material-handling equipment to be used, skills of warehouse personnel, and cost efficiency. For example, popular (high demand/usage) materials should be located in the most accessible areas, and heavy materials should be stored on the lower levels, which are close to the point of use.

- **Step 5: Implement the selected system design**—This step involves acquiring the chosen alternative, training warehouse personnel, and then managing the installed system within the budget constraints. Once the installation has been completed, the installed system should be validated through periodic audit processes.

In particular, the audit of the material-handling system may follow the 20 principles recommended by the 1996 College Industry Council on Material Handling Education (CICMHE). These principles are as follows (Tompkins, 1996, p. 158):

- **Orientation principle**—Study the system relationships thoroughly prior to making preliminary plans in order to identify existing problems, physical and economic constraints, and to establish future targets and goals.
- **Planning principle**—Establish a plan to include basic requirements, desirable options, and the consideration of contingencies for all material-handling activities.
- **Systems principle**—Synchronize and coordinate those handling activities associated with receiving, inspection, storage, production, assembly, packaging, and shipping.
- **Unit load principle**—Handle product in as large unit load as practical.
- **Space utilization principle**—Utilize all warehouse cubic (both horizontal and vertical) space.
- **Standardization principle**—Standardize handling methods and equipment wherever possible.
- **Ergonomic principle**—Consider human factors for designing material-handling equipment and procedures to make interaction with the people using the system easier.
- **Energy principle**—Minimize energy consumption of the material-handling systems and material-handling procedures.
- **Ecology principle**—Minimize adverse environmental effects when selecting material-handling equipment and procedures.
- **Mechanization principle**—Automate the handling process where feasible to increase efficiency and economy in the handling of materials.
- **Flexibility principle**—Use methods and equipment that can perform a variety of tasks under a variety of operating conditions.

- **Simplification principle**—Simplify handling by eliminating, reducing, or combining unnecessary movements and/or equipment.
- **Gravity principle**—Utilize gravity to move material wherever possible, while respecting limitations concerning worker safety, product damage, and loss.
- **Safety principle**—Provide safe material-handling equipment and methods that follow existing safety codes and regulations.
- **Computerization principle**—Consider computerization in material-handling and storage systems, when circumstances warrant, for improved material and information control.
- **System flow principle**—Integrate data flow with the physical material flow in handling and storage.
- **Layout principle**—Find an optimal operational sequence and layout for all viable systems, and then select the alternative system that works best.
- **Cost principle**—Compare the benefits of alternate equipment solutions and handling methods on the basis of cost efficiency as measured by expense per unit handled.
- **Maintenance principle**—Prepare a plan for scheduled preventive maintenance and repairs on all material-handling equipment.
- **Obsolescence principle**—Develop a long-range and economically viable policy for replacement of obsolete equipment and methods with special consideration to after-tax life cycle costs.

Order Picking

Among various material-handling activities, order picking makes up the largest expense category. In fact, order picking has long been identified as the most labor-intensive and costly activity for most warehouses given that the cost of order picking is estimated to be as much as 55% of the total warehouse operating expenses (de Koster et al., 2007). Four types of order picking methods are used in today's warehouses (Ackerman, 1999):

- **Discrete picking**—This is the most commonly used order-picking method in which a single order picker picks an entire order from start to finish, one order at a time. In this picking, all the items on one order are picked before the next order is picked. Thus, the order picker's job is simplified and the rehandling or repacking of picked items is not needed. Generally, discrete order picking is best for a small organization with low order counts, but the order picker is forced to travel throughout the warehouse and thus his or her picking efficiency is decreased while his or her travel distance and cycle time are increased (Modern Materials Handling, 2011).

- **Batch picking**—In this picking, orders are grouped into small batches containing many common SKUs. Thus, an order picker will pick all orders within the batch in one pass and then deliver them to the staging area where they are separated into individual orders. Batch picking is usually associated with pickers with multi-tiered picking carts/totes moving up and down aisles picking batches of usually four to 12 orders. Batch picking is also very common when working with automated material-handling equipment such as carousels. Batch picking is efficient for small-item, multiple-order picking with a significant distance, because it can reduce repeat visits and the subsequent wasted travel time. Batch picking is also useful when you have a high concentration of SKUs over a larger area. In this picking, however, order pickers may have to travel the entire picking area, and two order pickers may block each other while picking (Parikh and Meller, 2008). Batch picking is common in food, fashion, and e-commerce industries.
- **Zone picking**—In this picking, each order picker is assigned to a particular area ("zone") of the warehouse. A warehouse is divided into several pick zones (e.g., a fast-moving item zone and a slow-moving item zone), and order pickers are only allowed to pick the items in that zone. Therefore, orders are moved from one zone to the next (usually on conveyor systems) as they are picked (also known as "pick and pass"). Zone picking is suitable for orders with many different types (requiring different equipment or expertise) to reduce the order picker's travel time. In this picking, order pickers can gain great familiarity with the items and locations in their assigned zones, thereby increasing both picking frequency and the accuracy of picks (Modern Materials Handling, 2011). However, this method requires many individuals to handle a condensed pick (Ackerman, 1999).
- **Wave picking**—In this picking, orders are grouped by a common characteristic, such as specific carriers, routes, and/or destinations (retail stores). Unlike with zone picking, all zones are picked at the same time, and the items are later sorted via the conveyor sortation process and consolidated into individual orders/shipments. Wave picking is the quickest method for picking multi-item orders. However, the sorting and consolidation process can be time consuming and confusing. Wave picking is effective in large warehousing facilities that have thousands of SKUs or high volume such as grocery items and footwear. This picking method is also frequently used to exploit full truckload (FTL) freight discounts.

Warehouse Productivity

As an important link in the supply chain, warehouse operations influence the overall efficiency of the supply chain. Therefore, systematic efforts have to be made to enhance warehouse productivity with respect to cycle time, order fulfillment, asset utilization, and labor management. These efforts begin with the establishment of warehouse productivity measurements. The warehouse productivity measurements are classified into two categories: external and internal productivity measurements (Ackerman, 2000). The external productivity measurements are composed of the following elements (e.g., Ackerman, 2000; Frazelle, 2002):

- The frequency of customer complaints
- Out-of-stock occurrences
- Backorder rate
- Warehouse order cycle time, which represents the elapsed time from when an order is released to the warehouse floor until it is picked, packed, and ready for shipping
- Dock-to-stock time, which represents the elapsed time from when a receipt arrives on the warehouse premises until it is ready for picking and shipping
- The speed and accuracy of documentation
- The efficiency of handling special requests
- Compliance with loading, marking, and tagging rules

The internal productivity measurements include these (Ackerman, 2000; Frazelle, 2002; Coyle et al., 2003):

- Inventory accuracy, measured by the percentage of warehouse locations without inventory discrepancy
- Picking accuracy, measured by the percentage of order lines picked without errors
- Putaway accuracy, measured by the percentage of items put away correctly
- Shipping accuracy, measured by the percentage of order lines shipped correctly
- The length of equipment downtime
- Storage density, measured by the percentage of warehouse cubes occupied
- The frequency of accidents and personal injuries in the warehouse
- Employee (voluntary) turnover rate
- Dock utilization rate, measured by turnaround time to load and unload
- Warehouse labor productivity, measured by the ratio of actual output in receiving, picking, and storing, and shipping to the amount of labor hours spent
- Product damage rate, measured by the ratio of the total number of items handled with damage to the total number of items handled in the warehouse
- Warehouse throughput, measured by the amount of items moved through the warehouse system in a given time period

Warehouse Security and Safety

Parallel with the growth of the logistics industry, total annual inventory loss in the United States resultant from cargo- and warehouse-related theft has been on the rise since 2006 and increased by 17% between 2010 and 2011 (CargoNet, 2011). Theft of goods in the warehouses will continue to rise unless proper theft-prevention guidelines are developed and management vigilance is increased. There are two types of theft: *mass theft*, where thieves either hijack a truck or break into a warehouse, and *pilferage* (called "mysterious disappearance"), which may involve collusion between truck drivers and warehouse employees who load "extra" products on the truck when shipping or receiving less than the full quantity of material off an inbound shipment. They may also load the orders obtained through their fellow thieves employed as salespeople (D.M. Disney and Associates, Inc., 2007). In particular, pilferage involving collusion among truck drivers and warehouse employees is difficult to control because a paperwork trail is seldom left, and it often involves the clandestine removal of small amounts of merchandise over a long period of time (Ackerman, 2000). There are several reasons for frequent inventory loss at the warehouse (Brandman, 1989):

- The abundant storage of marketable and valuable products in the warehouse that are subject to potential employee theft
- The difficulty in detecting internal theft that is camouflaged as part of day-to-day operations
- A lack of pre-employee screening due to an increased labor crunch in the warehousing industry
- A heavy reliance on physical deterrents such as security alarms, guards, and closed-circuit television equipment, which cannot protect warehousing property from internal theft

To prevent the aforementioned mass theft and pilferage, the following loss prevention measures should be taken (Brandman, 1986; Ackerman, 2000):

- Enhance loss-prevention awareness by establishing formal security rules and regulations involving employee integrity, package inspection, locker inspection, unauthorized use/consumption of company equipment or inventory, disposal of company property and trash, falsification of company paperwork and record, and periodic loss-prevention meetings.
- Conduct random tests on loadings for outbound trucks as a form of surveillance.
- Cycle count inventory periodically and carefully check inventory discrepancies as a way of controlling stock in the warehouse.
- Use RFID tags on the high-value inventory.

- Conduct undercover investigations when the following signs of inventory theft are evident:
  - Unexplained inventory shortages, overages, or discrepancies
  - Customer complaints of incomplete or missing shipments
  - An unexplained decrease in a particular customer's purchasing patterns
  - Disappearance of invoices, receipts, or purchase orders
  - Discovery of inventory cases containing goods that differ from those described on the affixed labels
- Conduct external security audits as a guard against physical, personnel, and procedural security measures.
- Conduct rigorous pre-employment screening tests, including personal interviews, reference checks, background reviews, drug/alcohol screening, and psychological tests.

Warehouse Automation

Given rising labor costs and the increasing complexity of warehousing operations resultant from multitasking processes, warehouse automation can drive a company toward greater productivity and a subsequent increase in profitability. Examples of automated warehouse systems include powered conveyor storage/sortation systems, automatic storage and retrieval systems (AS/RS), automatic case or item-pick systems, automated electrified monorail (AEM), automatic guided vehicle systems (AGVS), pick-to-light/put-to-light systems, computerized stock locator systems, wireless communication terminals for fork trucks, batch order picking systems, computerized shipping/manifesting, and barcode-verified replenishment. The typical features of warehouse automation are optimizing and controlling movement of ordered materials along a warehouse conveyor, tracking orders, electronically transmitting information, reducing paper processes, verifying items and counts for all orders and replenishments, preventing inventory shortages and out-of-stocks, initiating replenishments/restocks, providing warehousing managers a real-time view of workflow throughout the warehouse, and accommodating multinational shipping requirements. Regardless of its features, warehouse automation requires careful planning to be successful. This careful plan needs to follow the following four steps (http://logistics.about.com/od/increaseproductivity/a/whse_automation.htm):

1. Create the process map of the proposed automation process. This process map may include the examination of customer order patterns and volume fluctuations to see if the proposed automated system has the capacity to handle those patterns and fluctuations.

2. Identify routine and repetitive warehousing operations that can be a candidate for automation. For example, cartons and containers conforming to a standard size (length, width, height, shape) may enhance the efficiency of automated material handling.
3. Conduct a sensitivity analysis to determine the impact of additional volume during peak periods and develop a flexible process for handling that additional volume. For example, if a carton is not properly labeled or is damaged, can it be handled by the proposed automated system?
4. Assemble a team that would undertake the warehouse automation project. Every team member should provide valuable input into the automation project, especially the team members closest to the work. They have lived through many challenges in the warehouse, and their input is invaluable. Assign a project manager who will track and report on progress.

Once warehouse automation is fully in place, it can bring a number of managerial benefits. These benefits include increased throughput, better resource utilization, reduced human error, labor cost savings, savings in handling and storage costs, improved data accuracy, improved inventory/order accuracy, and the less risk of human injury. On the other hand, warehouse automation requires a huge startup investment and involves a significant learning curve. Therefore, the rapid automation of warehousing operations without a proper transition plan can spell disaster.

Warehouse Workforce Planning

Regardless of the extent of warehouse automation, warehousing operations are labor intensive by nature. In particular, growing customization and e-fulfillment operations require individual picking of online orders. Considering the labor intensity of warehousing operations, warehouse workforce planning has a profound impact on the effectiveness and efficiency of warehousing operations, which in turn affect the profitability of warehousing operations. Typically, warehouse workforce planning aims to recruit, retain, and nurture talented warehouse employees, including warehouse supervisors, order pickers, forklift operators, and inventory clerks. As the growth of the warehousing industry recently accelerated, warehouse workforce planning has become a daunting task. For example, the compound annual growth rate (CAGR) of the warehousing industry is estimated to be 9.67% during the period of 2013 through 2018 (PR Newswire, 2014). Reflecting such trends, the number of wage and salary jobs in the warehousing industry grew around 11% from 1998 through 2008 (U.S. Department of Labor, 2009). However, the pool of qualified labor forces has constantly lagged behind the growth of the warehousing industry. This imbalance created a chronic shortage of qualified warehouse workforces and undermined the warehousing productivity. As a matter of fact, Min (2004) observed that warehouse employee shortages adversely affected customer service, absenteeism, logistics productivity, employee morale, and profit margin. According to a survey conducted by the Warehousing Education and Research Council (WERC), nearly half (44.1%) of the surveyed respondents indicated an average annual turnover rate in excess of 20% among entry-level warehouse positions, whereas median employee turnover in the U.S. was 8.4 % (Bureau of Labor Statistics, 1999; Wilson, 2000). Facing such labor scarcity, warehousing managers should consider taking the following steps to attract potential warehouse employees (Ackerman, 2000; Min, 2004):

- Develop a labor requirement plan based on short-term demand forecasts.
- Use local labor recruiters and reward them for their recruitment efforts.
- Increase the chances for active employee involvement in new employee recruitment and warehousing operational changes to enhance the sense of community among employees.
- Establish flexible working schedules for nontraditional workforces such as housewives, students, retirees, immigrants, and the handicapped.
- Direct job advertisement toward nontraditional workforces.
- Provide long-term career advancement opportunities built upon warehousing jobs, including blue-collar jobs.

Warehouse Management Systems

With the increasing demand for impeccable customer service, today's warehouses are pressured to raise their goals for inventory accuracy, timely delivery service, individualized order fulfillment, flexible value-added services, and responsiveness to special customer requests. To meet these high performance goals, the warehouse needs to find a way to eliminate any waste, streamline its operations, and improve efficiency in every aspect of warehousing activities. One of the most proactive ways of doing so is the use of a warehouse management system (WMS) that is designed to speed up order turnaround time, improve inventory accuracy, provide instant order status information, manage warehouse space, and enhance labor productivity (Min, 2006). WMS has revolutionized the way to schedule, plan, and fulfill orders, track inventories, and ensure the on-time delivery of the right products. WMS has recently expanded to include the additional features that can handle light manufacturing, transportation management, order management, and complete accounting systems. In general, WMS is referred to as a "real-time" inventory-tracking, resource management, and communication system that links corporate-level production, purchasing, scheduling, and logistics activities through improved supply chain visibility (see Adams et al., 1996, and Alexander Communications Group, 2003, for a conceptual foundation of WMS). WMS updates the inventory level on a real-time basis and fills customer orders in the distribution environment by enhancing inventory visibility.

The key objectives of the WMS include the following:

- Eliminate order fulfillment errors by product identification and continuous cycle counting.
- Send and receive critical customer/warehouse information with minimum lead time through electronic transmission.
- Maximize labor productivity by managing and prioritizing tasks.
- Maximize space utilization by selecting a proper storage location.
- Reduce inventory and handling requirements through a continuous flow of information.

Since its inception in the 1970s, WMS has gained popularity due in part to an ongoing trend in warehouse automation and the subsequent growth in supply chain solution software. For example, the U.S. market size for automated material handling systems alone stood at $10.7 billion in 2012 and was expected to grow 4–5% a year (Doering and Hausladen, 2013). The market of supply chain solution software such as WMS is projected to grow to $9.8 billion by 2018 from $1.646 billion in 2001 (ARC Advisory Group, 2003; Trebilcock, 2014). According to Ramaswami (2001) and Clearwater International (2014), WMS is one of the most popular (most frequently used) among supply chain solution software, including customer relationship management (CRM), enterprise resource planning (ERP), and a transport management system (TMS). Indeed, WMS accounted for approximately 56% of the sales revenue of the supply chain solution software in 2007 and its market reached more than $1.1 billion in 2013 (Cooke, 2008; Clearwater International, 2014; Trebilcock, 2014). Following this trend, a number of firms successfully implemented WMS and reported numerous benefits resultant from WMS.

For example, Kimberly-Clark enhanced product flow and eliminated "deadheading" (an empty return trip without a picked order), which accounted for more than half of the order picking time, by replacing the manual stock location system with WMS. As such, it could reduce labor cost by 10% to 30%. Similarly, the Disney Store improved product flow to stores, better managed its workload during the peak period, and reduced order processing time by optimally determining whether incoming products should be cross-docked or entered into inventory with the help of WMS (Maloney, 1999). With the addition of new material-handling equipment and an updated real-time WMS, Nike more than doubled its throughput from 100,000 to 250,000 units per eight-hour shift, increased overall productivity from 40–45 shipped units per labor-hour to 73 units, and reached an order accuracy of 99.8% (Maloney, 2000). Thanks to WMS, Nestle optimized its storage facilities, improved stock control and accuracy, and maximized efficiency across the entire distribution operation at its central finished goods depot (Central Systems and Automation, 2000). The U.S. Social Security Administration also reported that it achieved a warehousing space reduction of 60,000 square feet, yielding a cost saving of $300,000, and a reduction in contracted order-picking staff thanks to the efficiency created by WMS (Harps, 2000). In general, WMS can lead to a 25% gain in warehouse productivity, near perfect (100%) inventory accuracy, a 10% to 20% improvement in floor space utilization, a 15% to 30% reduction in safety stocks, and significant improvement in customer services (Autry and Bobbitt, 2003; Smith, 2003).

Despite this rosy picture, WMS implementation is still a daunting task, as illustrated by the reported WMS failure of Adidas. Problems with WMS once brought Adidas' main warehouse almost to a standstill with its retailers getting few or no Adidas shoes. In fact, only 17% of participants in a survey commissioned by the Warehousing Education and Research Council regarded their WMS implementation as a real success (Barnes, 2004). To make matters worse, WMS can be very expensive to implement. WMS software alone can typically cost a company somewhere between $100,000 and $500,000 (Autry and Bobbitt, 2003; Jones, 2014). To elaborate, the typical license fee for the WMS software ranges from $50,000 to $250,000 per site, whereas the implementation service fee ranges from $5,000 to $75,000. In addition, the software vendor typically charges 15% to 20% of the license fee for routine maintenance (Supply Chain Systems Magazine Staff, 2001). Other pitfalls of WMS include a lack of standards/compatibility, limited functionalities, frequent updates/modifications, and employee resistance to the changes brought by WMS (Mulaik and Cooper, 2000; Wilson and Drea, 2002).

Key Features of the Warehouse Management System

As summarized in Table 6.1, WMS typically performs a variety of warehousing tasks essential to daily warehousing operations (see eSync International, 2001; Alexander Communications Group, 2003; Barnes, 2004; and Min, 2006 for typical WMS capabilities). Examples of such tasks follow:

- **Order preparation, entry, and scheduling**—These tasks include EDI order entry, order validation, order routing, and carton size selection.
- **Inbound appointment scheduling**—Allows traffic managers to enter the information (e.g., carrier name, anticipated arrival time, supplier, inbound purchase order number, product number, SKU and quantity of each line, and trailer number) about the trailers coming to the warehouse to deliver inbound receipts

and automatically recommend appointment times and dock doors (Mulaik and Cooper, 2000).

- **Pre-receiving**—This task estimates the anticipated receipt times and schedules inbound carriers.
- **Receiving**—Creates inventory records, prepares inventory for stocking, utilizes advanced shipping notice, and assigns a list of docks to incoming trailers.
- **Putaway/storage**—Handles materials from the point of receipt to a storage area and manages the time needed to find open stock locations.
- **Cross-docking**—Bypasses the putaway/storage process to facilitate the process of combining shipments destined for the same destination.
- **Inspection**—Examines damage and checks the discrepancy between the shipping order and the load.
- **Wave planning**—Chooses which orders you want to process now, prints pick lists, and sorts items into groups that should be picked in a similar fashion.
- **Picking**—Relieves inventory from stocking locations to fill customer orders.
- **Cycle counting**—Monitors inventory levels and checks the inventory record continuously.
- **Inventory adjustment**—Accommodates adjustments due to product returns and cycle counting.
- **Inventory re-warehousing**—Performs stock consolidation and rotation tasks (Alexander Communications Group, 2003).
- **Manifesting**—Develops a list of all cargoes that pertain to a specific shipment, grouping of shipments, or piece of equipment.
- **Performance reporting**—Produces performance measurements against established standards for space utilization, order fulfillment, total throughput, and loss/damage, while creating an audit trail for warehousing activities.
- **Workload management and labor planning**—Estimates, tracks, and prepares reports on labor requirements (e.g., personnel hours to complete the day's expected activities) while managing workflow throughout the warehouse.
- **Automated data collection**—Compiles updated information regarding inventory location and lot sequence through radio-frequency (RF) readers, barcode scanners, mobile computers, and wireless local area networks (WLANs).

The scope of these functionalities may expand with the advent of new WMS technology. Although WMS functionalities can encompass most of warehousing activities, many firms would like to streamline WMS for their unique and customized warehousing functions; that is to say, many firms preferred a tailor-made WMS over a standard WMS. The five most popular areas of WMS applications include putaway/storage, picking, receiving, inventory adjustment, and cycle counting. Notice that all these represent primary warehousing functions that, in a typical warehouse, need to be handled on a daily basis. On the other hand, WMS is seldom used to perform secondary warehousing functions such as performance reporting, inspection and quality control, and workload management/labor planning (see, e.g., Autry and Bobbitt, 2003; Min, 2006).

Table 6.1. Typical WMS Features

Functions	WMS Features
Pre-receiving/receiving	• Purchase order update
	• Advanced shipping notice
	• Scanning
	• Self-checking of receipt information
Inspection and quality control	• Notification of inspection requirements
	• Confirmation of product inspection
Storage and putaway	• Dedicated, random, or hybrid storage
	• Reserve storage
	• Cross-docking
	• Shelf-life monitoring
	• Lot and serial number tracking
	• Operator location
	• Operator-directed putaway
	• Re-warehousing
Inventory management	• FIFO/LIFO
	• Inventory relocation/consolidation
	• Inventory quarantine, allocation, and release
	• Cycle counting
	• Stock rotation
	• Lot/serial number tracking
Order entry/preparation/picking	• Order planning/scheduling
	• Forward pick area replenishment
	• Kitting/pick and pack
	• Order/pick confirmation

Shipping	• Carrier scheduling
	• Yard management
	• Load sequencing
	• Manifesting
	• Bill of lading generation
	• Trafficking
Performance reporting	• Customer service reports
	• Inventory/order accuracy reports
	• Space utilization reports
	• Labor productivity reports
	• Item activity reports
Automated data collection	• Fixed inventory location
	• Random inventory location
	• Reserved inventory location
	• Lot sequence
	• Safety stock level

Once the aforementioned specific features of a particular WMS are well matched to the functional requirements of warehousing activities, as displayed in Figure 6.4, the WMS will be selected and then implemented (Mulaik and Cooper, 2000). The typical WMS implementation procedure is composed of the following steps (Miesemer, 2001):

1. Initial user training
2. System configuration
3. Pilot testing of the configured WMS
4. The analysis of gaps between the current system and the required system
5. Acceptance testing
6. On-site system testing
7. Enhancement design, approval, and further development
8. Customization

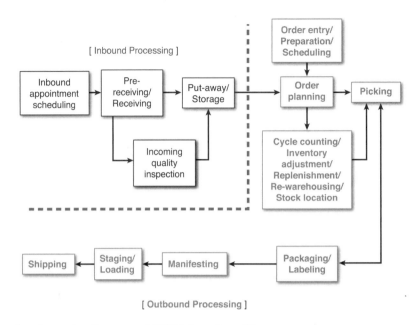

Figure 6.4. *A schematic diagram of WMS functionalities*

Integration of WMS with Other Supply Chain Execution Systems

The success of WMS often rests with the quality (e.g., accuracy, timeliness) of data that is fed by other supply chain execution systems (or supply chain solution software) such as enterprise resource planning (ERP), electronic data interchange (EDI), a transportation management system (TMS), barcoding, radio frequency identification (RFID), and a voice-directed system. Consequently, it is not uncommon that WMS has been integrated into ERP, EDI, and TMS. For instance, erroneous data introduced by ERP might allow an unnecessary order to be released to the warehouse, and then labor hours would be wasted by the WMS. Shipping activities cannot be performed and outbound transportation carriers cannot be scheduled unless the order is picked and released by the warehouse. A standalone WMS cannot provide visibility into how long it takes a warehouse employee to pick a set of orders or how long it takes to drive a forklift truck down an aisle in comparison to the benchmark performance measure (Schnorbach, 2005). Therefore, incorporation of RFID into WMS can create synergy for managing labor in the warehouse, because it can pinpoint inventory locations and direct order pickers to those locations. For instance, data synchronization via a supply chain execution system such as RFID can lead to a 3%–5% reduction in out-of-stocks, a 0.2%–0.7% reduction in logistics costs, and a 0.5% reduction in inventory (Blanchard, 2004). After adopting RFID, Wal-Mart reported $6.7 billion of labor cost savings, $575 million savings by eliminating employee theft and administrative error, and $300 million savings from more efficient tracking of its pallets (IDTechEx, 2005). Similarly, because the advanced shipping notice (ASN) via EDI can enhance the efficiency of WMS in planning inbound shipments and receiving activities, the WMS interface with EDI makes sense. Although less than half of the surveyed WMS adopters integrated WMS with either RFID or EDI, there is a great potential for the increased interface between WMS and RFID/EDI.

In transition, barcoding is still the most cost-efficient way of identifying products/cartons and complying with customer requests for compliance labeling. The WMS benefits can be further maximized by incorporating emerging communication technology such as a voice-directed system. The WMS aided by a voice-directed system allows the warehouse employees to operate in hands- and eye-free environments as they perform order picking tasks, which consequently helps reduce picking errors and labor fatigue (Alexander Communications Group, 2003). For example, an integration of a voice-directed system with WMS enabled a leading U.K. grocery retailer called Somerfield to reduce its warehouse employees' picking errors by 60%–70% and late picks by 50% (The Newsletter for Warehouse Management & Control Systems User, 2003). According to the survey conducted by the Accenture consulting firm, the integration of a voice-direct system with WMS improved picking accuracy by 40% on average and enhanced warehouse productivity by somewhere between 12% and 20% (Warehouse Management and Control Systems, 2003). As a matter of fact, all (100%) of the surveyed WMS users embedded a voice-directed system within their WMS to exploit such a benefit. In particular, more labor-intensive environments such as e-fulfillment and the grocery and food industry may take advantage of this technology due to its efficiency in enhancing labor productivity.

Handling Returned Products and Reverse Logistics

Product returns cost businesses more than $100 billion a year and caused an average profit loss of 3.8% (Petersen and Kumar, 2010). Wall Street Journal once reported that as much as one third of all online sales got returned (Banjo, 2013). According to Forrester Research, returns end up costing U.S. merchants as much as $20 billion a year (Kharif, 2014). As such, a well-planned and organized return process is one of the few opportunities that firms can exploit to improve their warehousing efficiencies and save unnecessary costs. In other words, reverse logistics involving a systematic return process should not be regarded as just an unavoidable cost of doing business or a necessary evil, but as a way to better utilize the company's assets and create shareholder value. Indeed, Guide et al. (2003) have observed that handling product returns is not considered a company's core "value-creating" business and therefore most companies often passively accept product returns. Such passive handling of returned products in fast clock-speed industries (e.g., personal computers) could erode the commercial value of products rapidly because the reverse logistics processes could slow the customer response time and shorten product life cycles.

In a broader sense, reverse logistics refers to the distribution activities involved in product returns, source reduction/conservation, recycling, substitution, reuse, disposal, refurbishment, repair, and remanufacturing (e.g., Stock, 1992). By nature, reverse logistics involving the product return process is more complicated than forward logistics operations due to the presence of multiple reverse distribution channels (direct return to manufacturers versus indirect return to repair facilities), individualized returns with small quantities, extended order cycles associated with product exchanges, and a variety of disposition options (e.g., repair versus liquidation), as summarized in Table 6.2 (Shear et al., 2003; Min, 2006).

Table 6.2. Comparison between Reverse Logistics and Forward Logistics

	Reverse Logistics	**Forward Logistics**
Quantity	Small quantities	Large quantities of standardized items
Information tracking	Combination of automated and manual information systems used to track items	Automated information systems used to track items
Order cycle time	Medium to long order cycle time	Short order cycle time
Product value	Moderate to low product value	High product value
Inventory control	Not focused	Focused
Priority	Low	High
Cost elements	More hidden	More transparent
Product flow	Two-way ("push and pull")	One-way ("pull")
Channel	More complex and diverse (multi-echelon)	Less complex (single or multi-echelon)

In particular, if your company has seen the following signs recurring, it needs to consider organizing and planning reverse logistics activities by developing a viable reverse logistics strategy (e.g., in-housing versus outsourcing, closed-supply chain network design, and reverse logistics facility location):

- Your customers are returning too many products at an alarming rate (e.g., more than 40% "no-fault" returns). There are noticeable high returns on specific products.
- Product returns are sprawling through the existing space of your warehouses.
- Your company needs to force its existing employees to work overtime or hire extra workforce to handle returned products.

- There is no transparency (e.g., reports, accounting records) on what has been going on with returned products.
- Your company never received a credit for returned products and took a substantial loss from the return process.
- Your customers increasingly complain about improperly or untimely processed returns.
- Your company has never had a clearly written return policy.

Types of Returns

In a broad sense, returns can be categorized as either controllable or uncontrollable, depending on their sources. Examples of controllable (or preventable) returns include the following (Stock, 2002):

- Damaged products
- Defects
- Mis-picks
- Refused shipments due to late delivery
- Unsalable consignments
- Overstocks due to poor forecasts
- Product recalls

On the other hand, some of the returns that cannot be controlled or avoided include these:

- End-of-life and seasonal returns
- End-of-lease
- Obsolete products
- Customer rejects for no reason
- Environmental disposals

The returns can be further subdivided into at least six different categories: consumer returns, marketing returns, damage returns, asset returns, product recalls, and environmental returns (Rigers and Tibben-Lembke, 2006).

Returns Management Strategy

The streamlined management of returns can improve a company's profitability by an estimated 10% to 15% (Norek, 2003). Therefore, the company needs to develop a viable return management strategy. Good examples of this strategy include the following:

- **Asset recovery**—The asset recovery option is primarily concerned with sending the product back to as close to its original condition and into the original channel as quickly as possible. This option can be divided into the following strategies (Norek, 2003):
  - Selling as new
  - Repairing or repacking and then reselling as new
  - Repairing or repacking and then reselling as used
  - Reselling at a lower value to a discount store
  - Selling by the pound to a scrap and salvage vendors
- **Outsourcing return management**—Overwhelmed by the volume of returned products, many e-tailers have outsourced their return management process to third-party logistics service providers (3PLs) such as GENCO, USF, and Ingram Micro Logistics, which specialize in reverse logistics operations. The use of the return specialist 3PLs can help e-tailers to save investment costs associated with the purchase of special equipment, hiring of additional personnel, and the installment of reverse logistics information systems. However, prior to hiring 3PLs for return management processes, the company needs to assess both the hard and soft costs of returns. Hard costs are composed of various costs associated with stocking returned items, handling, shipping, processing, refunding, repacking, disposal, lost value, and reselling. Soft costs are negative publicity, tarnished brand, lost customer goodwill, and lost bargaining power with customers (Morrell, 2001). Also, the outsourcing return process may reduce revenue-generating opportunities such as the sale of the recycled, remanufactured, and scrapped products.
- **Gatekeeping**—Successful gatekeeping usually allows companies to control and reduce the rate of returns without deteriorating customer service (Rogers and Tibben-Lembke, 2006). In general, gatekeeping refers to the screening of defective and unwarranted returned merchandise at the entry point into the reverse logistics process (Rogers and Tibben-Lembke, 1999). For example, gatekeeping would allow the warehousing employee to detect a returned product that was sold by another retailer and whose serial number does not match the packaging and then reject that returned product instantly. In doing so, gatekeeping can avoid unnecessary costs associated with unwarranted returned merchandise.
- **Recycling, refurbishment, and remanufacturing**—All returns cannot be handled in the same manner due to sheer volume and different product characteristics (Murphy, 2004). Many of the products that have been returned are repaired, refurbished, or remanufactured to the level where they can be salvaged for revenue-generating sales/resale opportunities. For example, Hewlett-Packard often sends returned computers in bulk to recycling centers to salvage any metals or plastic parts that have some commercial value before disposing of

them. As such, recycling is the reuse of parts and materials that were part of another (e.g., virgin) product or subassembly (Thierry, 1995). Another option of salvaging returned products includes refurbishment or remanufacturing, which can save reverse logistics costs by 40% to 60% (Rogers and Tibben-Lembke, 2006).

- **Supplier "zero-return" program**—This program provides the customer (e.g., retailer) with some discounts off the invoice amount in exchange for not sending any returns except when products are the supplier's defects. In this program, the customer may dispose of the product, donate it to charity, or sell it to alterative channels such as "dollar" stores or flea markets (Rogers and Tibben-Lembke, 2006). By passing the reverse logistics responsibility onto the customer, this program saves costs for the supplier. However, this program will work well only if the customer's warehouse knows the exact cost of returns processing and/or the revenue associated with reverse logistics activities. If the costs are lower or the revenues are higher than expected, the customer will be better off in this program. Otherwise, the customer will be reluctant to enter this program (Stock, 2004).

Chapter Summary

The following are key lessons learned from this chapter:

- The traditional role of a warehouse as a mere storage facility has been changing over the years. The emerging role of a warehouse includes "flow-through" or "postponement" of value-adding logistics activities. To elaborate, the concept of flow-through becomes a catalyst for cross-docking, which helps reduce order cycle time and improve customer services. Postponement is the art of putting off value-added services such as labeling, packaging, kitting, and assembling products when they are about to be shipped to end customers.
- This decision comes down to the selection of a particular type of warehousing among three potential choices: private, public, and contract. Each one of these choices has pros and cons, so the warehousing manager must carefully weigh those pros and cons prior to finding the warehousing type most suitable to his or her company.
- Without understanding and estimating warehousing costs, a company cannot control these costs. Unfortunately, warehousing costs often include some hidden elements that are either invisible or unquantifiable. These hidden costs can include pilferage, loss, damage, security, worker safety, compliance with rules and regulations, and labor management. Therefore, warehousing cost data should be collected, stored, traced, and updated periodically. To do so more effectively, these cost data should be properly categorized into elements such as real-estate

- costs, equipment costs, labor costs, administrative expenses, and hidden miscellaneous costs.
- Because the warehouse is a vital link to other supply chain processes, it should be designed in such a way that it is well connected to customer bases, supplier bases, and other supply chain partners' facilities. Such a design involves centralization/decentralization of warehouse networks as well as determination of the optimal number, size, and location of warehouses.
- Once the warehouse is established, its internal layout should be carefully planned to maximize the utilization of given vertical/horizontal space and to improve work/material flows in the warehouse. The typical warehouse layout plan follows one of four types: u-shape, straight-thru, modular, and multistory.
- Because warehouse productivity has a profound impact on the overall supply chain efficiency, it should be monitored by assessing warehousing performances periodically. Therefore, warehousing performance metrics should be identified and developed. The most commonly used warehousing performance metrics include inventory accuracy, order accuracy, order-fill rate, warehouse labor productivity, storage density, and warehouse throughput.
- To cope with the increasing complexity of warehousing operations, many firms have begun to use warehouse automation tools such as an automatic storage and retrieval system (AS/RS) and information technology such as a warehouse management system (WMS). In particular, the use of WMS is on the rise due to its ability to improve inventory tracking, movement, storage utilization, and labor management.
- With the increase in online sales and retailers' loose, liberal return policies, a growing number of products sold to customers are returning to the warehouse. In the past, these returns were not handled efficiently and therefore became major sources of warehousing messes. To deal with product returns, proper return management strategies should be developed. These include asset recovery, outsourcing return management, gatekeeping, recycling, refurbishing, remanufacturing, and supplier zero-return programs.

Study Questions

The following are important thought starters for class discussions and exercises:

1. Explain the changing role of warehouses in various organizational settings (e.g., retail stores such as Walmart, grocery stores such as Kroger, third-party logistics providers such as UPS Supply Chain Solutions, automobile manufacturers such as Toyota, and building/construction materials suppliers such as Owens-Corning).
2. Discuss the pros and cons of the different types of warehouses.

3. How would you estimate total warehousing costs? What comprises warehousing costs? What are hidden warehousing cost elements? Develop a detailed work sheet for warehousing cost elements.

4. What are value-added services? Why are they increasingly important? How do those services reshape warehouse operations and layouts?

5. What are major barriers for warehouse productivity? How do blocked lanes, poor layout, uneven workload, poorly trained and motivated employees, and disorganized workforce scheduling impact warehouse productivity?

6. Which factors should be considered for selecting the warehouse site? Which of those factors should be weighed more heavily than others?

7. What are the pros and cons of warehouse strategy? Why are most of today's organizations migrating toward the warehouse centralization strategy? (What may have prompted the warehouse centralization strategy?)

8. Explain the differences among four different forms of warehouse layout plans. Also, list the examples of those forms in various warehouse settings.

9. What are potential measures that would help enhance warehouse security and safety? Discuss how 9/11 impacts the importance of warehouse security and safety. Also, discuss how the warehouse can work as the security buffer in case of supply chain disruptions.

10. Illustrate various warehouse automation tools that are useful for warehousing operations.

11. Identify the major sources of (voluntary) warehouse employee turnover and assess its impact on warehouse productivity. Also, discuss ways to reduce warehouse employee turnover.

12. Discuss how a warehouse management system (WMS) can be applied to eliminate order fulfillment errors, enhance labor productivity, and utilize given warehouse space. Also, explain the key functionality of WMS. What are the selection criteria for WMS software? What are the challenges for implementing WMS and integrating it with other supply chain execution systems such as enterprise resource planning (ERP) and a transport management system (TMS)?

13. What are the sources of product returns? What makes reverse logistics differ from forward logistics? How would you formulate viable return management strategy or reverse logistics strategy?

Case: One Bad Apple and Thousands of Headaches

This case was inspired by a story of Mr. Ken Ackerman and Mr. William G. Sheehan as part of the UPS Center workshop on lean warehousing at the University of Louisville, which was held on February 28 and March 1, 2002.

First Week of the Work

Pedro Lopez, a newly hired warehouse executive, recently got a high-paying job from a fast-growing e-tailer, Jungle.com, in his hometown of San Antonio, Texas. In this tough job market, he feels very lucky to land a new job with a substantial pay raise from his previous job in the small regional warehouse located in Birmingham, Alabama. Today is his fifth day at work and he's excited to play a leadership role in the enormous mega-warehouse, which handles tens of thousands of products sourced from all over the world, including his parents' mother country, Mexico.

As soon as he walks into his new office this morning, he sits down with his predecessor, Harold Maxwell, who decided to step down from his warehouse director position several months ago and retire three years earlier than planned for deteriorating health reasons. Harold looks like a nice, courteous, and soft-spoken gentleman for whom many workers may want to work. However, Pedro is somewhat surprised to notice that he looks tired and concerned. After Harold congratulates Pedro on his new job, he begins to debrief the current warehousing situation and then express a lot of concerns about the relatively high employee turnover, ranging from 35% to 50%. Although high employee turnover in the warehousing sector is not unusual, as reported by Dr. Hokey Min of BGSU, Harold sounds pretty glum and frustrated when he has to tell Pedro about the declining productivity and rising cost per order ($1.75 from $1.25) of this warehouse in San Antonio. Harold had primarily managed the warehouse in San Antonio, a distribution hub serving the growing Southwest and Pacific Coast markets. Another warehouse located in Louisville, Kentucky serves the Midwest and East Coast markets. Except for product lines, these two warehouses are nearly identical. Both are about a similar size (approximately half-million square feet) and layout. They also use comparable equipment, have identical work rules, and are unionized. When it comes to labor productivity and financial performances, however, the similarities stop. San Antonio warehouse's throughput per man hour is 7% lower and its shrinkage is calculated at .95%, as compared to Louisville warehouse's shrinkage rate of .18%. Harold also hints that San Antonio warehouse's recent struggle might have something to do with a few whiners and disgruntled workers who constantly complain about their compensation and benefits. Plus, some of them do not seem to be happy with cramped and uncomfortable working conditions inside the outdated warehouse.

Trouble-Maker

According to Harold, Carlos González, a supervisor of the warehouse crew, has been the root of the Harold's headache. Carlos was a former soccer player who had once played for a semi-professional soccer club in Argentina and worked as a fashion model thanks to his handsome looks. However, he was cut and released from the soccer club after he missed quite a few practices for no particular reason and often argued with his coach for his lack of playing time and demotion to a backup player position. For the past six years prior to his recent promotion to supervisor, he had worked every day as an order picker at the San Antonio warehouse. For the most part, Carlos was an outstanding order picker and a very charismatic person with a lot of charm. In fact, he often set the pace for the rest of the crew by picking an average of 120 ordered items per hour, and over the course of the past several year had become their "informal" leader and the "mentor" for new employees. Such a performance and leadership potential warranted his promotion to warehouse supervisor.

On the other hand, Carlos is known as a "party animal" on the weekends. He frequently reports for work on Monday mornings groggy and irritable—and the day before yesterday was no exception. A couple of days ago, Carlos showed up two hours late for his shift. Pedro approached Carlos as he was standing on the dock, noticing that he appeared unsteady on his feet and slurred of speech. When Pedro asked, "What's wrong with you this morning?" Carlos shot back angrily, "None of your damned #@$% business." Pedro was almost speechless, but after gathering himself he told Carlos to go home and sober up. Also, Pedro verbally warned Carlos that he would consider taking serious disciplinary actions against him, including a permanent dismissal from his current job.

Yesterday, Carlos failed to show up for work. Pedro noticed, however, a definite undercurrent running through the warehouse. There was none of the good-natured banter that usually filled the lunchroom. A pallet of frozen eggs was dropped and five bottles of orange juice were spilled on the floor. Someone apparently hit and dented one of the rack uprights with a lift truck. No one came forward and reported the accident. At the end of the first shift, three inbound trucks remained unloaded and were sitting idle at the receiving docks. Normally, they would have been emptied an hour before the shift ended. Because the first shift crew had to go home, Pedro and one of the "lumpers" (a laborer who unloads cargo) he hired stayed to unload the trucks.

Stirring the Pot

Today, Carlos is still "AWOL" (absent without official leave) and the situation in the warehouse remains messy—the same as the day before. Four electric lift trucks have low batteries, and when their operators go to charge them, they discover that the replacement batteries are also low. The operators complain that the chargers were not working properly. During the rest of the day, Pedro observes that two lift trucks appear to be running at half speed. Operators stop frequently to check their lift trucks for "unknown mechanical problems." By shift's end, the day's scheduled work is not finished, and Pedro has to hold everyone over two hours—at overtime rates—costing the company an extra $25,000.

This evening, Pedro calls his immediate boss, Charlie White, and asks for his advice in handling the worsening situation. Unfortunately, Charlie does not seem to have any idea what needs to be done other than saying "cost, cost, cost and nothing matters." Although Pedro suspects that Carlos is behind the warehousing mess and ongoing sabotage, he has no clear evidence. Earlier this afternoon, Pedro received an email from his company's corporate legal counsel informing him that the union wanted to arrange a meeting to discuss Carlos's situation. The union believes any disciplinary action was unwarranted in this situation and wants a formal apology to Carlos for the threat of disciplinary actions. The union leader also reminds Pedro that the company's documentation was insufficient for any potential disciplinary actions, including Carlos's suspension without pay.

To elaborate, the union leader reminds Pedro that a manager who has not followed the specific disciplinary and termination policies in place could face a wrongful discharge suit. Also, Pedro is reminded that wrongfully discharged employees could claim reinstatement, back pay, compensatory damages, and attorneys' fees. To his dismay, after a long conversation with his boss, Pedro finds that his company has no employee grievance guidelines and performance measurement systems in place. With his hands tied and unbearable stress stemming from this incidence, Pedro feels sick to his stomach and suffers severe chest pains. To make matters worse, when Pedro goes to his new, sporty BMW 325i this evening, he finds two front tires slashed and obscenities scratched into the paint of his trunk lid. Pedro is convinced that this bad apple can bring down his warehouse, so he knows that he cannot downplay and ignore the ongoing problems with Carlos. However, he still cannot figure out what will be the best course of action to take to deal with this situation without hurting his own professional career, warehouse workflow, productivity, worker morale, and working environment. Knowing that he will have a sleepless night this evening, he decides to pull up his old training manual from his file cabinet that he kept after he joined the manager training seminar about a decade ago. The manual shows a number of potential remedies in case a manager runs into a troublemaker such as Carlos. It suggests the following (see Demand Media, http://everydaylife.globalpost.com/handle-sneaky-troublemaker-work-5696.html):

- *Keep a private log of the troublemaker's mischievous conduct.* If the offender's misconduct escalates and you are forced to take actions against that individual, you need documented evidence to support your concerns. According to the career website *The Grindstone*, proof of sneaky or unacceptable behavior at work requires a legitimate paper trail. However, you need to avoid documenting small or irrelevant details such as frivolous or overly sensitive issues. Focus on specific examples of when the worker cheated, misbehaved, or purposely violated the company's policy and guidelines.

- *Report "unacceptable" behavior to upper management.* The company may have confidential procedures for reporting misconduct, but even if your company does not, you will likely need to address the issues with upper management. When you report your subordinate's deceitful, unprofessional, or unethical behavior, you must rely on your documented evidence and stick to the facts. Also, you should explain to your boss why the accused individual's behavior is detrimental to your company's productivity. You do not want upper management to perceive that you take the matter personally and are trying to slander his professional reputation. If the offenses are serious, the manager will have to decide whether terminating the troublemaker is the best course of action. If criminal activity such as fraud, money laundering, or embezzlement are not involved, upper management might be able to put the worker on probation, giving him the chance to rectify his untrustworthy behavior, notes *The Grindstone*.

- *Confront the troublemaker, if you have the authority to do something about his behavior.* If his misconduct does not affect your company's business activities or other employees, a firm rebuke may set him on the right path and no other actions may be necessary. If the troublemaker is still defiant and reluctant to change his behavior, you may give him at least two gentle warnings in a calm and easy tone. Afterward, you should make sure that both parties develop an improvement plan with the time limits and the proviso that, if measurable goals are not met, he will be either suspended without pay or out of work. Your documented warning with a paper trail may help him realize that he had better change his behavior before he endures the serious disciplinary action.

Discussion Questions

Step into Pedro Lopez's shoes and list a number of next steps you need to take to resolve the current problems.

1. What do you plan to do tomorrow?
2. If Carlos shows up, what will you say to him? Can you articulate to Carlos which of his actions and words are causing problems for the company? (If so, please specify.) Is it a good idea for you to ask Carlos why he said the things and acted the way he did? (If so, why? If not, why not?) With your best effort, if Carlos still slides through, what would you then do?
3. What do you suggest to prevent the recurrence of situations such as this?
4. Why do you think this incident took place? What mistakes do you think were made in hiring and treating a troublemaker such as Carlos?
5. What company policies (e.g., hiring policy, performance standards) should be developed, changed, or reviewed?

Bibliography

Ackerman, K.B. (1999), *Warehousing Fundamentals: A Self-Study Program for Managers and Lead Personnel*, Columbus, Ohio: Searl Education Service.

Ackerman, K.B. (2000), *Warehousing Profitably*, Columbus, Ohio: Ackerman Publications

Alexander Communications Group (2003), *Warehouse Management and Control Systems*, New York, NY: Alexander Communications Group.

Andersen Consulting (1998), *Warehouse Systems and the Supply Chain: A Survey of Success Factors*, Oak Brook, IL: Warehousing Education and Research Council.

Arc Advisory Group (2003), "Logistics Software Market Grows Despite Economy," The Newsletter for Warehouse Management and Control Systems Users, 5, 6.

Armstrong & Associates (2004), *Who's Who in Logistics? Armstrong's Guide to Global Supply Chain Management*, 12th Edition, Stoughton, Wisconsin: Armstrong & Associates.

Autry, C.W. and Bobbitt, L.M. (2003), "WMS and Warehouse Performance," WERC Watch, Dec., 1–8.

Autry, C.W., Griffis, S.E., Goldsby, T.J. and Bobbitt, L.M. (2005), "Warehouse Manage-ment Systems: Resource Commitment, Capabilities, and Organizational Performance," *Journal of Business Logistics*, 26, 165–183.

Banjo, S. (2013), "Rampant Returns Plague E-retailers: Sellers Suggest Sizes and Redirect Discounts to Break Bad Habits," *Wall Street Journal*, December 23, 2013, http://www.wsj.com/news/articles/SB10001424052702304773104579270260683155216, retrieved on March 15, 2015.

Barnes, C.R. (2002), "Benefits from Barriers to New Warehouse Technology," The Newsletter for Warehouse Management and Control Systems Users, 4, 3.

Barnes, C.R. (2004), *Warehouse Management Systems: Assessment and Selection*, Oak Brook, IL: Warehousing Education and Research Council.

Bolten, E.F. (1997), *Managing Time and Space in the Modern Warehouse*, New York, NY: American Management Association.

Ballou, R.H. (1999), *Business Logistics Management: Planning, Organizing, and Controlling the Supply Chain*, Upper Saddle River, NJ: Prentice-Hall

Brandman, B. (1989), *Implementing a Security Program for Reducing Warehouse Losses*, Oak Brook, IL: The Warehousing Education and Research Council.

Bureau of Labor Statistics (1999), http://stats.bls.gov/news.release/tenure.toc.htm, U.S. Department of Labor, Washington, DC.

CargoNet (2011), *United States Cargo Theft Report: First and Second Quarters 2011*, Unpublished Report, Jersey City, NJ: The Cargo Theft Prevention and Recovery Network.

Clearwater International (2014), *Global Supply Chain Management Software Market Report 2014*, Unpublished Report, Aarhus, Denmark: A Clearwater International TMT Team.

Cooke, J.A. (2008), "The Race Isn't Over…," DC Velocity, January, 54–56.

Coyle, J.J., Bardi, E.J., and Langley Jr., C.J. (2003), *The Management of Business Logistics: A Supply Chain Perspective*, 7th Edition, Mason, Ohio: South-Western.

de Koster, R., Le-Duc, T., and Roodbergen, K.J. (2007), "Design and control of warehouse order picking: A literature review," *European Journal of Operational Research*, 182(2), 481–501.

D.M. Disney and Associates, Inc. (2007), "Loss Prevention," http://www.dmdisney.com/loss_prevention.htm.

Doering, J. and Hausladen, R. (2013), "Warehouse and Distribution Solutions," Presented at the Swisslog 2013 Investor Day sponsored by Swisslog, May 29, 2013.

eSync (2001), "Justifying Warehouse Management Systems," Supply Chain Forum White Paper Series, eSync Toledo.

eSync International (2001), "Warehouse Management Systems Perspective '01," White Paper, Toledo: eSync International.

Frazelle, E.H. (1986), "Material Handling: A Technology for Industrial Competitiveness," Unpublished Material Handling Center Research Report, Atlanta, GA: Georgia Institute of Technology.

Frazelle, E.H. (2002), *World-Class Warehousing and Material Handling*, New York, NY: McGraw-Hill.

Gilmore, D. (2011), "Supply Chain News: Inventory Performance 2011," Supply Chain Digest, July 22, 2011, http://www.scdigest.com/assets/FirstThoughts/11-07-22.php?cid=4759, retrieved on March 15, 2013.

Harps, L. (2000), "What's Working in WMS," *The Newsletter for Warehouse Management and Control Systems Users*, 2, 1–8.

Harrington, L. (2005), "Inventory Velocity: All the Right Moves," Inbound Logistics, http://www.inboundlogistics.com/cms/article/inventory-velocity-all-the-right-moves/, retrieved on October 10, 2010.

Hill, J.M. (2002), "Warehouse Management Systems: Perspective," Supply Chain Forum White Paper Series, Toledo, OH: eSync.

Hill, T. (1994), *Manufacturing Strategy: Text and Cases*, Burr Ridge, IL: Richard D. Irwin.

IDTechEx (2005), "Cost Reduction in Retailing and Products Using RFID," http://www.idtechex.com/research/articles/cost_reduction_in_retailing_and_products_using_rfid_00000205.asp, retrieved on March 15, 2015.

Jones, R. (2014), *Warehouse Management Systems by the Numbers*, Unpublished Report, Dallas, TX: Total Logistics Solutions, Inc.

Kharif, O. (2014), "Merchants Try to Trim Many Unhappy Returns," *Bloomberg BusinessWeek*, November 24–30, 19–22.

Maloney, D. (1999), "The Magical World of Disney Distribution," *Modern Material Handling*, December, 32–37.

Maloney, D. (2000), "Nike Just Does It," *Modern Material Handling*, January, 42–47.

Maltz, A. (1980), *The Changing Role of Warehousing*, Oak Brook, IL: Warehousing Education and Research Council.

Miesemer, K.D. (2001), *Starting Up A World-Class DC: A Roadmap to Success*, Oak Brook, IL: Warehousing Education and Research Council.

Min, H. (2004), "An Examination of Warehouse Employee Recruitment and Retention Practices in the USA," *International Journal of Logistics: Research and Applications*, 7(4), 345–359.

Min, H. (2006), "The Applications of Warehouse Management Systems: An Exploratory Study," *International Journal of Logistics: Research and Applications*, 9(2), 111–126.

Min, H., and Melachrinoudis, E. (2001), "Restructuring a Warehouse Network: Strategies and Models," In *Handbook of Industrial Engineering*, 3rd Edition, edited by G. Salvendy, New York, NY: Wiley Interscience, pp. 2070–2082.

Modern Materials Handling (2011), "Order Picking Basics," *Modern Materials Handling*, March 8, http://www.mmh.com/article/order_picking_basics/, retrieved on October 15, 2014.

Morrell, A.L. (2001), "The Forgotten Child of the Supply Chain," *Modern Materials Handling Online*, http://www.manafacturing.net/mmh/index.asp?layout+articleprint&articleID=CA82287, May 15, 1–6.

Mulaik, S. and Cooper, C. (2000), "Warehouse Management Systems," Presented at the Executive Development Workshop sponsored by the UPS Center for Worldwide Supply Chain Management at the University of Louisville

Murphy, J.V. (2004), "New Solutions Help Companies Find the Return in Returns Management," *Global Logistics and Supply Chain Strategy*, 8(9), 44–50.

Newsletter for Warehouse Management and Control Systems Users (2003), "What's Working in WMS," *The Newsletter for Management and Control Systems Users*, 5, 7–8.

Norek, C. (2003), "Throwing it into Reverse," *DC Velocity*, January, 54–56.

Parikh, P.J. and Meller, R.D. (2008), "Selecting Between Batch and Zone Order Picking Strategies in a Distribution Center," *Transportation Research Part E: Logistics and Transportation Review*, 44(5), 696–719.

Petersen, J.A. and Kumar, V. (2010), "Can Product Returns Make You Money?," *Sloan Management Review*, 51(3), 85–89.

PR Newswire (2014), "Warehousing and Storage Market in the US 2014-2018," http://finance.yahoo.com/news/warehousing-storage-market-us-2014-125000209.html, retrieved on March 15, 2015.

Rogers, D.S. and Tibben-Lembke, R.S. (2006), "Returns Management and Reverse Logistics for Competitive Advantage," *CSCMP Explores*, 3, 1–15.

Rogers, D.S. and Tibben-Lembke, R.S. (1999), *Going Backwards: Reverse Logistics Trends and Practices*, Pittsburg, PA: Reverse Logistics Executive Council Press.

Schnorbach, P. (2005), "Managing the Labor Supply Chain," *WERC Watch*, March, 6–8.

Shear, H., Speh, T. W., and Stock, J. R. (2003), "The Warehousing Link of Reverse Logistics," Presented at the 26th annual warehousing and research council conference, San Francisco, CA.

Smith, B. (2003), "Justifying a WMS," *The Newsletter for Warehouse Management and Control Systems Users*, **5**, 1–6.

Speh, T.W. (1999), *Warehouse Inventory Turnover*, Oak Brook, IL: Warehousing Education and Research Council.

Stock, J. R. (2002), "Warehousing's Role in Reverse Logistics," Presented at the 25th annual Warehouse Education and Research Council (WERC) conference, April 29.

Stock, J. R. (2004), *Product Returns/Reverse Logistics in Warehousing: Strategies, Policies and Programs*, Oak Brook, IL: Warehousing Education and Research Council (WERC).

Thierry, M. (1999), "Strategic Issues in Product Recovery Management," *California Management Review*, 37(2), 114–135.

Tompkins, J.A., White, J.A., Frazelle, E.H., Tanchoco, J.M.A., and Trevino, J. (1996), *Facilities Planning*, 2nd Edition, New York, NY: John Wiley & Sons, Inc.

Trebilcock, B. (2014), "The Market for Supply Chain Management Software Continues to Expand, Highlighting the Importance of Software in Today's Supply Chains," *Supply Chain 247*, http://www.supplychain247.com/article/2014_top_20_global_supply_chain_management_software_suppliers, retrieved on March 17, 2015.

U.S. Department of Labor (2009), "Career Guide to Industries," www.stats.bls.gov/oco/cg, retrieved on January 14, 2010.

WERC Sheet (2003), "In Through the Out Door," *WERC Sheet*, February, 1–3.

Wilson, G. S. (2000), "Are You Finding the Right Warehousing Employees?," *WERC Watch*, December, 1–13.

Wilson, G.S. and Drea, J. (2002), "Growing Use of Technology in the Warehouse," *WERC Watch*, 2002, February, 1–14.

7

Transportation Planning

"Life is like riding a bicycle. To keep your balance, you must keep moving."
—Albert Einstein, in a letter to his son Eduard (February 5, 1930)

Learning Objectives

After reading this chapter, you should be able to:

- Understand the basic concept of transportation.
- Comprehend the roles of transportation in the national economy and the supply chain.
- Understand the Central Place Theory and its conceptual background.
- Grasp the evolving transportation regulations and regulatory reforms and their impact on transportation practices and supply chain efficiency.
- Understand the rationale behind transportation regulations and deregulations.
- Identify specific carrier management issues.
- Select a right carrier (transportation service provider) and a proper mode of transportation that minimizes costs and best meets the shipper's service needs and requirements in deregulated transportation environments.
- Understand the role of intermodalism in globalized transportation activities.
- Exploit various technological tools and equipment to enhance transportation productivity.
- Cope with the challenges of recruiting, hiring, and retaining productive transportation employees, including drivers, dispatchers, and traffic managers.
- Leverage the transport management system to enhance transportation productivity.
- Understand ways of controlling transportation costs through effective and efficient freight rate negotiation, yield management, and revenue management.
- Plan terminal operations effectively and efficiently.

Transportation as a Vital Link in the Supply Chain

To satisfy the daily needs of life, people consume resources. These resources include basic daily necessities such as water, food, and cloths. Because these resources are not readily available everywhere on the surface of Earth, they need to be moved to different locations. This movement is called *transportation*. Therefore, transportation is essential for sustaining human life. Its role is analogous to blood, which carries vital oxygen to the entire human body. In other words, transportation plays a vital role in connecting one side of the supply chain to another by carrying vital resources to the entire supply chain. Given its significance to the supply chain, transportation should be better planned, managed, and leveraged by the supply chain professional. However, transportation gets more complicated as human society evolves. Part of transportation complexity stems from its increased geographic coverage, diverse modal options, changing regulatory/deregulatory transportation laws/rules, limited fuel resources, and multiple stakeholders. To ease transportation complexity and unravel a transportation puzzle, one should start understanding the fundamental functions, evolution, and surrounding environments of transportation.

Speaking of fundamental transportation functions, transportation creates spatial (or place) utility by reducing distance gaps among suppliers, producers, and consumers. For example, transportation allows raw materials and parts/components needed for production to be shipped to producers' locations. Likewise, transportation allows finished goods made by producers to be shipped to consumers' locations. Thus, transportation creates value by changing the locations of raw materials, parts/components, and finished goods to the places where they are needed or consumed. Transportation also can create temporal (time) utility by making the needed resources available at the same location at the same time. For example, assembly at the production facility requires a variety of parts/components and workers at the same time. Transportation allows them to arrive at the same production location at the same time, thus avoiding production delays (waste of time). These fundamental functions of transportation have been greatly expanded by gradual changes in business environments and management philosophy for the last few decades. Those changes include the following:

- Advancements in information technology such as geographic positioning systems (GPS), satellite tracking systems, and radio frequency identification (RFID) that help track shipment status during transit
- Government intervention in transportation practices that intends to control the extent of competition in the marketplace and enhance transportation safety

- Unstable fuel prices, which necessitates the increasing use of alternative fuel
- Globalization of business activities that stretch the entire supply chain and then increasingly require intermodal transportation with frequent modal transfers
- Just-in-time (JIT) delivery services, which require faster, uninterrupted transportation within limited delivery time windows

To better cope with these changes, transportation planners or supply chain professionals must formulate a flexible transportation strategy that can answer the following transportation questions (see Temple, Barker and Sloane, Inc., 1982, for a typical transportation strategy):

- What kind of transportation services are needed to fulfill the supply chain missions? Included in this question are the following:
    - Should the firm use its own transportation equipment and personnel to handle its transportation activities?
    - If the answer to this question is yes, which modes of transportation, or mix of modes should be used? How should the firm finance (lease or buy) transportation equipment? How should the firm hire and manage transportation personnel?
    - Which carriers should be selected?
    - Which mode of transportation services is most fuel-efficient? Which alternative fuel is available for each mode or carrier?
- How should required transportation services be purchased? Included in this question are the following:
    - To what extent should transportation services be outsourced? How often should outsourced contracts be renewed?
    - When and under what circumstances should intermediaries (e.g., brokers) be used?
    - How should the terms of contract be negotiated?
- What resources are needed to support the firm's transportation activities or improve the firm's transportation efficiency? Included in this question are the following:
    - What information should be transmitted to the firm's supply chain partners and how much information should be collected, analyzed, and stored for future usage?
    - Which information technology should be acquired and developed to support transportation decisions? What level of investment should be made to adopt and maintain such technology?

- What kind of human resources are required to support transportation activities? How should those resources be hired, utilized, and managed?
- How should the transportation revenue be generated to create financial resources?

Discussions in the following sections will center around the potential answers for these questions.

Central Place Theory

The process of exchanging something of value between two parties (e.g., producers and suppliers, producers and consumers, producers and distributors, or distributors and consumers) that are distant from each other often triggers transportation activities. Although this process can take place anywhere in between the respective locations of those two parties, its location can either boost or undermine transportation activities. For example, if produced goods are to be exchanged at the location (i.e., a trading area or a marketplace) further away from consumer population centers, producers will lose their consumer-drawing power due to larger spatial gaps between them and their consumers and thus transportation activities from consumers and producers will be limited. However, if that remote location of the marketplace can offer a diverse range of product assortments due to its scale (size), it may still draw consumers from longer distances. In other words, the location and size of the trading area can dictate transportation activities. This concept can be succinctly explained by the Central Place Theory, which was originally introduced by Christaller (1933) and further developed by Lösch (1954). Generally speaking, the Central Place Theory is a geographical concept that seeks to explain the development patterns, number, size, and location of human settlements such as cities and towns around the world. It helps explain reasons why people gather together in cities and towns to exchange their goods, services, and ideas. Extending its theory, it can also explain why distribution centers and retail establishments are located in certain areas and how people get engaged in transportation activities to reach those specific trading areas.

To elaborate, this theory can be described from two different perspectives: producers and consumers. From the producer's perspective, the producer will gravitate toward the trading area where he or she can find a larger number of potential consumers with higher disposable incomes. Because travel distance to such a trading area can create spatial gaps and thus increase the cost of travel, distance will also affect the producer's decision to trade his or her products and services in that trading area. From the consumer's perspective, the consumer will be drawn into the trading area where he or she can find particular products that he or she wants or has access to a wide variety of products to choose from. Despite the availability of products and services that the consumer wants, distance for shopping trips to the trading area can be an obstacle because it will increase his or her cost of travel and thus increase the total cost of a product. Presuming that the consumer can move freely in any direction, the total cost to a consumer for a product will be the summation of product price and consumer travel distance multiplied by cost-per-unit distance of transportation. In addition to distance, there may be other influencing factors such as modes of transportation available, natural barriers (e.g., rivers and lakes), road conditions, traffic congestion, product characteristics (e.g., perishability), and taxes. The spatial and temporal dynamics of the Central Place Theory will be further complicated by the evolving technological advances such as the Internet, which allows consumers to buy products and services online without making a trip. Also, rising energy costs can change the travel behavior of consumers. Furthermore, today's time-conscious lifestyle may limit the maximum distance ("breaking point") that consumers are willing to travel for shopping.

Transportation Regulations and Deregulations

Due to the heavy influence of transportation on our daily lives, national defense, and national economy, transportation draws more attention from the federal government than any other logistics function. Therefore, it comes as no surprise that the federal government has been deeply involved in transportation affairs and has extensively controlled transportation activities. This means that neither transportation carriers nor shippers are completely free from various forms of government intervention. The most well-documented interventions are transportation regulations that set many ground rules for transportation activities and then control both economic and safety issues associated with transportation activities. Although transportation regulations are evolutionary with constant changes and fine-tunings, they aim to ensure that adequate transportation services are provided to any potential users and protect those users from excessive price and unfair practices. In a broad sense, transportation regulations are broken down into two forms: economic regulations and non-economic regulations. Economic regulations are designed to correct market imperfections, whereas non-economic regulations focus on safety enhancement and environmental protection. To elaborate, economic regulations are intended to combat monopolies by setting the maximum ceilings of freight rates and preventing discriminatory pricing; obviate destructive competition by stabilizing prices and controlling transportation market entry/exit; improve the utilization of transportation resources in the presence of "externality" (e.g., air pollution, traffic congestion). Economic regulations often play the following roles:

- They govern the entry of new firms into a particular mode.
- They govern the expansion of operating authority of established carriers and services (or abandonment/shrinkage of operations).
- They govern the exit from the transportation business.
- They rule freight rates, mergers/acquisitions, and financial arrangements and accounting practices.

On the other hand, non-economic regulations address public safety, health, and environmental issues by mandating seat-belt usage, restricting driving hours, prohibiting smoking on airplanes, imposing vehicle registration taxes, and setting vehicle emission standards.

Given the increasing complexity, transportation regulations are developed, promoted, administered, and enforced by many different government agencies. These agencies include the U.S. Department of Transportation (DOT) as a principal agency and its wings, such as the Federal Highway Administration, Federal Aviation Administration, Federal Railroad Administration, Maritime Administration (MARAD), National Highway Traffic Safety Administration, Office of Intermodalism, Federal Transit Administration, and National Transportation Safety Board (NTSB). In addition, there are several independent regulatory commissions, such as the Surface Transportation Board (STB), Federal Maritime Commission, and Federal Energy Regulatory Commission (FERC). The U.S. Coast Guard also regulates marine safety under the jurisdiction of the U.S. Department of Homeland Security. The specific roles of these agencies and commissions are summarized in Table 7.1.

Table 7.1. Major Transportation Agencies and Their Roles

Agency	Role
Federal Aviation Administration	Involved in the development and enforcement of safety regulations and operation of the nation's airway system
Federal Highway Administration	Administers the federal financial aid program for states for highway construction and rehabilitation
Federal Railroad Administration	Develops and administers railroad safety regulations
Federal Maritime Administration (MARAD)	Involved in the promotion (subsidies) of the nation's merchant marine
National Highway Traffic Safety Administration	Involved in prescription and implementation of motor vehicle safety standards (e.g., mandatory seat belts use, speed limit)
Office of Intermodalism	Facilitates projects that cross modal lines
National Transportation Safety Board (NTSB)	Investigates every civil aviation accident, determines probable causes of transportation accidents, and promotes transportation safety (is connected to the DOT but is somewhat autonomous/independent)

Surface Transportation Board (STB)	Controls many of the economic aspects (e.g., entry and abandonment of service, ratemaking, consolidations, mergers and acquisitions) of interstate surface transportation such as rail, pipelines, and trucks
Federal Maritime Commission (FMC)	Oversees the services, rates, practices, and agreements of foreign and domestic water carriers
Federal Energy Regulatory Commission	Regulates the interstate movement of oil and natural gas by pipelines
U.S. Coast Guard	Protects maritime economy and environments, defends maritime borders, and ensures the safety of navigation in ocean and inland waterways

Historical Backgrounds of Transportation Regulations

For the last century, transportation has been one of the most regulated sectors of the logistics industry. The origin of transportation regulation can be traced to the English common law that intends to protect the rights of shippers and passengers. The need to protect those rights originated from the evolution of the U.S. railroad system. In the mid-nineteenth century, the U.S. railroad system was rapidly expanded to meet the increasing transportation demands of western parts of the U.S. This initial railroad system was virtually "regulation free" and thus grew at an astonishing rate with the expectation that the western part of the U.S. would be developed continuously without limit. After the Civil War, however, the railroad industry began to discover that railroad capacity eventually outstripped the growth of the U.S. economy and forced that industry to practice discriminatory pricing to survive in a tough competitive environment. Ironically, despite aggressive rate reductions in major intermarket lines where more than one railroad competed, most of the rural areas of the U.S. were served by single railroad lines only and therefore shippers and passengers in those rural areas had to endure discriminatory pricing. This unfair practice in the early railroad industry became the impetus for great government interventions in transportation activities (Fair and Williams, 1975).

Indeed, the federal government felt a sense of urgency in regulating the railroad industry after the Windom Committee in 1884 reported that the U.S. railroad industry became out of control with exorbitant freight rates, inadequate equipment, and unfair/instable rate structures. Backed by U.S. Supreme Court rulings in 1877 and 1886, the federal government established the powerful Interstate Commerce Commission (ICC) and promulgated the Interstate Commerce Act that allowed it to regulate transportation activities. This Act was written in four parts to control the various modes of transportation available during late nineteenth and early twentieth centuries (Fair and Williams, 1981):

- 1887—Railroads and Pipelines
- 1935—Motor Carriers
- 1940—Domestic Water Carriers
- 1942—Freight Forwarders

The rationale for the government's early regulatory actions was twofold: (1) assurance of the role of the U.S. transportation system as a public utility, which protects the general public's economic and safety interests; (2) a catalyst for social and economic developments across the U.S. This rationale had not been changed for nearly a century until the U.S. government witnessed a near-disastrous decline of railroads and the crisis of inner-city transportation in 1970s. The era of a hundred years of transportation regulation can be broken down into three periods (Fair and Williams, 1981):

- Restrictive regulation from 1887 to 1917
- Positive and extended regulation from the end of World War I to 1970
- Relaxation of regulation and government assistance to railroads in the post-1970 period

Notable acts that were enacted during the first period (so-called Gilded Age and Progressive Era of 1887–1917) included the Interstate Commerce Act, the Elkins Act, and the Hepburn Act. The Interstate Commerce Act (ICA) of 1887 made the railroad industry the first transportation industry subject to federal regulation and prevented railroad monopolies by outlawing special rates and railroad rebates, forbidding long-haul/short-haul discrimination, requiring "just and reasonable" rate changes, and prohibiting pooling of traffic and markets (*U.S. News and World Report,* 2010). Congress further addressed the weaknesses of this Act by introducing the Elkins Act of 1903, the Hepburn Act of 1906, and the Mann-Elkins Act of 1910. The Elkins Act strengthened the government's control of anti-rebate initiatives by making it illegal to receive rebates as well as to give them. The Hepburn Act enabled the ICC to put a cap on rate charges, to determine adequate accounting procedures, and to alter unfair rates to ones it deemed "just and reasonable." The Mann-Elkins Act empowered the ICC to suspend proposed rate increases pending an investigation of the potential effects (http://www.answers.com/topic/interstate-commerce-act).

In the second period, the government extended its control over the transportation system by proliferating regulations and regulation bodies, most of which were mode-specific. One of the most notable acts during this period was the Urban Mass Transportation Act of 1964, which intended to slow down the deterioration of inner-city mass transit systems. More importantly, the U.S. Department of Transportation was formed in 1967 as the central government agency to coordinate national transportation policy. Its main mission is to ensure a fast, safe, efficient, accessible, and convenient transportation system that meets the vital national interests and enhances the quality of life of the American people, today and into the future.

During the third period, a growing number of legislators began to view the past regulatory efforts with increasing skepticism after those efforts often adversely affected competitive incentives, protected inefficiencies, and caused administrative delays and costs. The total cost of regulation grew rapidly in the late 1970s and led to a greater dissatisfaction with the government's regulatory policy among the legislators. This dissatisfaction created a new momentum toward a more free-market approach to transportation. In the next section, the details of such an approach will be discussed.

Landmark Deregulatory Acts and Their Impact on the Transportation Industry

As the free-market approach gained more support from legislators, a series of sweeping changes in federal transportation policy started in the late 1970s. One of the earliest landmark regulatory reforms was the enactment of the Railroad Revitalization and Regulatory Reform Act (so-called 4R Act) in 1976. This act focused primarily on limiting the ICC's authority to control rates. Especially, the ICC was forbidden from proscribing any rates that equaled or exceeded the variable cost of providing the service and then made a contribution to the "going concern value" of the railroad. Furthermore, a rate could not be considered "unreasonable" unless it was proven that the railroad was operating a service without serious competition from other railroads or other modes of transportation (Fair and Williams, 1981). This act was followed by another regulatory reform called the Air Cargo Deregulation Act of 1977, which deregulated the domestic air cargo business by allowing it to establish any rate and not restricting the aircraft size. Other regulatory reforms that lessened government controls over the airlines, trucks, railroads, and other related transportation industries followed suit. Some notable and important reforms include the Airline Deregulation Act of 1978, the "TOTO" decision in 1978, the Motor Carrier Act of 1980, the Staggers Rail Act of 1980, and the Shipping Act of 1984.

To elaborate, the Airline Deregulation Act of 1978 no longer controlled the market entry by allowing an airline to enter one new market per year, created a zone of reasonableness for either an increase or a decrease in rates within 5% range, and allowed the airline to freely abandon any routes for noncompensatory markets. This act intended to reduce airfares, stimulate demand and airline employment, create operating efficiencies, and improve airline profits. The "TOTO" decision in 1978 removed backhaul restrictions on private carriers and then allowed private carriers to secure common carrier certificates or contract carrier permits on their empty backhauls. Going a step further, the Motor Carrier Act of 1980 created more freedom for the trucking industry so that it could enhance operating efficiencies and developed more innovative pricing strategies. More specifically, the Motor Carrier Act of 1980 eased restrictions on market entry, allowed for a round trip with a backhaul, permitted intercorporate hauling for private carriers, no longer restricted contract carriage, created a zone of flexibility for rate determination, allowed more licensed brokers, and limited the impact of the rate bureaus, which used to establish and publish "official rates" for common carriers. In a sense, this act was intended to be an anti-inflation and pro-shipper measure as opposed to a pro-carrier measure.

Similar to the intent of the Motor Carrier Act of 1980, the Staggers Rail Act of 1980 significantly relaxed government controls over rate making, contract terms, and management rights. This act created a zone of "reasonableness (rate freedom)" for rate increases, thereby allowing railroads to increase rates according to a railroad cost index (adjusted to inflation) published quarterly, with a maximum of 18% in the first four years after its enactment. This act also limited the power of the rate bureaus, returned to management freedom of choice for many day-to-day operations, and sped up/simplified mergers and abandonments. Focusing on the deregulation of water carriers, the Shipping Act of 1984 permitted pooling practices for cargoes and then allowed a group of ocean carriers to maintain shipping conferences that could establish rates (e.g., intermodal through rates) and decide on which ports to serve. It also allowed greater flexibility for discounted tariffs and contracts, while allowing shippers' associations to obtain volume rates. This act, however, required the service contract to be filed with the Federal Maritime Commission (FMC). Faced with the increasing complexity of global maritime transportation and the constantly evolving maritime industry with mergers and consolidations, the Shipping Act of 1984 was replaced by the Ocean Shipping Reform Act (OSRA) of 1998. The OSRA provided shippers and ocean carriers with greater choice and flexibility in entering into contractual relationships with shippers for ocean transportation and intermodal services and allowed shipping conferences to negotiate/enter into confidential service contracts with one or more shippers and to do so independently of the other members of the shipping conferences (Wang, 2006).

Regardless of their specific provisions, these deregulatory acts aimed to restructure the three "pillars" of transportation regulation: control of entry and exit; control of rates and earnings, and control of services (see Sampson et al., 1990, for the three pillars of transportation regulation). These deregulatory acts had profound impacts on the transportation industry. These impacts can be summarized into four points:

- Less control of entry
  - Changes in market entry requirements from "fit, willing, and able" carriers to "fit" carriers (with proper equipment, insurance coverage, safety inspection).
  - Increase in the number of motor carriers. (Before deregulation, 18,000 licensed truck carriers were in the market. However, after deregulation, 48,000 licensed truck carriers competed in the market in 1992.)
- Rate and service flexibility
  - Deep discounting became a common practice among carriers.
  - Labor conflicts increased in the aftermath of more mergers/bankruptcy.
  - Negotiated rates became more prevalent than ever before. (There was an increased use of brokers who negotiated the price per truckload with the shipper and with the truckload operator.)
  - Contract carriers became increasingly utilized due to allowances for special individual contracts between carriers and shippers.
- Improved logistics efficiency
  - A greater number of carriers began to exploit freight consolidation through the hub-and-spoke system for air and the terminal for trucks so that they could improve their transportation efficiency.
  - Piggybacking became popular (before deregulation, car-loadings accounted for 8%, but after deregulation, card loadings accounted for 16.4% in 1987).
  - "Do-it-yourself" transportation (e.g., backhauling via private carriers) became common to save transportation costs in very competitive environments.
- Unstable carrier earnings
  - To survive in more competitive environments, carriers tried to find their niches in the transportation market.
  - Transportation cost savings became the top priority for many struggling carriers at the expense of deteriorated service quality. Also, the safety, maintenance, and age of carriers received less attention than before deregulation due to mounting cost pressures.
  - In an effort to save labor costs, many carriers began to employ a "two-tier" wage system, where new employees earned less in a given work category.
  - Labor unions began to lose their clout.

Hours of Service Regulations

Though non-economic transportation regulations did not receive as much attention as economic transportation regulation discussed earlier, there has been a growing awareness of safety regulations such as hours of service (HOS) regulations as fatal crashes involving large trucks gradually increased for the last few years after bottoming out in 2009. In fact, large truck involvement in fatal crashes rose from 3,211 (involvement rate of 1.11%) in 2009 to 3,802 (involvement rate of 1.42%) in 2012. In particular, between 2011 and 2012, fatalities in truck crashes showed a 5 % increase in the number of occupants of other vehicles killed and a 9 % increase in the number of large-truck occupants (mainly drivers) killed (National Highway Traffic Safety Administration, 2014). Reflecting the growing concern over traffic safety, the Federal Motor Carrier Safety Administration (FMCSA) introduced a series of HOS regulations that aimed to reduce serious truck accidents on the highway resulting from fatigued driving. Despite its good intent to help ensure truck drivers get adequate rest and perform safe operations, HOS regulations may lead to substantial cost increases (e.g., compliance cost, cost of hiring more drivers) for common carriers, which have already been hit hard by a lack of qualified drivers and declining shipping demands (Min, 2009). As such, HOS regulations have been a controversial topic in the trucking industry (especially for long-haul carriers). The key elements of HSO regulations for property-carrying commercial motor vehicles (CMVs) are composed of three limits (Federal Motor Carrier Safety Administration, 2014):

- **11-hour driving rule**—After driving a maximum of 11 hours, a driver must have at least 10 consecutive hours off duty before he or she can drive again.
- **14-consecutive hour duty rule**—A driver is not allowed to drive the CMV for more than 14 consecutive hours (including lunch breaks and other off-duty rest breaks) after coming on duty. After the 14th hour, the driver must have at least 10 hours off duty before the driver is allowed to resume his or her driving.
- **60-hour/7-day and 70-hour/8-day rules**—Under the 60-hour/7-day limit, a driver may not drive the CMV after having been on duty for 60 hours in any period of seven consecutive days. Under the 70-hour/8-day limit, a driver may not drive after having been on duty for 70 hours in any period of eight consecutive days. A driver may restart a 7/8 consecutive day period after taking 34 or more consecutive hours off duty. He or she must include two periods from 1 a.m. to 5 a.m. home terminal time, and may only be used once per week, or 168 hours, measured from the beginning of the previous restart.

Driving schedules illustrated in Figures 7.1a and 7.1b show examples of HOS rule compliance and violation, respectively.

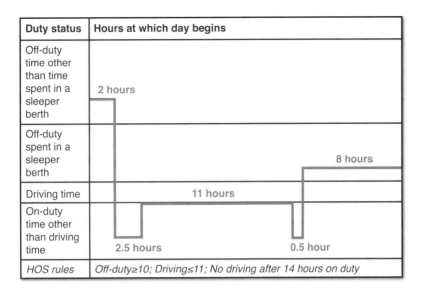

Figure 7.1a. *The illustrative example of a driver schedule in compliance with HOS rules*

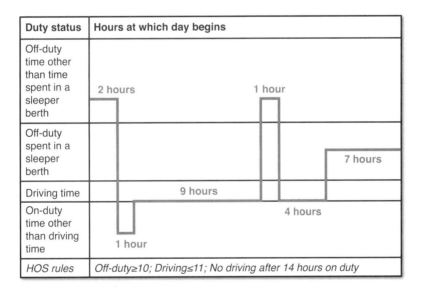

Figure 7.1b. *The illustrative example of a driver schedule violating HOS rules*

Legal Forms of Transportation

Transportation means can be legally classified into four different forms: private carriers, common carriers, contract carriers, and exempt carriers. Private carriers are those carriers that perform transportation of their own goods and related services for the furtherance

Chapter 7: Transportation Planning 233

of their primary business unrelated to transportation using their own transportation equipment (e.g., trucks, ships, aircraft). For example, if a retail supermarket chain owns and operates its own truck fleets to deliver grocery items such as cereals, meats, fruits, and vegetables to their local franchise stores while picking up fresh produce from farmers across the United States, those truck fleets are considered to be "private carriers." Notice that private carriers are rare in railroads, with an exception of steel companies. On the other hand, common carriers are those that perform the licensed, regulated transportation services to the general public over the specific routes identified in their scope of operating authority. The common carriers are required to fulfill duties to serve, deliver, charge reasonable rates, and avoid undue or unjust discrimination (Stephenson, 1987). These duties are specified as follows:

- **Duty to serve**—Must make a standing offer or holding out to serve all comers within the carrier's ability
- **Duty to deliver (carrier liability)**—Must deliver the goods in the same condition as received with reasonable dispatch and to the right party with the following exceptions (limitations):
    - Acts of God (disaster)
    - Acts of a public enemy (belligerent actions)
    - Acts of public authority (confiscation)
    - Inherent nature or vice of goods (defects)
    - Acts of the shipper (improper loading and packaging)
- **Duty to charge "reasonable" rates**—Keep the rates of existing carriers at "reasonable" rates. Regulatory bodies determine whether a rate is "reasonable."
- **Duty to avoid "undue" or "unjust" discrimination**—Treat all customers, products, and geographic locations the same when similar circumstances are present.

Contract carriers are another form of for-hire carriers that offer transportation services for those shippers with whom they have held a continuing, long-term contract with respect to rates, commodities, shipping volume, and service duration. Unlike common carriers, contract carriers do not serve the general public and do not issue a single bill of lading contract. Therefore, contract carriers can discriminate between small and large shippers and charge different freight rates to different shippers because freight rates are often set by carrier/shipper negotiations (Stephenson, 1987). Exempt carriers are special for-hire carriers that are free from economic regulations and are not subject to the same insurance filing requirements as common and contract carriers, but they are subject to the same safety regulations. They are often used to haul unprocessed or unmanufactured commodities including agricultural commodities (e.g., peanuts, sweet potatoes), seafood (e.g., fresh fish), live stock, or newspapers (Logistics Platform, 2010).

However, grains, soybeans, and sunflower seeds are not exempted from regulation (Stephenson, 1987).

Carrier Management

Carrier management is vital to the supply chain success because it directly affects a company's logistic costs and the subsequent bottom line. Although the primary tasks of carrier management are to set appropriate prices for transportation services and meet the customer's (shipper's) service needs given available capital and human resources, their full tasks can encompass carrier/modal selection, carrier routing/scheduling, dispatching, staffing, pick-up/delivery arrangement, financing, rate negotiation, documentation, shipment tracking, claims handling, carrier quality assurance, carrier safety, and billing/payment. Therefore, carrier management is complex by nature. This complexity requires careful planning and preparation. To make wise carrier management decisions, supply chain professionals should address a variety of managerial issues, as shown in Table 7.2. Among these issues, carrier selection may be one of the most important but challenging issues to tackle. Given its importance, we will discuss details of the carrier selection process.

Table 7.2. Carrier Management Decisions and Related Issues

Decisions	Managerial Issues
Selection	• Outsourcing versus in-housing • Carrier selection • Modal choice • Size and composition of fleets • Leasing versus purchasing fleets • Standardized versus nonstandardized fleets

Marketing	• Line of business
	  • Product type (e.g., grocery, liquid, automobile)
	  • Bulk versus non-bulk
	  • High, medium, or low value
	  • Fragile or sturdy
	  • Toxic or nonhazardous
	  • Packaged or containerized
	• Geographical scope (regional, national or international)
	• Alliances

Decisions	Managerial Issues

Capacity planning	• Long term versus short term
	• Transportation equipment
	• Maintenance equipment
	• Staffing
	    • Employee (e.g., driver) recruitment and retention
	    • Unionized versus non-unionized
	    • Full time versus part time
	• Number, size, and location of terminals
	• Information technology
	    • Computer equipment
	    • Shipment tracking
	    • Advanced shipping notice
	    • Satellite tracking systems, transport management system (TMS), RFID
	    • Electronic Data Interchange (EDI)
	    • Delivery information acquisition device (DIAD)

Network design	• Routing/scheduling • Dispatching • Lines of communication • Ship direct or consolidation • Backhauling
Service offerings	• Service gradients • Same day, next day, second day, or others • Just-in-time delivery services • Expedited services • Intermodal services • Value-added services • Shipper/carrier relationship management • ISO certification • Claims/damage control • Safety measures

Decisions	Managerial Issues
Pricing	• Rate negotiation • Deep discounting during the off-peak periods • Accessorial charges • Yield management • Revenue management • Freight terms • Incoterms • Zone billing
Financing	• Lease financing • Asset ownership • Debt financing

Put simply, carrier selection is the process of choosing the "right" transportation service provider that best meets the shipper's service needs and requirements. The main objectives of carrier selection are to minimize transportation cost, minimize in-transit inventories (i.e., minimize delivery/pick-up delays or forbid early shipments), and minimize transit variability. Thus, carrier selection was made on three decision criteria: cost, time (speed), and dependability (reliability). These three primary criteria can be further extended to include other criteria and then subdivided into subcategories, as listed here:

- Cost
  - Cost per mile
  - Fuel cost
  - Driver pay
  - Maintenance cost
  - Taxes/toll
  - Insurance
  - Vehicle registration fee
- Time
  - Transit time
  - Service response time
- Dependability
  - Consistent on-time pickups/deliveries
  - Fulfilled promises
  - Avoidance of work stoppage (strikes)

- Accessibility
  - Network (door-to-door) coverage
  - Convenient schedules
  - Frequency of service
  - Minimum/maximum weight limit
  - Commodity restrictions
- Control and liability
  - Ability not to lose or damage shipments
  - Claims settlement practices
- Other service elements
  - Consolidation services
  - Flexibility to meet changing shipper needs
  - Availability of specialized equipment
  - Accessorial services
  - Driver performance
  - Department of Transportation (DOT) safety ratings
  - Green initiatives (e.g., carbon footprint reduction)

Under the preceding selection criteria, a carrier choice can be made with respect to four legal forms (private carriage, common carriage, contract carriage, and exempt carriage) and three different shipment sizes. Based on shipment sizes, the following three carrier choices are available.

- **A parcel carrier**—Usually handles small packages and freight that can be broken down into units less than 150 U.S. pounds.
- **A LTL carrier**—Collects freight (100 to 100,000 pounds) from various shippers and consolidates that freight onto enclosed trailers for line-haul to the delivering terminal or to a hub terminal where the freight will be further sorted and consolidated for additional line-haul.
- **A TL carrier**—Moves freight loaded into a semi trailer, which is typically between 26 and 53 feet and therefore requires a substantial amount of freight to make such transportation economical. TL carriers transport freight at an average rate of 47 miles per hour (including traffic jams or queues at intersections).

Once the proper carrier is selected, the next step to take is to find the "right" mode of transportation and the "right" combination of the chosen transportation modes. The available transportation modes are trucks, rail, water carriers (including barges), and air carriers. In the next section, these modes will be discussed in great detail.

Surface Transportation

Despite the recent economic downturn, demand for surface freight transportation has risen steadily over the last two decades and forecasts show continued growth at least over the next several decades. In 2001, the Bureau of Transportation Statistics reported that more than 3.18 trillion ton-miles of freight were moved over the nation's domestic transportation system, up almost 22% from the 2.61 trillion ton-miles of freight moved in 1990—an annual growth rate of 2%. During the period from 1990 to 2001, surface transportation modes such as trucks and rails continued to capture a larger portion of the domestic freight market as measured in ton-miles, as shown in Figures 7.2 and 7.3. As shown in Figure 7.2, trucking market share rose from 28% in 1990 to 33% in 2001 and then 41% in 2010. Although rail ton-mile market share grew and represented the largest portion of the intercity freight market at 47% in 2001, up from 40% in 1990, it began to slide and then decreased to 29% in 2010 (U.S. Department of Transportation, 2010). By 2018, trucking's total tonnage share is expected to rise to 70% from 69% in 2006. In the meantime, total rail tonnage share will continue to taper offer and dwindle to 14.7% of domestic tonnage in 2018 (O'Reilly, 2008).

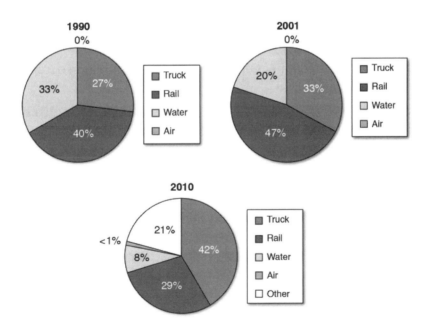

Figure 7.2. *The market share of transportation modes with respect to domestic ton-miles during 1990, 2001, and 2010 (source: Bureau of Transportation Statistics, 2004 and 2011)*

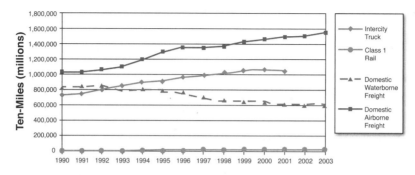

Figure 7.3. The market share trend of transportation modes during 1990 and 2003 (source: Bureau of Transportation Statistics, 2004)

As evidenced by the sizable market share, trucks and rails are clearly the two most dominant modes of transportation in the United States. Their dominance, however, has not necessarily translated into success. For example, the trucking industry has suffered from increasing business failures for the last decade, although trucking failures began to decline in 2011 (Truck Gauge, 2011). To make matters worse, the bankruptcies in the highly fragmented U.S. trucking industry peaked in 2008 and then escalated again in 2010 as higher diesel fuel prices and excess capacity squeezed already thin profit margins further and lenders started to tighten their noose on the sector (Card, 2009; Ananthalakshmi, 2010). In particular, many small trucking firms or owner-operators will be more vulnerable to intense competition created by transportation deregulation and overcapacity in the trucking industry.

To overcome the economic downturn and survive in an increasingly competitive market, the trucking industry has to deal with the following challenges.

- **Labor issues**—High driver turnover, demographic shifts for employee bases, and a growing need for ergonomic transportation equipment
- **SKU proliferation**—Difficulty in standardizing loads
- **Continued decline in order size and increase in order frequency due to greater online retail sales**—Limited freight consolidation opportunities and freight cost increases due to frequent small shipments under JIT delivery schedules
- **Rising energy costs**—Increased pressure for fleet productivity and the use of alternate fuels
- **Load planning**—The need to build loads that improve unloading efficiencies at destination
- **Uneven order volume**—Forecasting challenges and an increased difficulty in capacity planning

- **Dynamic pricing**—Real-time rate bidding/negotiation and flexible accessorial charges, including fuel surcharges

To cope with these challenges, trucking companies may consider using the following strategies:

- Eliminate or reduce "partial" trips.
  - Consolidate outbound shipments.
- Reduce empty mileage.
  - Use intercorporate hauling.
  - Reduce empty backhauls.
  - Use optimal routing/scheduling.
- Limit allocation of private fleet capacity to the most "profitable" trips, substituting for-hire carriers for the remainder.
- Make pre-scheduled deliveries.
  - May help reduce emergency/rush delivery.
- Make preventive vehicle maintenance a priority.
  - Set up in-house maintenance shops.
- Use fuel-saving devices.
  - Consider larger tanks, radial tires, tractor air shields, turbo charges, synthetic lubricants, high-flow air intake systems, aerodynamic drag reduction devices, and so forth.
  - Employ trailers with double-deck levels.
- Boost driver performance.
  - Link driver pay to productivity.
  - Use flexible pay systems such as hourly pay, mileage-basis pay, and trip-basis pay.
  - Install electronic trip recorders.
  - Have dispatchers and industrial engineers accompany drivers occasionally.
  - Reduce driver waiting time during loading/unloading.
  - Use relay driving to reduce layover expenses.

Although trucks are capable of carrying any type of shipments to anywhere ("door-to-door" services), they are often more expensive and slower than railroads for longer-distance (e.g., more than 500 miles) shipping. In some, instances such as hauling waste freight, rail shipment can be as cost-efficient as trucks for distances as short as 19 miles (TSSA, 2010). As a viable alternative to trucking, rail has emerged as the preferred mode of surface transportation for cost-conscious shippers. Indeed, U.S. freight railroads serve every sector of the U.S. economy, encompassing manufacturing, wholesale, retail, and agricultural sectors, and they move various shipments, ranging from lumber to vegetables, coal to orange juice, grain to automobile, and chemicals to scrap irons. As such, the rail freight role is not confined to bulk freight such as coal. The U.S. railroad industry has spent more than $20 billion a year to maintain, renew, and expand rail tracks and equipment, and that spending has helped it build the most efficient and lowest-cost rail network (Association of American Railroads, 2008). Based on the average freight revenue ton-mile (a surrogate for freight rate), rail tended to cost 10% to 20% less than trucking for the last decade (Bureau of Transportation Statistics, 2013). For example, the average freight revenue per ton-mile of Class I rail was $2.24, as compared to that of trucking, which was estimated to be $26.6 in 2001 (RITA, 2007). Rail can be classified into the following different types in terms of their freight revenue, distance travelled, and service coverage (Association of American Railroads, 2008):

- **Class I railroads**—Those that operate in many different states and concentrate on long-haul and high intercity traffic, with annual revenue exceeding $350 million
- **Regional railroads**—Those that handle line-haul movement with at least 350 miles and generate annual revenue between $40 million and the Class I threshold
- **Local line-haul railroads**—Those that operate less than 350 miles and earn less than $40 million per year
- **Switching and terminal (S&T) railroads**—Those that primarily provide switching and/or terminal services and pickup/delivery services within a specified area for one or more line-haul railroads, regardless of revenue

Water Carriers

Excluding coastal areas (Atlantic, Pacific, and Gulf Coast) and the Great Lakes, the United States contains approximately 29,000 miles of navigational waterways, including 25,000 miles (40,000 km) of navigational inland waterways (Sampson et al., 1990). Much of the commercially viable inland and intercoastal waterways are concentrated in the Mississippi River System, made up of the Mississippi, Missouri, Ohio, Tennessee, and Cumberland rivers. Thanks to great access to these navigational inland waterways, water carriers became an integral part of waterborne trade and domestic passenger/freight traffic in the U.S. As of 1997, domestic water carriers moved about 650 million tons of cargo, valued at over $73 billion annually, through these inland waterways in the U.S. (U.S. Army Corps of Engineers, 2000). The volume of waterborne freight traffic gradually increased to 956 million tons in 2008 (U.S. Army Corps of Engineers, 2009). The growth of traffic volume via inland waterways may be attributed to the water carrier's ability to convey a relatively large volume of bulk commodities for a long distance. For example, a single 15-barge tow is equivalent to about 225 rail cars and 870 tractor-trailer trucks; therefore, a barge can carry much larger volumes of bulk commodities such as coal, petroleum, steel, chemicals, grain, and minerals than other modes of transportation. Also, water carriers such as barges are far more fuel-efficient and consequently more environmentally friendly than other modes of transportation. For example, a gallon of fuel allows one ton of cargo to be shipped 59 miles by truck, 202 miles by rail, and 514 miles by barge (U.S. Army Corps of Engineers, 2000). Furthermore, water carriers are almost free from any traffic congestion, which often increases the travel time of trucks. Considering these benefits, water carriers should be regarded as a viable alternative to popular surface transportation.

Despite numerous merits, water carriers are not without their drawbacks. Those drawbacks include the substantial involvement of federal, state, and local governments in promoting and subsidizing water carriers; the dependence of inland waterways on locks and dams; the influence of weather (especially winter freezes) on some inland waterways; and the increased environmental concerns over dredged materials extracted from inland waterways (Stephenson, 1987; Wood and Johnson, 1996). In particular, more than 50% of the locks and dams operated by the U.S. Army Corps of Engineers are over 50 years old and reaching the end of their design lives (U.S. Army Corps of Engineers, 2000). Therefore, without sizable investment in the improvement and upgrade of water transportation infrastructure, inland waterways may cause severe congestion at locks and dams and consequently offset the advantages of water transportation. A lack of government funds to maintain and upgrade water transportation infrastructure may also lead to higher waterway user fees and the subsequent increase in water carrier costs. Another challenge of water carriers is the globalization of business activities that require the increasing use of ocean carriers (oceangoing ships) under multiple nations' jurisdictions. Typical services offered by ocean carriers are liner and charter services (WCL Consulting, 2006):

- Liner services
  - Regularly scheduled services between major ports.
  - Typical port calls of 1–3 days to unload imports and load exports using shore-based or on-board equipment (e.g., cranes) in a 24-hour per day operation.
- Charter services
  - Vessels chartered/leased by importers with shipload quantities.
  - Lower overall freight costs and greater control over scheduling.
  - Charter agreements (charter parties/fixtures) can be time based (e.g., in months) or voyage based.

- May include crew and equipment or vessel only (bareboat charter).
- Also known as "tramp" steamers (because they operate on irregular schedules and go anywhere the cargo is available).

Air Carriers

Although human fascination with flying has existed for centuries, the so-called "golden age" of aviation that saw rapid advancements in aviation technology started in 1918 during the World War I period. Due in part to its relative young history and expensive operations, air carriers have not been heavily used for freight transportation and instead primarily used for passenger transportation. Over four-fifths of the air transportation industry revenues used to come from passenger transportation (Sampson et al., 1990). However, this trend is gradually reversing, with increasing traffic associated with global trade. For instance, the U.S. shipped around 30% of its exports by value using air carriers during January and March of 2010 (Mcclnis, 2010). Indeed, international airfreight traffic increased by 25% in 2010 compared to 2009, despite disruptions caused by the eruption of the volcano in Iceland (Concil, 2010). Generally, air carriers engage directly, by lease or other arrangements, in providing domestic and international air transportation services, operating airfields, and furnishing terminal services. In a broad sense, air carriers are classified into air freight (cargo) carriers and air passenger carriers. Air freight carriers are dedicated to the transportation of cargo and are sometimes divisions or subsidiaries of larger passenger airlines (http://en.wikipedia.org/wiki/Cargo_airline). United Parcel Services (UPS), FedEx, and National Air Cargo are examples of air freight carriers. The air passenger carriers that also carry some cargoes can be further subdivided into four different types (U.S. Environmental Protection Agency, 1998):

- **Commercial air carriers**—This category encompasses air carriers and air taxi flights. Air carriers are airlines holding a certificate issued of public convenience and necessity under Section 401 of the Federal Aviation Act of 1958 authorizing them to perform passenger and cargo services. Air carriers operate aircraft designed to have a maximum seating capacity of more than 60 seats, to have a maximum payload capacity of more than 18,000 pounds, or to conduct international operations. According to annual operating revenues, the Civil Aeronautics Board (CAB) subclassified scheduled commercial air carriers in 1981 as follows:
    - Majors (greater than $1 billion)
    - Nationals ($100 million to $1 billion)
    - Large regionals ($20 million to $100 million)
    - Medium regionals (Up to $20 million)

- **Air taxi carriers**—Those carriers for which departure time, departure location, and arrival location are specifically negotiated with the customer or by the customer's representative and are conducted with airplanes or rotorcraft having a seating configuration of 30 or fewer seats.
- **Commuter air carriers**—Noncertified small regionals that perform scheduled service to smaller cities and serve as feeders to the major hub airports. They generally carry 60 or fewer passengers.
- **General carriers**—Those carriers that are not commercial or military. General carriers encompass all facets of aviation, such as corporate/executive transportation, instruction, rental, aerial application, aerial observation, business, pleasure, and other special uses.

When it comes to fast shipment of time-sensitive or valuable orders or receipt of a critical component for your production, an air carrier may be the best mode of transportation. However, an air carrier is arguably the most expensive mode of transportation, and its pricing is very complex. An air freight rate often depends on many factors: air carrier nationality, origin and destination airports, freight zones, shipment weight and dimensions, packaging of shipment, type of commodity shipped, and type of buyer of the services. It should also be noted that the air freight rate is based on transportation from origin airport to destination airport, excluding any ground transportation or port service. In addition to this pricing complexity, air freight services can be easily disrupted by unexpected terrorist attacks, inclement weather conditions, and maintenance mishaps. Therefore, a careful plan should be made before choosing an air carrier as the primary mode of transportation.

Intermodalism

As discussed earlier, each transportation carrier (each mode of transportation) has its pros and cons. In an effort to make up for some shortcomings of each mode of transportation, there has been a growing need for combining and mixing multiple modes of transportation without separating each. Such a need can be translated into *intermodalism,* which involves transportation of cargo or passengers using multiple modes (truck, rail, air, ship) under a single freight bill. For example, a train can haul much heavier loads such as automobiles at much lower cost than a truck, but is not designed to offer "door-to-door" delivery services. What's more, its schedule is far more limited than a truck. In this situation, shippers can use rail for long-haul and then switch to trucks for the shorter distances at either end of the trip. This scheme allows shippers to take advantage of the best attributes of both trucks and rails at the same time. Intermodal mix is not just limited to the combined use of trucks and rails (called "piggyback," or more formally Trailers On Flat Cars [TOFC]). Depending on modal combinations, intermodal transportation can take the following different forms: piggyback (truck and rail mix), birdyback (truck and air carrier mix), and fishyback (truck and water carrier mix).

Regardless of its form, intermodal transportation brings a lot of benefits. These benefits include reduction in cargo handling due to the use of standardized shipping containers for intermodal transfers, shipping flexibility in hauling a wide variety of products due to the combined use of diverse modes of transportation, improved cargo security with less damage (including spoilage) and loss during transit resulting from the use of protective intermodal shipping containers, faster shipment and lower freight than the use of a single mode of transportation, and documentation cost savings via a single bill of lading. Thanks to these benefits, intermodal transportation began to grow continuously for the last several decades, as shown in Figure 7.4. Although there is a slight decline in 2009 due to economic recession, intermodal traffic rebounded nicely in 2010. As a matter of fact, intermodal traffic in the U.S. was up 21.2% from 2009 to 227,985 trailers and containers (Truckinginfo, 2010). In particular, intermodal volume on U.S. freight railroads for the week ending May 29, 2010, reached its highest level since November 2008. Intermodal traffic totaled 225,111 trailers and containers, up 35.5% from 2009 and 10.3% from 2008 (Association of American Railroads, 2010).

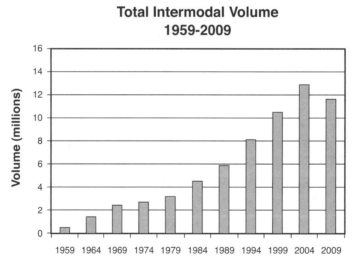

Figure 7.4. *Intermodal traffic volume (1959–2009) (source: Intermodal Association of North America, 2009)*

Despite the increasing popularity of intermodal transportation, it has some drawbacks. These drawbacks include limited capacity and rising rates during peak seasons, higher packaging costs for containerizing shipments for intermodal transfers, uneven modal capacity, inconsistent standards for intermodal shipping containers and trailers, and potential congestion and the subsequent delay at the intermodal terminal. Therefore, the pros and cons of intermodal transportation should be carefully weighed before choosing it.

Transportation Documentation

Because transportation carriers often haul valuable goods, the movement and transfer of those goods should be traced and documented properly. Various documents are needed to verify the acceptance and receipt of goods before/after shipments and clarify the ownership of those goods. These documents include a bill of lading, a carrier's certificate, delivery receipts, freight bills, shipping manifests, shipper's declaration for dangerous goods, a shipper's certification for live animals, and other documents needed for freight claims. Among these, a bill of lading (B/L) is one of the most important documents. A typical B/L is illustrated in Figure 7.5. Generally, a B/L is a document issued to a consignor (shipper) by a carrier listing and acknowledging that specified goods have been received onboard as cargo for conveyance to a named place for delivery to the consignee, who is usually identified (http://en.wikipedia.org/wiki/Bill_of_lading). A B/L can be classified into four different categories:

- Straight (consignment) bill of lading
  - Is used when goods have already been paid for.
  - Is non-negotiable.
  - Contains terms of the sale, including the time/place of title transfer.
  - A carrier delivers the shipment to its consignee on presentation of an identification under this B/L.
  - Is not safe.
- Order bill of lading
  - Is typically used in letter of credit transactions where goods have not been paid for in advance.
  - Is negotiable.
  - Ownership of goods under this B/L can be transferred from one party to another by signature (endorsement) and delivery of this B/L.
  - Consignor retains original until bill is paid.
- Bearer bill of lading
  - Under this B/L, delivery will be made to whomever holds the B/L.
  - Is negotiable.

- Surrender bill of lading
    - The importer does not pay the bank until the maturity of the draft under the relative credit.

Depending on the use of a particular mode of transportation, a B/L can be further subclassified into an inland B/L, ocean B/L, through B/L, and air waybill (or airway B/L).

Bill of Lading

SHIPPER NUMBER: ______________
CARRIER NAME: ______________
TRAILER NUMBER: ______________
SEAL NUMBER: ______________
BILL DATE: ______________

TO	*FROM*
Consignee (Destination)	Shipper
Street	Street
Destination	Origin
City/State/Zip	City/State/Zip
Delivery Date	Shipping Date
Route:	Special Instructions:

FOR PAYMENT, SEND BILL TO	SHIPPER'S INSTRUCTIONS
Name	
Company	
Street	
City/State/Zip	

NUMBER OF SHIPPING UNITS	HAZARDOUS MATERIAL	DESCRIPTION OF ARTICLES SPECIAL MARKS & EXCEPTIONS	TOTAL QUANTITY	WEIGHT	CLASS OR RATE	CHARGES

REMIT C.O.D.		C.O.D. AMOUNT: $	C.O.D. FEE PREPAID ☐ COLLECT ☐
TO: ADDRESS:		(Signature of Consignor)	TOTAL CHARGES $
FREIGHT CHARGE TERMS	*CHECK BOX IF PREPAID* ☐ *COLLECT* ☐ *OTHERS* ☐		
Shipper Signature		Carrier Signature	
Pickup Date:		Pickup Date:	

Figure 7.5. *An example of a blank B/L form (source: Samplewords.com, 2010)*

Transportation Pricing

Prior to a series of regulatory reforms that took effect in the early 1980s, a transportation price (or a freight rate) was often determined by the fixed rate published by rate bureaus. Deregulations, however, facilitated more competition among carriers and provided greater flexibility in changing transportation price structures. Ever since, transportation pricing has become more dynamic and complicated. The transportation price depends on a shipping distance, a dimensional shipping weight/volume, the form of cargo, the mode of transportation, accessorial charges, and the quality of services in terms of speed, accessibility, reliability, frequency (intervals between departures), rhythm (ratio of maximum intervals between departures) and punctuality. Among these, the most dominant pricing factors are: a distance, a weight, and a class that categorizes the type of cargo. Generally, there are three ways to determine the transportation price. These are published tariffs, negotiated contracts, and open-market arrangements (Tyworth et al., 1987).

To elaborate, published tariffs are official transportation industry publications that contain fixed freight rates, rules, and other useful information needed to purchase common carrier services (Tyworth et al., 1987). The tariffs include freight classifications (numbers assigned to different shipments based on their handling characteristics), passenger fare tables, routing guides, and so forth (Sampson et al., 1990). These tariffs are classified into rate tariffs and supporting tariffs. Rate tariffs quote freight rates for line-haul (door-to-door or terminal-to-terminal) movements, while supporting tariffs (or services tariffs) provide supporting information about the correct application of rates, route characteristics, and additional charges for accessorial services such as switching, storage, refrigeration, stop-offs, and diversion in transit (Tyworth et al., 1987; Sampson et al., 1990). Rate tariffs are further subdivided into class rates and commodity rates. Class rates are standard freight rates that list the prices of shipping every product from and to everywhere and are often the highest rates that carriers would quote. The class rates can only be used in conjunction with a specific freight classification scheme usually given in cents per 100 pounds and are primarily distance-scale rates although they are not in direct proportion to the distance hauled (Sampson et al., 1990). This classification scheme includes *Uniform Freight Classification*, used primarily by railroads, and *National Motor Freight Classification*. This classification scheme has nothing to do with the shipment's origin, destination, or routes. Instead, it evaluates the shipments according to the following five composite characteristics (Lieb, 1994; Wood and Johnson, 1996):

- *Density*, as measured by shipping weight per cubic foot and the value per pound in comparison with other articles;

- *Stowability*, which includes excessive weight or length;
- *Extent of handling* (including loading and unloading), which includes special care and attention necessary to handle the goods;
- *Liability*, as measured by value per pound, susceptibility to theft, liability to damage, perishability, propensity to damage other freight with which it is shipped and propensity to spontaneous combustion or explosion;
- *Freight characteristics*, which encompass trade condition, competition with other commodities transported, quantity offered as a single consignment, and value of service required.

In contrast with class rates, commodity rates cover only the specific commodities and the specific shipments named in the tariff and do not rely on the classification. Thus, commodity rates are generally discounted and used for repeated shipments in large quantities. Since a shipper is legally entitled to use the lowest published rate, the commodity rate takes precedence over the class rate (Sampson et al., 1990). In addition to aforementioned rate-making procedures, a transportation price can be ultimately determined by either one of two contrasting pricing principles: (1) *cost of service pricing* (cost-based pricing); (2) *value of service pricing* (differentiated pricing). In cost of service pricing, freight rates should cover the total summation of the variable cost (e.g., fuel cost, operating cost) of a particular shipment, a proportionate, fair share of the fixed costs (e.g., equipment cost), and a "reasonable" profit margin. On the other hand, in value of service pricing, freight rates are set based on what the tariff will bear and thus can be settled at any level above the variable cost or the floor price (Stephenson, 1987).

Freight Rate Negotiation

After a series of deregulatory acts were put into place, carriers were given more freedom to set their freight rates. Given this rate flexibility, a growing number of shippers look for bargain or discounted prices through negotiation. This rate negotiation opportunity was further enhanced by the introduction of the Negotiated Rates Act of 1993, which was later modified slightly in 1995. This Act stipulates the following rules:

- Shippers must contract for a series of shipments rather than individual shipments under a bill of lading. Also, they must keep evidence of the written contract (agreement) and rates for at least one year.
- The contract carrier must agree to meet the "distinct needs" of the shipper (e.g., dedicated equipment).
- The contract must contain the agreed-upon rates.
- Concealed damage claims should be addressed in the contract and negotiated "up front" (subjecting the contract carrier to common carrier liability).

- A broker is allowed to represent a shipper, but is not empowered to assume a shipper's rights and role in creating or executing a bill of lading. A broker whose name is listed on the bill of lading as a "carrier" willfully misrepresents itself in law and creates immediate liability for cargo loss and/or public liability.

As rate negotiation between the shipper and the carrier becomes a daily routine, it has a far-reaching effect on transportation costs and the subsequent bottom line (profitability) of the affected companies. Therefore, a successful rate negotiation strategy has to be developed based on careful strategic plans. These plans usually start with a full understanding of the rate-making process and the required services. Before coming to the negotiation table, both shippers and carriers should accurately estimate the cost of conveying freight from origin and destination points as well as the cost of handling freight. In addition, they should study the market trends that often shape the current rate. For example, for the first three years of the twenty-first century, shippers had the upper hand in freight rate negotiations. In 2003, the tables turned. Driver shortages, capacity constraints, and fuel cost increases tipped the scales in favor of carriers. For most of 2003 to 2005, shippers had to endure double-digit rate increases, particularly in the truckload sector. In 2006, the economy reversed its direction again, transportation capacity increased, and rate increases became more reasonable. This trend was changed again in late 2007, when the severe economic recession started and shipping demand subsided. Given volatile market conditions, the following rate negotiation strategy might make sense:

- Negotiate a modest rate increase.
    - Depending on economic downturns and upturns, both shippers and carriers may adjust their rates without deal-breaking dramatic rate increases or reductions. In other words, rate negotiation should factor into the long-term viability of both shippers and carriers without sacrificing one party. Indeed, many shippers are increasingly stressing a collaborative relationship with their carriers and considering sharing risk.
- Take a hard line with carriers and seek a freeze or rollback on rates.
    - For shippers with significant volumes, they can secure a freeze or rollback. The downside to this strategy is that if the economy tightens, the carriers may allocate their equipment to those shippers with more attractive yields. This may jeopardize the carrier's available capacity and thus adversely affect the shippers' service requirements.
- Conduct a freight bid to leverage volume discounts and sign a multiyear rate agreement.
    - Under this scenario, a shipper can place its freight for bid to a selected group of existing and new carriers. The shipper can then negotiate with a short list of the selected carriers that can supply the service and capacity at the best

rate. To secure the carriers' capacity over a longer run, the shipper can negotiate a multiyear agreement with modest rate increases and SLAs (service-level agreements).
- The downside to this strategy is that if the economy softens, the shipper would be bound by a multiyear agreement with little negotiation room for favorable rates in subsequent years. However, this strategy offers the best opportunity for cost certainty, secured capacity, and good service. Establishing a multiyear relationship with the core carriers fosters a stronger, more collaborative bond between the carriers and shippers. The carriers that have a multiyear commitment on volumes and rates are more likely to work more diligently on behalf of the shippers.
- Use freight derivatives to mitigate risk against rate spikes.
  - This strategy is intended to hedge against sharp rate increases in the volatile marketplace where the uncertainty and risk are very high. This strategy provides peace of mind for both shippers and carriers due to the guaranteed rate.

Revenue/Yield Management

Revenue management (also known as yield management) is an approach that aims to maximize revenue streams or profits given fixed but perishable resources (e.g., airline seats) through a thorough understanding of market dynamics. This approach has become popular among the commercial airliners ever since American Airlines successfully used it in 1985 to fill a sufficient number of seats without selling every seat at a discounted price. This approach helped American Airlines defray at least its fixed operating expenses. Once fixed operating expenses were covered, they could sell fewer remaining seats at higher prices and thus could maximize their revenues and profits. Typically, revenue management boosted incremental revenue gains by 3%–7% and profit gains by about 50%–100% when it was implemented successfully (Wikipedia, http://en.wikipedia.org/wiki/Yield_management). Revenue management is particularly relevant to the following situations (Salerno, 2010):

- When there is a fixed amount of resources available for sale.
- When the resources to sell are very perishable.
- When customers are willing to pay a different price for using the same resources.
- When fixed costs are relatively high as compared to variable costs. In other words, the less variable costs to absorb, the greater chance the larger amount of revenue could be generated.

The airline industry perfectly fits into these conditions and practices of revenue management more frequently than other carriers. Nowadays, revenue management principles are not confined to the airline industry because they have spread to other carriers as well. For instance, Amtrak in the railroad industry began to adopt the revenue management philosophy to fill inventory seats for its long-distance intercity train operations in 1991. Revenue management can take three different approaches: allocation of capacity by different classes (e.g., first class, business class, and economy class in airlines); price bidding based on the estimated opportunity cost; and direct price adjustment for an individual sale, where the price increases after every sale because there is one less unit to sell. Revenue management is also affected by a multitude of factors, such as demand volatility, seasonality, price sensitivity, uncertainty, marginal variable costs, and market segmentation (see McGill and Van Ryzin, 1999, for the details of revenue management elements). Revenue management is particularly useful for determining how to do the following (Cross, 1997):

- Discount rates with discretion to build market share
- Uncover hidden demand, which allows opportunistic pricing
- Understand customer tradeoffs between price and other service attributes
- Increase revenue without sales promotions or improving service quality
- Identify lost revenue opportunity
- Focus on profit growth

Transport Management Systems

With mounting transportation costs resulting from unstable fuel prices and higher labor costs, there is a growing need for streamlining all facets of transportation operations, identifying any cost-saving opportunities, and elevating supply chain visibility through transparent transportation planning. A transport management system (TMS) fits the bill for such a need. Generally, *transport management system* refers to a software application that helps to manage all aspects of inbound and outbound transportation activities including order processing, load planning, carrier selection, vehicle routing/scheduling, freight consolidation, claims management, freight bill payment, and auditing. In response to increasing complexity and globalization of transportation activities, TMS has recently expanded to include additional features that can handle real-time asset management, yard management, event management with shipment tracking, shipping demand forecasting, palletization, import/export documentation, multimodal operations, hazardous material transportation, reverse logistics, and performance management (e.g., service quality control). The typical TMS functions listed by Conover (2001) are graphically displayed in Figure 7.6.

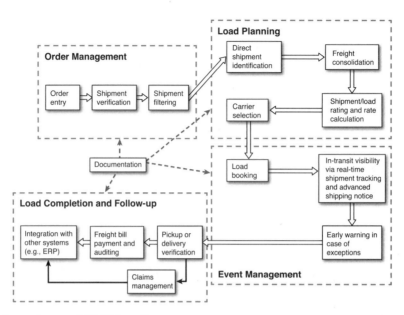

Figure 7.6. *A roadmap of TMS functions*

If properly executed, a TMS brings numerous managerial benefits. These benefits include:

- Cost savings. TMS users reported 5%–30% transportation cost savings and about 80% of those users had one or two years of the payback periods (Conover, 2001).
- Increased customer satisfaction through optimal route configurations/scheduling, delivery confirmation, and timely claims processing.
- Improved supply chain visibility through real-time shipment tracking and advanced shipping notice.
- Reduced paperwork through automation.

Terminal Operations

Put simply, a (transportation) terminal is any facility where freight or passengers originate, terminate, or are consolidated and then transshipped to other locations. For example, travelers usually go to bus terminals, railroad stations, or airports to reach their final destinations. Likewise, all the (freight or passenger) traffic flows involve the movement between one terminal and another. The typical roles of a terminal are threefold (Rodrigue et al., 2009):

- A point of interchange for continuous traffic flow (e.g., flight change at the hub airport)
- A point of transfer for transshipment
- A point of convergence for assembly, consolidation, or break-bulking

Also, a terminal can offer a variety of accessorial services such as pickup/delivery, loading/unloading, weighing/reweighing, sorting, coopering, switching (including floatage and lighterage for barge operations), inspection, storage, and elevation. In a nutshell, a terminal is an essential link in the supply chain because it can dictate the effectiveness and efficiency of traffic flows. Therefore, to ensure the supply chain success, terminal operations have to be optimized. Because terminal operations involve the management of terminal facilities (e.g., buildings, runaways, hangars, depots, piers, quays, wharfs), equipment (e.g., lift trucks, cranes), and employees (e.g., lumpers, stevedore gangs, dockers, baggage handlers), various costs associated with such management should be estimated and controlled before value-added processes performed by the terminal are considered. These costs include the cost of constructing or leasing terminal facilities, the cost of maintaining terminal facilities and equipment, the cost of loading/unloading and handling freight or passengers, the cost of processing terminal-related information, as well as labor costs, utility expenses, and administrative costs (including billing and paperwork).

These costs often depend on the size, number, location, and infrastructure of a terminal. In particular, the terminal location decision is one of the most important parts of terminal operations. Although terminals are typically located where traffic flows intersect or where traffic starts or ends, terminal locations are affected by a host of factors, including main transportation routes, feeder transportation routes, frequency of direct/indirect shipments, proximity to clusters of shippers and carriers, access to major transportation arteries (e.g., highways, railway lines, seaways, rivers), available lands, taxes, subsidies, utilities, weather/climate, local wages, externalities (e.g., community support), and room for expansion. In addition to these factors, other factors come into play for the location of terminals according to a particular mode of transportation. For example, in the case of motor carriers, traffic congestion on the designated truck routes should be factored into the terminal location. In the case of rail, urban blight and deterioration near rail yards and shops can affect rail terminal operational environments. In the case of air, noise and safety concerns for the surrounding community often affect the establishment and expansion of airports due to noise abatement rules (Sampson et al., 1990).

Chapter Summary

The following are key lessons learned from this chapter

- Transportation creates place (spatial) utility by moving people and/or goods from a point of origin to a point of destination. Thus, transportation allows people to get access to daily necessities such as water, food, and cloths without producing those at their place of living. Transportation also allows people to exchange their goods, service, and ideas by bringing them all together at the same location. In other words, transportation is a lifeline essential for human survival. Due to its role in connecting the dots in the supply chain, transportation is a vital link in the supply chain. This means that transportation efficiency will mirror supply chain efficiency. As such, the first step toward supply chain efficiency is the development of a successful transportation strategy.

- Considering the lasting impact of transportation on the national economy, the federal government gets deeply involved in transportation activities. Going a step further, the government often controls and dictates some of the transportation activities through regulatory acts. At the infancy of transportation's evolution in the United States during the nineteenth century, the transportation industry (especially railroads) was heavily regulated. This trend continued until the late twentieth century (i.e., the late 1970s and early 1980s), during which time the transportation industry began to mature and therefore did not need government intervention. The era of deregulation was represented by several landmark regulatory reforms, including the Airline Deregulation Act of 1978, the Motor Carrier Act of 1980, the Staggers Rail Act of 1980, and the Shipping Act of 1984. These reforms dramatically altered transportation practices by providing transportation carriers with a lot more freedom to enter/exit the market and set transportation price.

- In 2007, the United States spent $1.469 trillion for various transportation activities. This transportation spending accounted for 10.6% of the gross domestic production (GDP) in the United States (U.S. Bureau of National Statistics, 2008). With unstable energy prices and stretched supply chains across the world, there is no sign of a decline in transportation costs. To better control transportation costs, a growing number of firms are seeking for cost-saving opportunities. These include use of alternate fuel, adoption of information technology (e.g., satellite tracking systems, global positioning systems), better shipping forecasts, streamlined transportation operations, delivery/pickup coordination, and rate negotiation.

- Carrier selection is one of the most important transportation decisions. Specific carrier selection criteria include speed, reliability, accessibility, control and liability, security, environmental initiatives, and technological advances (e.g., RFID, EDI capability). In addition, a carrier's geographical coverage, range of services, loading capacity, and financial stability can be factored into the carrier selection decision.

- Intermodalism is on the rise due in part to globalization of transportation activities. Intermodal options include piggyback, fishyback, and birdyback. Generally, intermodal transportation is faster and less expensive because it exploits the best attributes of different modes of transportation. For example, while a truck has the lower cost function for relatively short distances (somewhere between 500 and 700 kilometers of the point of departure, or roughly 500 miles), its cost function climbs faster than both rail and water as it travels longer distances (e.g., 1,500 kilometers, or roughly 935 miles). Thus, the combined use of truck for last-miles and rail for line-haul movement may help the company speed up the delivery process and save transportation cost.

- A transportation price can be set in many different ways. Two key factors for transportation pricing are the cost of transportation operations and the cost of value-added services. To determine an optimal price that is fair to both carriers and shippers, transportation cost elements should be identified and estimated. Transportation cost elements typically include the cost of vehicle ownership, vehicle operations, repair and maintenance, fuel, shipment weight, travel distances, loading/unloading, type of shipment, insurance, toll, vehicle registration, and taxes. Thanks to deregulation, the transportation price (freight rate) can be negotiated between the carrier and the shipper. To get negotiation leverage, both parties should understand market dynamics in terms of supply and demand and then carefully conduct cost analysis.

- To cope with the increasing complexity of transportation operations, many firms have begun to use a transport management system (TMS). A TMS would be useful for order management, load planning, routing configurations, shipment tracking, delivery/pickup confirmation, and claims management.

Study Questions

The following are important thought starters for class discussions and exercises.

1. Explain the importance of transportation to our daily lives and national economy.
2. Distinguish different forms of transportation carriers and then discuss the pros and cons of the different forms of transportation carriers.
3. Which factors should be considered in choosing a particular carrier? Which of those factors should be weighed more heavily than others?
4. Discuss the pros and cons of the different modes of transportation.
5. Which factors have accelerated the evolution of intermodal transportation?
6. What are the benefits of intermodal transportation?

7. How would you estimate total transportation costs? What comprises transportation costs? What are hidden transportation cost elements (such as accessorial charges)? Develop a detailed work sheet for transportation cost elements.
8. Who are major transportation agencies? Explain the specific roles of each transportation agency.
9. Discuss the history of transportation regulation in the United States.
10. What are the impacts of transportation deregulation on the transportation industry? How did transportation reforms reshape the transportation industry and alter transportation practices? Who are the proponents or the opponents of transportation deregulation?
11. Explain how class and commodity rates are determined.
12. What is a bill of lading (B/L)? Why is it needed?
13. How can you improve transportation productivity?
14. How has information technology been used by carriers as means of improving their productivity? Which information technology is popular among the carriers?
15. What are the most viable transportation strategies in times of worldwide financial crisis? How do you come up with "blue-ocean" strategy for transportation carriers?
16. What are potential measures that would help enhance transportation security and safety? Discuss how 9/11 impacts the importance of transportation security and safety. Also, discuss how the transportation industry can work as a security buffer in case of supply chain disruptions.
17. How do you file freight claims when there are in-transit loss and damage? What are the proper steps to take for filing freight claims?
18. Under what circumstances could mergers/acquisitions in the transportation industry (e.g., airline, rail) be considered successful in improving transportation efficiency?
19. Identify the major sources of truck driver turnover and assess their impact on transportation productivity. Also, discuss ways to reduce truck driver turnover.
20. Discuss the role of a transport management system (TMS) in enhancing transportation productivity. Also, explain the key functionality of TMS. What are the selection criteria for TMS software? What are the challenges for implementing TMS and integrating it with other supply chain execution systems such as enterprise resource planning (ERP) and a warehouse management system (WMS)?
21. What impact did the oil spills in the Gulf Coast and Alaska have on the transportation industry?

22. What factors have driven many motor carriers out of the trucking business? (What has caused high trucking failures for the last decade?)
23. Why did many airlines offer discriminatory pricing on airfare?
24. Estimate the price sensitivity of airline passengers.
25. Discuss ways to increase the profitability of the fiercely competitive U.S. airline industry.
26. Why do many transportation carriers expand their business scope? (For example, many trucking firms began to offer warehousing or third-party logistics services in addition to their traditional trucking services.)
27. What are the environmental implications of transportation activities? How can we reduce the carbon footprint resulting from transportation activities?

Case: Louis Cab On Demand

This case is a fictitious story of a fast-growing ride-sharing taxi service.

Limited Public Transportation

As of 2014, the Louisville metropolitan area had a population of 1,284,195 people. Since 2000, it has had a population growth of 10.52 % thanks to its thriving shipping, warehousing, and healthcare industries. This growth includes Oldham, a fast-growing suburb of Louisville, Kentucky, that draws more commuters but allows local government to regulate taxicab operations. This regulation deteriorates public transportation services by limiting the number of cabs in operation (i.e., in a form of "regulated monopoly"). Until two years ago, taxi service to Oldham was primarily offered by Bill Gabriel, his wife, and their two sons, operating five cabs in a manner that rarely generated complaints from local residents. However, just two years ago, Bill, who began to experience some health problems after his company struggled to survive financially, decided to quit his taxi business and moved to Ocala, Florida. His company's cabs, garage, radios, and office were sold when they left, but the new owner continued to generate a sufficient amount of profit in tough economic times with skyrocketing fuel prices.

A few other competing firms have gone into business, operating whenever they want to and charging whatever they can get away with. Given that there is limited public transit service available in the Louisville metro area, some of its suburban counties such as the fast-growing Oldham County began to subsidize additional bus routes to accommodate public transit needs within those counties. However, such bus services still would not meet the need for more individualized cab services, because the given bus routes could not cover all the residential areas in suburban counties. Also, public transit would cost the taxpayers money, whereas the cab operations—such as they are—are self-supporting. For example, Oldham's mayor and city council have begun receiving numerous complaints about Oldham's lack of cab service or the poor quality of current cab service. There have been constant complaints. The local newspaper has run a brief series on cab problems in Oldham and has printed several hundred letters received from past passengers with complaints about rude driver behavior, reckless driving, and inadequately maintained vehicles. Other sources of frequent complaints include no cab service during inclement (rainy/snowy) weather; cabs rarely responding to phone calls in a timely manner; drivers refusing to give any change by claiming that they do not carry small bills and often arguing over fare/tip; a growing number of immigrant drivers who can barely speak English; no cab service during the Kentucky Derby and March Madness games; drivers who refuse to pick up drunken passengers for religious reasons; drivers who are reluctant to carry handicapped passengers and/or their guard dogs for unspecified reasons; drivers who ask for excessive tips for loading or unloading luggage; drivers who refuse to accept credit cards and only take cash for fares by not equipping their cabs with credit card scanners; and drivers who take lengthier routes for visitors to increase fares.

Service Dilemma

Although local cab operators seem to be aware of customer complaints and acknowledge that they are valid, their excuses are as follows:

- There is not enough profit earned to hire "quality" drivers or to maintain "first-rate" cabs due to rising fuel costs and high driver turnover.
- Restrictions imposed by local laws and regulations (e.g., consumer protection law, public safety rules, license requirements, fair competition) limit cab operators' business options and thus undermine their efficiency.
- There will always be a shortage of cabs during inclement weather such as rainstorms and blizzards because of unusual demand surge. In a recent survey of cab drivers, they said that the probability of being in a costly and dangerous accident is higher with precipitation. Most of them said that pedestrians tended to act crazy and unpredictable when it rained or snowed. Some of the studies showed that cab drivers drove 2.4% fewer miles with passengers during rainy days, thus making driving in the rain less profitable. As for the shortage of cabs during Derby week and other sporting events, no local residents (including drivers) want to work.
- Hauling intoxicated passengers is a losing proposition, even if the barkeeper often picks up the tab. If the drunk vomits in the cab, the cab company or sometimes the driver has to clean it up. Also, many drivers who recently

immigrated from Somalia, Pakistan, or other Middle-Eastern countries believe that serving drunks is a violation of their religious principles and therefore refuse to carry drunken passengers.
- Handicapped passengers or seniors with limited mobility require too many special services that many cab drivers are not properly trained to handle. Therefore, traditional cab operators are not equipped to handle handicapped passengers.

New Breed of Taxi Services

Recognizing a greater need for better taxi services, Louis Cab, Inc. (LCI) recently started an "on-demand" taxi company serving visitors and residents in the Louisville Metropolitan area, including Oldham. Louis Cab uses popular taxi-booking apps to allow its customer to order a luxury vehicle and track the vehicle's location using smartphones or mobile devices, while informing its customer of the estimated time of pickup. A few months after its inception, Louis Cab received rave reviews from the local newspaper and loyal passengers who praised its excellent services that distinguish it from its existing rivals. However, Louis Cab's rivals (traditional cab operators) filed petition to the local council and mayor claiming that Louis Cab violated local ordinance and regulations. They argued that Louis Cab's fee-paying transport services using unregistered/unlicensed private or rented vehicles violated city ordinance and therefore should be banned from local taxi services. Also, they added that Louis Cab's calculation of fares based on distance and time taken infringed upon their right to be the sole users of taxi meters in Louisville. After reviewing the exiting cab operators' petition, the local council sent a "cease-and-desist" letter to Louis Cab and then demanded that it cease its "on-demand" taxi services on the grounds that it was operating an unlicensed limousine dispatch. Since the local newspaper reported the local council's decision to ban Louis Cab, however, there has been a lot of public uproar over the council's hasty decision. Although other on-demand taxi service companies such as Uber, Lyft, and SideCar were frequently sued by traditional cab operators in San Francisco, Chicago, Seattle, and New York, there was a precedent that in September of 2013, the California Public Utilities Commission (CPUC) unanimously voted to create a new category of taxi services called "transportation network companies" to cover on-demand taxi services, thereby making California the first jurisdiction to recognize such services. Also, as the local election nears, local media and consumer rights advocates harshly criticized the local government and the mayor's office for their lack of vision and imagination, which got trapped in the past. More importantly, the council learned that a local political coalition began to organize its movement to put the ride-sharing issue to voters in a referendum and obtained 328,000 signatures in support. As political pressure mounts, the council members have called for an emergency meeting to further discuss this matter and take a more sensible course of action.

Discussion Questions

1. Is taxi service considered a common carrier? (If so, why? If not, why not?)
2. Based on the documented complaints made by traditional cab passengers, which of the carrier's duties may have been violated?
3. What are the main motivations behind the local ordinance and rules regulating local cab services?
4. What are the sources of controversies surrounding on-demand taxi services?
5. How would you resolve the dilemma involving the Louis Cab's on-demand, ride-sharing services?

Bibliography

Association of American Railroads (2008), "Overview of America's Freight Railroads," http://www.aar.org/PubCommon/Documents/AboutTheIndustry/Overview.pdf, retrieved on June 27, 2010.

Association of American Railroads (2010), "Intermodal Rail Traffic Hits New High," http://pragcap.com/intermodal-rail-traffic-hits-a-new-high, retrieved on June 30, 2010.

Bureau of Transportation Statistics (2013), *National Transportation Statistics*, Office of the Assistant Secretary for Research and Technology, Washington DC: United States Department of Transportation.

Card, M. (2009), *Trucking Failures*, Alexandria, VA: American Trucking Association.

Christaller, W. (1933), *Central Places in Southern Germany (Die zentralen Orte in Süddeutschland)*, translated by Carlisle W. Baskin, Englewood Cliffs, NJ: Prentice-Hall.

Concil, A. (2010), "Global Traffic Falls 2.4% in April—Volcano Dents Recovery," http://www.iata.org/pressroom/pr/Pages/2010-05-27-01.aspx, retrieved on June 29, 2010.

Conover, B. (2001), "Transportation Management Systems," presented at the UPS Center workshop, September 24, 2001, Louisville, KY: University of Louisville.

Cross, R.G. (1997), *Revenue Management: Hard-core Tactics for Market Domination*, New York, NY: Broadway Books.

Fair, M.L. and Williams, E.W. (1975), *Economics of Transportation and Logistics*, Dallas, TX: Business Publications, Inc.

Fair, M.L. and Williams, E.W. (1981), *Transportation and Logistics*, Dallas, TX: Business Publications, Inc.

Federal Motor Carrier Safety Administration (2014), *Summary of Hours of Service Regulations*, March 27, 2014 Updated, Washington DC: FMCSA, U.S. Department of Transportation.

Lieb, R.C. (1994), *Transportation*, 4th edition, Houston, TX: Dame Publications, Inc.

Logistics Platform (2010), "Logistics Glossary," http://en.logisticsplatform.com.ua/?module=static&id=16, retrieved on June 25, 2010.

Lösch, A. (1954), *The Economics of Location* (*Die räumliche Ordnung der Wirtschaft*), 2nd edition (First published in 1940), translated by W. H. Woglom and W. F. Stolper, New Haven, CT: Yale University Press.

Mcclnis, L. (2010), "Air Cargo Volume Rebound Quickly," *Reuters*, http://uk.reuters.com/article/idUKTRE65026820100601, retrieved on June 28, 2010.

McGill, J.I. and Van Ryzin, G.J. (1999), "Revenue Management: Research Overview and Prospects," *Transportation Science*, 33 (2), 233–256.

Min, H. (2009), "The Impact of Hours-of-Service Regulations on Transportation Productivity and Safety: A Summary of Findings from the Literature," *Journal of Transportation Management*, 21 (2), 49–64.

National Highway Traffic Safety Administration (2014), *Traffic Safety Facts 2012 Data*, NHTSA's National Center for Statistics and Analysis, Washington DC: U.S. Department of Transportation

O'Reilly, J. (2008), "Trends," *Inbound Logistics*, http://www.inboundlogistics.com/cms/article/trucking-looms-large-in-freight-forecast/, retrieved on January 15, 2010.

RITA-Research and Innovative Technology Administration (2007), "Average Freight Revenue Per Ton-mile Between 1960–2006," *Bureau of Transportation Statistics*, http://www.bts.gov/publications/national_transportation_statistics/html/table_03_17.html, retrieved on June 27, 2010.

Rodrigue, J-P, Comtois, C., and Slack, B. (2009), *The Geography of Transport Systems*, 2nd edition, New York, NY: Routledge.

Rogers, D.S. and Tibben-Lembke, R.S. (1999), *Going Backwards: Reverse Logistics Trends and Practices*, Pittsburg, PA: Reverse Logistics Executive Council Press.

Salerno, N. (2010), "What the Heck is Hotel Revenue Management, Anyway—The Hotel Marketer's Guide to Revenue Management," *Hotel Marketing Coach*, http://www.hotelmarketingcoach.com/What%20the%20heck%20is%20Revenue%20Management.htm, retrieved on July 2, 2010.

Sampson, R.J., Farris, M.T., and Shrock, D.L. (1990), *Domestic Transportation: Practice, Theory, and Policy*, Boston, MA: Houghton Mifflin Company.

Stephenson, Jr., F.J. (1987), *Transportation USA*, Reading, MA: Addison-Wesley Publishing Company.

Temple, Barker and Sloane Inc. (1982), *Transportation Strategies for the Eighties*, Oak Brook, IL: National Council of Physical Distribution Management.

TruckGauge (2011), "Trucking Failures Remain Very Low," http://www.truckgauge.com/2012/05/09/fleet-failures-stay-historically-low-in-q1/, retrieved on December 11, 2011.

Truckinfo (2010), "6/25/2010 Intermodal Traffic Reaches New High," http://www.truckinginfo.com/news/news-detail.asp?news_id=70836, retrieved on June 30, 2010.

TSSA (2010), "Freight on Rail—Dispelling Commonly Held Myths," http://www.tssa.org.uk/article-3.php3?id_article=1596, retrieved on June 27, 2010.

Tyworth, J.E., Cavinato, J.L., and Langley, Jr., C.J. (1987), *Traffic Management: Planning, Operations, and Control*, Prospect Heights, IL: Waveland Press, Inc.

U.S. Army Corps of Engineers (2000), "Inland Waterway Navigation: Value to the Nation," IWR Publication Office, http://www.iwr.usace.army.mil/docs/InlandNavigation.pdf, retrieved on June 25, 2010.

U.S. Army Corps of Engineers (2009), *Final Waterborne Commerce Statistics for Calendar Year 2008*, Institute for Water Resources, New Orleans, LA: Navigation Data Center, http://www.iwr.usace.army.mil/ndc/wcsc/pdf/finaltotal08.pdf, retrieved on June 25, 2010.

U.S. Bureau of Transportation Statistics (2008), *National Transportation Statistics*, http://www.census.gov/compendia/statab/2010/tables/10s1030.pdf, retrieved on July 5, 2010.

U.S. Department of Transportation (2010), "National Freight Transportation Trends and Emissions," Federal Highway Administration, http://www.fhwa.dot.gov/environment/freightaq/chapter2.htm, retrieved on June 27, 2010.

U.S. Environmental Protection Agency (2008), *Air Transportation Industry*, Unpublished Report, Washington DC: Office of Enforcement and Compliance Assurance, U.S. EPA.

U.S. News and World Report (2010), "The People's Vote: Interstate Commerce Act of 1887," http://www.usnews.com/usnews/documents/docpages.orig/document_page49.htm, retrieved on June 23, 2010.

Wang, D.-H. (2006), "The Ocean Shipping Reform Act: The Trans-Atlantic Cares," *Maritime Policy and Management*, 33(1), 23–33.

WCL Consulting (2006), *Global Supply Chain Overview (Consumer Goods): Ocean Carriers*, Unpublished White Paper, Long Beach, CA: World Class Logistics Consulting, Inc.

Wood, D.F. and Johnson, J.C. (1996), *Contemporary Transportation*, Upper Saddle River, NJ: Prentice Hall.

8

Sourcing

"Outsourcing and globalization of manufacturing allows companies to reduce costs, benefits consumers with lower cost goods and services, causes economic expansion that reduces unemployment, and increases productivity and job creation."
—Larry Elder

Learning Objectives

After reading this chapter, you should be able to:

- Understand the role of sourcing in supply chain management.
- Identify factors influencing in-housing or outsourcing (or "make-or-buy") decisions.
- Understand the rationale behind outsourcing and the key to the success of outsourcing operations.
- Understand what truly drives costs and how to control them.
- Differentiate between traditional costing and activity-based costing.
- Recognize and identify the distinction between value-adding and non-value-adding activities associated with sourcing.
- Assess the impact of sourcing on the company's bottom line.
- Discover, evaluate, and select potential sources of supply and then develop a supply base.
- Learn to build long-term supplier relationships.
- Learn to leverage the assistance of intermediaries for strategic sourcing.
- Learn to cope with unexpected supply risk and develop contingency plans to deal with such risk.
- Determine when to bid competitively or negotiate with suppliers.
- Understand the emerging trends of sourcing including global sourcing, e-purchasing, and online auctions.

In-Housing versus Outsourcing

One of the first steps for sourcing is the recognition of the need for outsourcing. In other words, sourcing is not needed if the company decides to make its own products and create its own services all by itself, without resorting to outside help or resources. The company can justify its decision to use its own resources if such a decision can bring numerous benefits that may outweigh the potential advantages of outsourcing. The choice between in-housing and outsourcing often hinges on the assessment of the following factors: costs (including start-up investments), added value, time (response time), production capacity, financial capacity, control over production schedules, quality control, workforce stability, technology transfer risk, production volume, know-how, and patent rights. For example, the in-house management of information technology (IT) requires the use of the company's own staff, computer equipment, software, and peripheral devices while bearing the costs of initial setup, software/hardware upgrades, maintenance, user training, and network development. On the other hand, outsourcing IT frees the company from the aforementioned hassles and costs, although it loses control over IT operations.

Likewise, the pros and cons of in-housing versus outsourcing should be carefully weighed before choosing either one of those two options. Regardless, one cannot ignore the fact that a growing number of companies tend to outsource their supply chain operations more than ever before. As a matter of fact, the outsourced materials expenditure in the recent years surpassed 60% of the typical company's total sales revenue as compared to less than 30% just after World War II (Baily et al., 2005). Likewise, in the IT sector, nearly 70% of small- and medium-sized businesses were known to outsource some or their entire web-hosting needs, according to NetStride Internet Solutions (2010). Following this trend, some Fortune 500 Companies such as Procter & Gamble and DuPont expanded their outsourcing operations. For example, in the past decade, Procter & Gamble has outsourced everything from IT infrastructure, data center operations, finance, accounting, and human resources to management of its offices/facilities from Cincinnati to Moscow. It wanted half of all new P&G products to come from outside in 2010, versus 20% several years earlier. In particular, Procter & Gamble intended to hand over 80% of its back-office IT functions to outsourcers such as Electronic Data Systems (EDS) and Affiliated Computer Services (ACS) (Information Age, 2006). DuPont also wanted to fix its unwieldy system for administering records, payroll, and fringe benefits for its 60,000 employees across 70 different nations, with data scattered among different software platforms and global business units. By awarding a long-term outsourcing contract to the Cincinnati-based Convergys Corp., the world's biggest call-center operator, which was hired to redesign and administer

DuPont's human resources programs, DuPont reduced its costs by 20% in the first year and 30% a year afterward.

Despite the increasing popularity of outsourcing, a firm should not jump onto the outsourcing bandwagon without carefully weighing the pros and cons of in-housing versus outsourcing, as summarized in Table 8.1.

Table 8.1. Factors Favoring In-Housing or Outsourcing

Factors Favoring In-Housing	Factors Favoring Outsourcing
• If the needed material and part can be less expensively obtained and/or made within the organization than outside the organization.	• If the organization has limited resources and financial capacity and therefore cannot afford to make additional investment of its capital in developing new products or markets.
• If the production and distribution schedules need to be controlled by the organization to maintain supply chain flexibility.	• If the organization would like to focus on its core competency and improve its overall supply chain efficiency by contracting out its costly and inefficient non-core business functions.
• If the organization excels at innovation and therefore needs to maintain its know-how or design secrecy without a risk of technology transfer.	• If the organization's existing personnel skills and technological know-how cannot be readily adapted to making a product or its parts within the organization.
• If a product or its part is vital and requires extremely close quality control.	• If patents or other legal barriers preclude the organization from making a product or its parts.
• If a product or its part can be produced on existing equipment and is of the type in which the firm has considerable manufacturing experience and expertise.	• If the anticipated demand for a product or its parts are either temporary or seasonal.
• If the organization does not need to make extensive start-up investment in facilities and equipment because it has already sufficient production capacity.	• If the anticipated demand for a product or its parts are small in volume.
• If requirements or demands for a product and its part are projected to be both relatively large and stable. Thus, the organization can create economies of scale for its own production.	• If the organization does not want to deal with potential labor-management conflicts and work stoppages (or labor strikes).
	• If the organization would like to have contingency plans in case of emergencies and unexpected supply chain interruption.

Principles of Outsourcing

Generally, *outsourcing* refers to the act of moving the company's "noncore" business activities and related decision responsibilities to outside providers. Its main goal is to enhance the organization's flexibility and competitiveness in rapidly changing marketplaces. Because outsourcing frees up the company's key resources, such as cash, personnel, time, equipment, and facilities, it is a popular way to make the company's supply chain operations lean. The business functions that are often outsourced include call center services, logistics services, janitorial services, payroll and secretarial services, information technology services (e.g., website development, web hosting, cloud support services, data entry, data warehousing), fabrication and assembly, audit and payment, bookkeeping and invoicing, tax management, sales and marketing, and health and safety compliances.

Once the outsourcing decision is made, the next step for outsourcing is to determine the scope of outsourcing. Depending on its scope, outsourcing can be implemented at four different stages (Sanders and Locke, 2005):

- **Out-tasking**—With out-tasking, a specific task with a narrow scope, such as the delivery of finished goods to retailers in a certain area, can be contracted out to an outside supplier (outsourcer), such as a trucking firm. In other words, out-tasking is characterized by the outsourcing of one or a few tasks that are considered primarily tactical and standardized.
- **Co-managed services**—Co-managed service is another form of outsourcing in which the scope of subcontracted work performed by the outside supplier is greater than out-tasking, but the outsourced tasks are still controlled by the customer. It is characterized by outsourcing of multiple tasks that are mostly tactical and partially standardized. An example of a co-managed service is the arrangement of vendor managed inventory (VMI) between the manufacturer and its supplier.
- **Managed services**—With managed services, an outside supplier is responsible for the design, implementation, and management of end-to-end supply chain solutions for the customer. Managed services often involve the customization of outsourced tasks. An example is the use of a third-party logistics provider for handling a full range of integrated logistics activities, such as staffing, equipment purchase/maintenance, facility management, software development, materials management, inventory management, and traffic/carrier management.
- **Full outsourcing**—In full outsourcing, the outside supplier takes full responsibility for the customized design, implementation, management, and often the determination of the strategic direction of the entire business function. Full outsourcing can make the company virtual because the outsourced tasks are completely in the hand of an outside supplier. An example is the outsourcing of the full spectrum of information technology (IT) services encompassing day-to-day execution, equipment purchase and maintenance, staff development/training, payroll, and strategic planning. Because the client heavily depends on its outside supplier and has no control over its operation, full outsourcing can be riskier than other forms of outsourcing.

If implemented appropriately, outsourcing can bring numerous managerial benefits:

- Increased operational efficiencies with lower total cost through reduction of investments in noncritical assets (e.g., maximization of IT investment)

- Increased speed to market by working with a partner with the expertise and capacity to bring new products and services to market quicker
- Enhanced opportunity for a firm entering new markets through the quick execution of an outsourced supply chain function by an outside partner (or outsourcer) who is more familiar with and specialized in those markets
- Mitigated risk of business failure by passing some of the risk on to the outsourcer
- Increased focus on the firm's core competencies and building on those focused world-class skills to directly add value to the firm's customers

Despite the aforementioned benefits, a poor outsourcing plan can cause more harm than good. In particular, one of the biggest challenges in the successful implementation of outsourcing is the management of a productive outsourcing relationship with the outsourcer (Lynch, 2000). The following steps can be taken to maintain such a relationship (Macronimous.com, 2011):

- *Sustain a good relationship between key management personnel.* The peer friendship and synchronized coordination with an outsourcer's key management personnel are important factors for sustaining long-term relationships with them. Also, maintaining a single point of contact with the outsourcer will help avoid confusion and duplicated communication efforts.
- *Develop well-defined, quantifiable performance metrics.* The outsourcing performance criteria must be quantifiable and must be established as criteria at the beginning of the contract. If the outsourced firm can compare the performance of the outsourcer with the pre-established performance measures, it can assess the clear benefits of outsourcing. The outsourcer would also know where it stands in meeting the outsourced client's expectations.
- *Setting up special boards or committees.* Successful outsourcing relationships involve setting up of special executive committees and boards that draw out the best strategies for the smooth and effective handling of the outsourcing relationship. Also, the prompt identification and resolution of outsourcing issues through the help of these committees and boards can preemptively tackle the issues and resolve conflicts.
- *Develop incentives and penalties.* The outsourcer is encouraged to meet or exceed customer expectations by establishing performance-base (merit-based) pricing. When performance exceeds the established performance standard, the incentives apply. On the other hand, when the outsourcer's performance falls short of that standard, the penalties would be imposed. This practice will help both parties understand the clear performance expectation and reward/penalty structure and thus prevent potential disputes over the outsourcing performances.
- *Arrange periodical review meetings.* For a successful outsourcing relationship, it is better to have frequent meetings for formal performance review. These

meetings can discuss what both parties are working towards and what the deliverables are, given a period of time.

- *Train outsourcer personnel.* The outsourcer personnel should be given ongoing training opportunities so that they can align their business goals to the outsourcing objectives of the outsourced client. The issues driving the client's needs have to be understood and the outsourcer's effort has to be related to those needs. The training agenda may include management skills, technology advancement, or anything that can improve the client relationship.
- *Understand the cultural differences.* Because both parties to the outsourcing relationship will have their own cultures, these differences have to be recognized and bridged. Organizing social events, education on company background, participation in each others' quality control programs are some of the ways to bridge the cultural gap.

Cost Analysis

To get the best possible value for each purchase, supply chain professionals need to ensure that the price they are paying is fair and reasonable. Such an assurance cannot be made without knowing the exact costs of products or services to be sourced outside. Therefore, cost analysis is one of the most important tools for sourcing. In a nutshell, cost analysis is an attempt to obtain the "lowest fair and reasonable" price of outsourced products or services. That attempt includes the breakdown and examination of current and anticipated costs associated with the purchase of those products or services. Cost analysis is particularly useful for answering the following questions related to sourcing:

- Are we paying more than what the product or the service is really worth?
- Are we being charged the same as another buyer would be charged?
- How does the current price compare to a price paid in the past for the same or similar goods or services?
- Is the quoted price a discounted price or a list price? Is there room for price haggling?

The ultimate purposes of cost analysis are to identify and eliminate unnecessary costs; to ascertain that the price being quoted is equitable for both the buyer and the supplier; and to provide the supply chain professional with an objective negotiation tool. Because cost often reflects managerial efficiency, its specific measures may vary from one company to another or one period to another. Also, depending on the method of cost analysis, its estimates may differ from one analysis to another. Considering this complexity and volatility, one should keep in mind the four key principles of cost analysis (Laseter, 1998):

- *Capture cost drivers, not cost elements.* Answer "what incurs costs," not just "what comprises cost."
- *Build commodity-specific cost drivers.* Inherent differences in products will cause different cost drivers.
- *Consider the total cost of ownership.* Few sourcing decisions should be based solely on the product or service's purchase price.
- *Start simply and then add complexity to the cost analysis only as needed.* Focus on the most important cost drivers.

With this in mind, the following subsections elaborate on the details of various cost analysis methods.

Traditional Costing

Traditional costing primarily focuses on what is spent in a given fiscal period instead of where and why costs incur (ExactCost, Inc., 2011). Because it is accounting oriented, it allocates costs to different categories, as summarized in Table 8.2. Traditional costing often allocates costs based on single-volume measures such as direct-labor hours, direct-labor costs, and machine hours. Although using a single volume measure as an overall cost driver seldom meets the cause-and-effect criterion desired in cost allocation, it provides a relatively inexpensive and convenient means of complying with accounting or financial reporting requirements (http://www.gale.cengage.com/pdf/samples/sp665568.pdf). However, because traditional costing uses the volume-based allocation for cost-assignment purposes, it tends to assign overhead (indirect) costs to products using an arbitrarily predetermined rate. This method can distort the true cost of making products, because it is hard to associate the specific volume of direct labor or direct material with the final output; therefore, a single pool of indirect costs assigned for the entire production unit (e.g., entire plant or department) can be either under- or overestimated. For example, traditional costing often allocates the cost of idle machine/equipment to products, although the manufacturer of those products did not consume any additional resources during idle time.

Table 8.2. Traditional Cost Classification

Cost	Features
Direct cost	• Costs accrued from the unit being produced.
	• Most direct costs are variable.
	• A reduction in the supplier's direct cost is generally worth more to the buyer than a major reduction in the supplier's profit.
	Examples: Direct material, direct labor, and purchasing cost

Cost	Features
Indirect cost	• Costs incurred in the operation of a production plant or process, but which normally cannot be related directly to any given unit of production.
	Examples: Rent, property taxes, equipment depreciation, data processing, utility, wages and salaries, and maintenance cost
Fixed cost	• Costs that tend to remain the same regardless of the number of units produced. • Costs decrease as a cost per unit when output levels are increased. • Incurrence of costs is a function of top-level management, not lower-level supervisors.
	Example: Land purchase and long-term leasing cost
Variable cost	• Costs that are expected to fluctuate in direct proportion to changes in the level of operational activities such as sales and production levels.
	Examples: Labor cost, hourly wage, and material cost
Semi-variable cost	• Costs that tend to change in proportion to changes in the level of operational activities, but *not* in direct proportion; partially variable and partially fixed.
	Examples: Heat, power, light, fuel, and salaries

Total Cost of Ownership

The total cost of ownership (TCO) is a management accounting philosophy that includes all supply chain–related costs expected to be incurred throughout the entire life of product. These costs are composed of three components (Burt et al., 2010):

- **Acquisition costs**—Costs associated with the purchase of a product or service
- **Ownership costs**—Costs associated with the ongoing use of a purchased product or service
- **Post-ownership costs**—Costs associated with the disposal and quality assurance of a product or service

The detail breakdown of these cost components is summarized in Table 8.3. TCO is useful for a firm that wants to control its supply chain costs, as detailed next:

- TCO allows the buyer to compare one supplier's total cost performance to others, regardless of the current price tag.
- TCO helps the buyer identify the cost of non-performance (e.g., defects, repair, maintenance, late delivery) and hidden cost (e.g., taxes, customs duties).
- TCO develops an awareness of non-price factors and enhances the suppliers' accountability for performance failures.
- TCO can capture the likely future revenue and expense streams from the purchases.

Table 8.3. Total Cost of Ownership Components

Cost Categories	Components
Acquisition costs	• Purchase price • Planning costs (e.g., order preparation) • Administrative costs (e.g., budgeting, bid specifications) • Taxes/tariffs/customs duties • Financing costs (e.g., interest)
Ownership costs	• Downtime costs • Inspection costs • Maintenance costs • Inventory carrying costs • Conversion costs • Non-value-added costs • Depreciation
Post-ownership costs	• Disposal costs • Clean-up costs • Environmental compliance costs • Repair/replacement costs • Product liability costs • Reverse logistics costs (e.g., product recall, recycling costs) • Cost of lost customer goodwill (e.g., lost sales)

Activity-Based Costing

As the focus of cost analysis shifts from the estimate of income to the extent of profit contribution, a growing number of firms are beginning to explore the opportunity to identify and eliminate incremental but avoidable cost elements. Such a shift in cost accounting has given birth to *activity-based costing (ABC)*, which generally refers to a cost accounting system that aims to identify money losers or winners by linking cost drivers directly to products or services that require activities consuming resources. ABC begins with the dissection of supply chain activities in terms of their causal relationship with cost objects and then specifies where non-value-adding activities exist. In other words, ABC helps the firm uncover the root causes of cost increases that do not contribute to profit increases. Therefore, the ultimate purpose of ABC is to increase profit margin by eliminating non-value-adding activities that cost money. Examples of non-value-adding activities include the use of nonstandard materials, components, and parts; red tape (too much paperwork); lengthy purchase approval processes; redundancy;

defects, rework, and scrap; split shipments, emergency/rush shipments, and damaged shipments; and an excessive number of suppliers. Based on the framework displayed in Figure 8.1, ABC makes the following fundamental premises (Hicks, 1999).

- Cost objects consume activities.
- Activities consume resources.
- The consumption of resources drives costs.
- The understanding of cost drivers is critical for managing overhead expenses and minimizing cost errors.

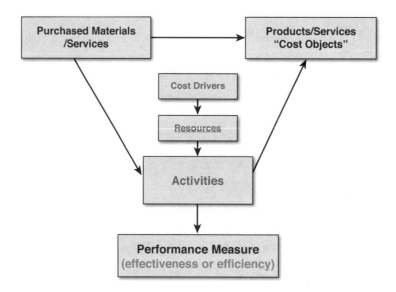

Figure 8.1. *The basic framework of activity-based costing*

These premises indicate that a company's outputs give rise to the need for various supply chain activities that incur costs. To make ABC more meaningful, these costs should be accurately defined and measured, while the cause-and-effect relationship between the company's supply chain activities and resultant costs should be examined carefully. Under these premises, ABC can be implemented by taking the following steps (Forrest, 1996):

1. Define activities.
2. Develop a "bill of activities" (or activity/process mapping).
3. Determine cost drivers for each cost component.
4. Calculate the total activity cost.
5. Estimate the cost driver (or activity allocation) rate (= total activity cost / total number of activities performed) for each cost component.

As summarized in Table 8.4, ABC differs from traditional costing and is designed to overcome the deficiencies of traditional costing (Cooper and Kaplan, 1988). These deficiencies include the following:

- **Poor cost classification**—Ambiguous distinction between fixed (uncontrollable) and variable costs
- **Oversimplified operation and production**—Poor reflection of varying material, rework, and scrap costs
- **Failure to identify alternate processes**—Over-costing automated operations (e.g., multiple machines run by a single operator)
- **Diverse aggregation of overhead costs**—Equal allocation of overhead costs among all departments
- **Ignorance of profit contribution**—Lack of evaluation of activities for profit contribution
- **Reliance on outdated cost figures**—No adjustment for inflation

Figure 8.2 illustrates how the cost allocation based on traditional costing can be translated into the ABC framework to overcome the preceding deficiencies (Pohlen, 1994, p. 1).

Table 8.4. Traditional Costing versus Activity-Based Costing

Category	Traditional Costing	Activity-based Costing
Principle	Allocate costs.	Trace costs.
Approach	Aggregation for categorization.	Disaggregation for details.
Objective	Report costs.	Manage costs.
Focus	"What is spent?" "How much was spent?"	"How the resource is utilized."
Basis	Volume driven.	Transaction driven.
Per unit cost	Same (no reflection of economies of scale, learning curve, or complexity).	Varies.
Feature	Can distort the overhead cost information by neglecting cost-varying activities.	Enhance the visibility of overhead expenses.
Philosophy	Satisfying.	Optimizing.

Figure 8.2. *Comparison of traditional costing to activity-based costing*

The advantages of ABC are as follows:

- Closer analysis of the cause of costs in production or service processes, which helps develop a better budget plan and more meaningful performance standards
- Identification of non-value-adding activities, which helps cost reduction

And here are the disadvantages of ABC:

- Difficulty in gathering relevant data because it no longer uses variable/fixed cost dichotomy and therefore relies heavily on historical data.
- Difficulty in breaking down some activities. For instance, it is difficult to discern value-adding activities from non-value-adding activities consistently.
- Declining employee morale because every activity performed by employees is under scrutiny.

Given these disadvantages, caution should be exercised when determining whether ABC is relevant to the company's cost analysis. Some of the checklists include the following (Innes et al., 1994):

- Is the product line or service mix "diverse" or "complex"?
- Do customers require "different" levels of services?
- Is overhead cost a "significant" element of total cost?
- Is overhead cost growing significantly?

Life Cycle Costing

Life cycle costing (LCC) is a cost management technique that determines the total discounted costs of owning, operating, maintaining, and disposing of an asset over its useful life. It uses a present value method to assess the value of an asset at the time of purchase, because not all the operating costs incur at the same time. LCC can be broken down into three cost components: pertinent costs of ownership, the period of time over which these costs are incurred, and the discount rate that is applied to future costs translated into present-day costs (Mearig et al., 1999). LCC is unique in that it estimates not only production costs, but also calculates the extent of revenue the product can generate as well as the extent of expenses it can incur at each stage of a supply chain throughout its entire product life cycle, as depicted in Figure 8.3 (http://www.businessdictionary.com/definition/life-cycle-costing.html). LCC, however, cannot account for such intangibles as personal preferences for a certain design, style, or color.

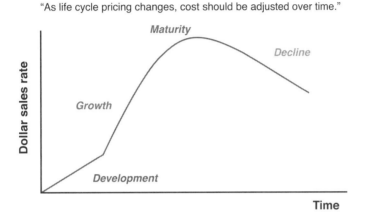

Figure 8.3. A life cycle curve of the product during its entire life

Value Analysis

Generally speaking, value analysis is a systematic, organized way to identify "unnecessary" supply chain activities used to create the intended functions of a product, service, or system. These unnecessary activities are the ones that can be eliminated, simplified, or reduced in scope without undermining those functions. Because these activities are often the sources of avoidable costs or waste, their identification and elimination help the firm improve its supply chain efficiency and the subsequent profitability. Put simply, value analysis is a better way to get the job done at the lowest possible cost without sacrificing quality. Its primary objective is to determine whether or not the maximum value is obtained for each dollar spent for the development of a product, service, or system. Herein, value is defined as the function (performance) of a product or service divided by its cost (Cooper and Slagmulder, 1977). This function can be further classified into primary (basic) and secondary functions. A product's primary function is the principal reason for its existence and usage. A product's secondary function is the auxiliary byproduct of the primary function, which becomes necessary to accomplish the primary function effectively. For example, the primary function of a cigarette package is to protect the cigarette, while its secondary function is to attract potential customers through its design appeal. By the same token, the primary function of a bicycle is to ride for transportation, while its secondary function is to exercise leg muscles for better physical fitness.

The concept of value can be categorized into four types

- **Use value**—Represents a property and quality that allows the product to accomplish its intended function
- **Cost value**—Represents a sum of labor, material, and overhead expenses required to produce the product
- **Esteem value**—Represents the pride (or prestige) of ownership
- **Exchange value**—Represents a property or quality of an item that makes it tradable for something else

To maximize the value of the product or service through value analysis, one should ask the following questions (Kaufman, 1990):

- What comprises a product (or service)?
- What is it for?
- What does it do?
- How much does it cost?
- What else will do the job?
- How important is the function of a product (or service)?
- What if the process changes?

With these questions in mind, value analysis can be conducted for a wide variety of supply chain areas, such as part/material design and specifications, production, assembly, standardization, packaging, transportation, warehousing, purchasing, and disposal of scrap, surplus, waste, or obsolete materials. For example, an engineer in an automobile manufacturer might conduct a value analysis on the motor casing and the process used to build it. He finds that three different sizes of nuts and bolts are used, with significant time taken to insert and tighten them. A redesign of the casing and the subsequent changes in the manufacturing process will allow this manufacturer to use only one size of bolt with threaded bolt holes, which removes the need for multiple nuts. The outcome is a significant savings in both material costs and assembly time (Sysque Quality, 2011). Similarly, an automobile part that uses too many fasteners to build it might be redesigned with a change in its shape so that it can use fewer fasteners and subsequently save production cost and time.

Typically, value analysis requires six different steps, as summarized in Figure 8.4. In particular, a value stream map at the goal-setting phase can play an important role in discerning value-adding activities from non-value-adding activities throughout the supply chain processes and thus helps the firm visualize the sources of waste (root causes of inefficiencies). A value stream map is an overarching visual technique that uses graphic symbols and charts to give managers and executives a broad picture of the entire supply chain processes with areas for potential improvements or lean transformation (Park, 1999).

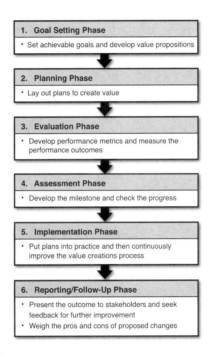

Figure 8.4. *Value analysis steps*

What Should Be Sourced

The fundamental role of purchasing in supply chain management includes the acquisition of *right* materials in the *right* quantity for delivery at the *right* time at the *right* place from the *right* source with the *right* service (before, during, and after the purchase) at the *right* price. Of these seven "rights," the determination of what materials should be sourced is one of the first steps of a purchasing process. Such determination starts with the identification of materials that are needed to support the organization's continuous supply chain operations. Those materials include commodities (e.g., iron, copper, coal, rubber, crude oil, lumber, paper, wheat, cotton, corn, meat, etc.), semi-finished parts/components, maintenance, repair and operating (MRO) supplies, and tools. In addition to these materials, capital equipment, real estates, and services (e.g., transportation services, construction services, legal services) can be bought from outside suppliers. Because the purchasing needs for materials, parts, supplies, and services differ from one organization to another and/or from one sourcing cycle to another, it is always difficult for supply chain professionals to forecast and then identify exactly what, how much, and when those needed items or services should be sourced. Therefore, the decision on what to buy should involve balancing a multitude of factors such as quality requirements, criticality of the need, budget constraints, substitutability, technical features, and shelf life. One of the systematic ways to make such a decision may be the use of a so-called "purchasing portfolio" tool introduced by Kraljic (1983). Using this tool, supply risk involved in the purchase of a certain material can be weighed against the profit to be generated by a purchase of that material. As such, a buying firm can determine which type of the material should be prioritized for sourcing using the company's portfolio in terms of supply risk and profit potential.

Who Can Be Supply Sources

Once the decision on what to buy has been made, the next step for sourcing is to identify potential supply sources, evaluate them, select the right source(s) among them, and then develop those chosen sources while nurturing the productive relationship with them for future partnerships. This step can be further broken down into five different stages, as described in Figure 8.5. The next subsections elaborate on these stages.

Discovery of Potential Supply Sources

Globalization of business activities along with the rapid emergence of low-cost countries such as China and India significantly increase potential supply bases. Although the expanded supply bases can be beneficial for buying firms due to a greater number of alternatives to consider, they often complicate the process of searching for the appropriate supplier. To make such a search process simpler, the buying firm should first explore and leverage the more reliable sources of information about the potential supply sources. These sources of supplier information can be classified into four different categories:

- **Published industry sources**—Available from publications and websites that contain information on new products and substitutions, suppliers' general backgrounds, and advertisements
- **Internal sources**—Available from the company's own historical files

- **Professional sources**—Available from the buyers' professional contacts
- **International sources**—Available from the World Trade Organization (WTO), foreign governments, and embassies

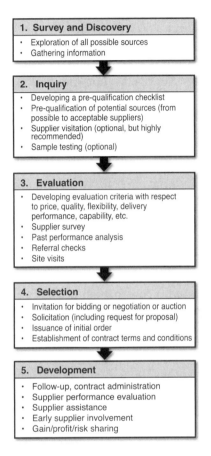

Figure 8.5. *Supplier selection steps*

Examples of these sources are listed next:

- Published Industry Sources
    - General industrial registers or directories that list the names of manufacturers, their addresses, products, and specialties as well as the addresses of their branches and affiliates (e.g., Thomas' Register of America Manufacturers, Conover – Mast Publishing Directory, McRae's Blue Book, directories published by chambers of commerce and industrial development agencies, and the website of the U.S. Securities and Exchange Commission [http://sec.gov])
    - Publications and websites of national and regional industry, trade, and professional associations (e.g., *Aviation Week* and *Iron Age*)
    - Trade journals (e.g., *Purchasing*, *Purchasing World*, and *Inbound Logistics*)

- Industry buyer's guide
- Consultant's guide (e.g., Who's Who in Logistics)
- Yellow pages—well-indexed source for local suppliers (www.yellow.com or www.worldpages.com)
- Manufacturers' catalogs that contain information on the specifications of and the location of suppliers for replacement parts and new equipment. Provides price information. Reference books with organized indexing (e.g., catalogues issued by Institute for Supply Management (ISM), American Society of Mechanical Engineers, Southern Supply and Machinery Dealers Association, and Automotive Jobbers Association)
- Internal Sources
    - Supplier files (e.g., Visual Search Microfilm Files, or VSMF)
    - Technical personnel who can identify technically qualified suppliers
    - Supplier information files
    - Internal mailing lists
    - Supplier catalogs
- Professional Sources
    - Sales representatives who can furnish not only valuable information on their products, but also other information of the trade
    - Professional colleagues/peers who can steer the search in the right direction based on referrals
    - Trade exhibits/tradeshows that offer a synopsis of an entire industry and its suppliers under one roof, permitting one-stop shopping and in-person comparisons of competitive offerings
- International Sources
    - U.S. Department of Commerce (e.g., World Trade Data Report)
    - Foreign Embassies and Consulates
    - International Chambers of Commerce, including Japanese External Trade Organization (JETRO), German-American Chamber of Commerce, and U.S.-Arab Chamber of Commerce

Inquiry of Potential Supply Sources

After preparing an exhaustive list of prospective supply sources through the discovery process, a purchasing manager has to narrow down the list to a manageable number of supply sources. This short list of supply sources should be further investigated to see if they are indeed worthy of serious consideration for upcoming sourcing contracts. In particular, if this list includes new suppliers with whom the purchasing manager has no experience in doing business, the following questions have to be addressed as part of the supplier prequalification checklists:

- Is the supplier capable of providing products and/or services of acceptable quality required by the buying firm on a consistent basis?
    - What is the supplier's quality management policy?
    - What are the supplier's environmental, ethics, and occupational safety policies?
    - What quality certifications (e.g., ISO 9000 certification) are held by the supplier?
    - How stable is the financial status of the supplier based on the last three years of audited financial statements?
    - How flexible is the supplier in meeting a variety of purchasing requirements? (e.g., does the supplier have room for growth or expansion?)
    - Does the supplier have a buyback or return policy?
- Is the supplier willing to complete a trial order? Are the test results of samples sent by the supplier favorable?
- Is the feedback of other buying firms who have the firsthand knowledge of the supplier mostly positive?

If all the answers to these questions are "yes," the purchasing manager may arrange a site visit with the supplier's plants and distribution centers. During the site visit, the purchasing manager needs to check and see if the supplier is duly registered and licensed; has adequate, well-maintained equipment and facility to manufacture and/or distribute required products; has a skilled workforce without any serious past labor-management issues; and has a contingency plan to cope with unexpected supply disruptions.

Supplier Evaluation

After developing a manageable list of prospective suppliers and gathering background information about them, the purchasing manager needs to evaluate these suppliers and compare them with each other in terms of their ability to provide *right* products and/or services with the *right* price at the *right* time. Such ability can be reflected in the host of supplier evaluation attributes summarized in Table 8.5. Although the importance of these attributes to supplier evaluation may vary from one organization to another and/or one purchase to another, the simultaneous consideration of these attributes will help the purchasing manager identify the strengths and weaknesses of each prospective supplier and then select the overall winner(s) of the purchasing contract. Among these attributes, some past studies (e.g., Min and Galle, 1991; Verma and Pullman, 1998) on supplier evaluation indicate that quality, price, and delivery services/performances are the three most dominant factors for selecting a particular supplier.

Table 8.5. Supplier Evaluation Attributes

Criteria	Attributes
Quality	• Quality of products • Warranty • Past records on the reliability of products • Quality certification, affidavits • Willingness to take the corrective actions • Willingness to accept a responsibility for defects or latent deficiencies • Prompt replacements of rejects
Price	• Competitive price • Accurate price quotation • No hidden costs • Correct billing/invoicing
Delivery services	• Delivery on schedule • Delivery per routing instructions • Delivery without constant follow-ups • Prompt responses to emergent and rush delivery requests • Good packaging • Geographical location
Production capacity and technical capability	• Adequate facility, equipment, and know-how • Adequate housekeeping (e.g., cleanliness, maintenance) • Good labor-management relations • Skilled labor • Technical ability to innovate • Information technology/communication system structure • Room for growth/expansion
Financial stability	• Credit rating (e.g., Dun and Bradstreet report) • Cash flow, liquidity, and profitability • Bank references
Environmental compliance	• Environment-friendly initiatives/policies • ISO 14000 certification

Supplier Selection

After the thorough evaluation of potential suppliers, a purchasing manager needs to select the most suitable supplier. However, the evaluation and the subsequent selection of suppliers are not simple tasks, especially when a relatively large number of supplier pools and many different attributes are considered at the same time. Such complexity calls for the use of a more systematic supplier evaluation and selection method. This method includes the categorical method, weighted-point method, cost-ratio method, analytic hierarchy process (AHP) method, and multiple attribute utility theory (MAUT) method. Each of these methods has its own pros and cons and therefore cannot be considered a panacea. With this in mind, these methods are discussed in greater detail next:

- **Categorical method**—In this method, a purchasing manager is required to keep and maintain a past performance record of all suppliers. The purchasing manager establishes a list of factors (with equal importance) for evaluation purposes and assigns a grade (plus, minus, and neutral) that measures supplier performance in each of the established areas. After an overall group rating, a positive evaluation in terms of a composite score may lead to increased business for a supplier, whereas a negative rating should result in discussions with the supplier to rectify the situation. This method is simple and straightforward; however, it may not be effective in handling a larger number of candidate suppliers (exceeding five). Also, this method may lead to many tie scores among the multiple suppliers.

- **Weighted-point method**—An application of the weighted-point method requires the assignment of an appropriate weight to each performance factor. The specific weight is a reflection of the purchasing manager's judgment about the relative importance of the specific performance factor. Then, a specific procedure or formula for measuring actual supplier performance must be developed for each individual factor. This results in a quantitative performance rating for each factor and then eventually results in a composite performance index for each supplier. Because the weighted-point method is quantitatively oriented, it is more objective than the categorical method, although their combined use may be meaningful. This method can be utilized by following the steps described next (Timmerman, 1986):

1. Establish a list of critical performance factors.
2. Assign weights to these factors to reflect their contributions to a supplier's overall performance rating; altogether, the weight must add up to 1.00, or 100% (= total performance).
3. Determine the procedure for measuring actual supplier performance on each factor.

4. Measure the actual performance of a supplier according to each factor—that is, develop performance ratings as a percentage of perfect performance (= 100%).
5. Cross-multiply the performance ratings with their respective weights to arrive at weighted ratings.
6. Add up the weighted ratings to compute the supplier's overall performance index.

An example illustrating how the weight-point method works is presented in Table 8.6.

- **Cost-ratio method**—The cost-ratio method relates all identifiable purchasing costs to the value of shipments received from individual suppliers (Zenz, 1994). Each cost ratio is assigned to a specific rating, subject to various performance criteria (e.g., quality, delivery, service). The lower the ratio of costs to shipments, the higher the rating for the supplier. Conversely, the higher the ratio of costs to shipments, the lower the rating for the supplier. The cost-ratio method is only used by a few progressive firms due to its complexity, which requires a computerized cost accounting system. This method typically follows the steps described here:

  1. Measure the costs for quality, delivery, and service separately.
  2. Divide them into the total purchases from the supplier, resulting in three cost-ratios (for quality cost, delivery cost, and service cost).
  3. Compute the supplier's overall cost-ratio by simply adding the three individual cost-ratios.
  4. Adjust the supplier's bid price for comparison purposes by multiplying it with a factor of (= 1 + overall cost-ratio).

  An example illustrating the cost-ratio method is presented in Table 8.7.

- **Analytical hierarchy process (AHP) method**—AHP is suitable for systematically selecting the most desirable suppliers with respect to multiple, confliction factors influencing the supplier selection decision. Indeed, AHP is one of the most popular methods for selecting the supplier as evidenced by its successful applications reported by Nydick and Hill (1992) and Min et al. (1994). Basically, AHP is a scoring method that helps the purchasing manager comparatively evaluate the strengths and weaknesses of each supplier through a series of pairwise comparisons. Also, unlike the weighted-point method, it does not require the arbitrary assignment of weights (relative importance) to each supplier selection attribute and allows the purchasing manager to conduct sensitivity analyses under various "what-if" scenarios.

- **Multiple attribute utility theory (MAUT) method**—MAUT was initially proposed by Min (1994) for supplier selection due to its ability to take into account both qualitative (e.g., quality assurance, perceived supply risk, communication barriers) and quantitative factors (e.g., quoted price) influencing supplier selection in the presence of risk and uncertainty. One of the important advantages of using the MAUT method includes its feature that allows the purchasing manager to see a performance "what-if" sensitivity analysis when the relative importance of supplier selection factors changes over time. In addition, unlike the weighted-point method or AHP, MAUT can handle a large number of potential suppliers (up to 500 prospective suppliers) and supplier selection attributes (up to 100 different attributes).

Table 8.6. An Example of the Weighted-Point Method

Supplier A for Part Number _____				
Factor	Weights	How Measured?	Supplier Performance (Past 12 Months)	Ratings
Quality	40	1% defective, subtract 5%	0.8% defective	40[100-(0.8×5)]/100 = 38.4
Delivery	30	1 day late, subtract 1%	Average 2 days late	30[100-(2×1)]/100 = 29.4
Price	20	Lowest price paid/price charged	$46/$50	20[(46/50)×100)]/100 = 18.4
Service	10	Good = 100%; Fair = 70%; Poor = 40%	Fair = 70%	10(70)/100 = 7.0
Total	100			93.2
Supplier B for Part Number _____				
Factor	Weights	How Measured?	Supplier Performance (Past 12 Months)	Ratings
Quality	40	1% defective, subtract 5%	1.5% defective	40[100-(1.5×5)]/100 = 37
Delivery	30	1 day late, subtract 1%	Average 3 days late	30[100-(3×1)]/100 = 29.1
Price	20	Lowest price paid/price charged	$46/$47	20[(46/47)×100)]/100 = 19.6
Service	10	Good = 100%; Fair = 70%; Poor = 40%	Poor = 40%	10(40)/100 = 4.0
Total	100			89.7

In this example, supplier A has to be selected because its composite rating of 93.2 is higher than that of supplier B (89.7).

Table 8.7. *The Example of a Cost-Ratio Method*

Cost Factor	Supplier A	Supplier B
Quality	+3% (minor quality problem)	-1 % (excellent quality = cost saving)
Delivery	+4% (more frequent delay)	+2% (less frequent delay)
Service	-2% (cost saving)	+1%
Overall cost-ratio (total penalty)	+5%	+2%
Original quoted (or bid) price	$150	$152
Adjustment factor	1 + 0.05 (5%) = 1.05	1 + 0.02 (2%) = 1.02
Total	$150 × 1.05 = 157.50	$152 × 1.02 = 155.04

In this example, supplier B should be selected because its cost-ratio of 155.04 is lower than supplier A's cost ratio of 157.50.

How to Split Orders

If the purchasing manager has chosen more than one supplier as the future sources of supply, he or she needs to make a decision as to how the orders should be split among these multiple sources. In dual-sourcing, it is customary that the purchasing manager splits orders between two suppliers using "70-30" approach. That is to say, 70% of the order volume will be awarded to a "big" supplier who can exploit its economies of scale, while the remaining 30% of the order volume will be awarded to a "little" supplier who may be regarded as a backup source providing competition. If the big supplier fails to perform up to the buying firm's performance standard, the allocated percentage of the order volume may be adjusted or completely reversed. In case that the order volume gets very large and therefore requires more than two sources of supply, the order split decision would be more complicated. Although there is no known systematic rule for splitting orders among more than two suppliers, one may consider the performance history, geographical location, capacity/size, past relationship, reputation, and stability of suppliers for splitting orders. For example, a relatively larger order volume will be awarded to a supplier who shows the best proven record of reliable quality and delivery services, whereas a relatively smaller order volume will be awarded to newer, unproven, or smaller suppliers whose stability or reliability may be questioned.

Supply Base Reduction

A *supply base* is the portion of the supplier network that is actively managed by the buying firm (Choi and Krause, 2006). Although a large supply base may provide the purchasing manager with many alternative sources of supply to choose from and plenty of backup sources in case of emergencies and unexpected supply disruptions, it would be costly and time consuming to manage all the suppliers for an extended period of time. In particular, it would be more challenging for the buying firm to develop a long-term partnership essential for sourcing stability with a larger supply base than with a smaller supply base due to the inherent complexity involved in a large number of suppliers. Considering this challenge, a greater number of companies such as John Deere, Toyota, Honda, and Chrysler have made a conscious effort to work with a fewer number of suppliers. In that effort, they were able to eliminate "problem" suppliers while developing more cooperative relationships with "prequalified" (or "preferred") suppliers who are more compatible with the buying firm in terms of their organizational culture, technical level, and quality commitments. The managerial benefits of a supply base reduction includes increased involvement of suppliers in new product development; less variation in product quality/specifications; better negotiated prices through volume discounts and economies of scale; lower supplier management costs; quick discovery of process flaws; and greater opportunities to share information with suppliers concerning inventory levels, fill rates, and demand levels (e.g., Ogden, 2003).

A key to the success of a supply base reduction strategy is to identify "problem" suppliers so that those can be weeded out. Some warning signs of a problem supplier include the following (Ostring, 2004):

- A supplier's country is experiencing drastic political, regulatory, or economic changes.
- A supplier has made changes in its key personnel.
- A supplier has been going through significant changes in ownership or has become a target of merger/acquisition in a short period of time.
- A supplier has prolonged labor issues.
- A supplier has outdated technology and limited budget for research & development (R&D) investments.
- A supplier's current liabilities exceed current assets.
- A supplier recently sold a significant amount of assets.
- A supplier suddenly requires significantly shorter payment terms.
- A supplier's credit rating is rapidly declining.
- A supplier's quality audit reveals many quality-related problems.

Supplier Certification

As discussed earlier, a supply base reduction starts with the identification of both "problem" suppliers and "prequalified" suppliers. Because we have already discussed how we can identify problem suppliers in the previous subsection, we still need to find a way to identify prequalified suppliers. The identification of prequalified suppliers often leads to supplier certification. Generally, a supplier certification program is a continuous incoming quality improvement process that recognizes suppliers who meet or exceed the buying firm's established quality, delivery performance, and price requirements consistently. This program begins with the supplier survey and audits to document the supplier's past performance history, long-term commitments to "built-in" quality control, and willingness to take corrective actions if the failure occurs. The supplier survey and audits will be followed by the extensive review of the supplier's supply chain process and the thorough inspection of the supplier's product samples. After these processes, a limited number of suppliers who showed the evidence of sound and effective quality control, cost control, reliable delivery performance, and technical support would be certified for a specified period of time (e.g., a year). Supplier certification can be renewed on a periodic basis.

Supplier certification can create "win-win" opportunities for both the buying firm and its supplier. To elaborate, a certified supplier will not be subject to routine testing/inspection and will be considered first for new business, whereas its buying firm saves time for selecting the right supplier and enjoys the opportunity to prevent the quality failure at the source. In order to reap the full benefits of supplier certification continuously, both the buying firm and its supplier must make sure that the following initiatives are in place for their supply chain activities (Grieco, et al., 1988):

- Preventive maintenance
- Supplier education and training
- Integration of design, engineering, and product testing into the certification program
- Early supplier involvement in new product design and development
- Coordination of supply chain activities between the buying firm and its supplier
- Cultural changes for total quality management (TQM)

Supplier Diversity

As of 2011, minority- and women-owned businesses account for more than $3 trillion in total sales revenue and employ approximately 25 million workers in the United States (Business Week, 2011). Indeed, the growth in minority- and women-owned businesses has outpaced the national average for the last decade. For example, the total number of African-American-owned businesses grew 45% between 1997 and 2002—more than four times the national average. During the same period, Hispanic-owned firms grew 31%, while women-owned firms shot up 20%. Also, the number of Asian-owned businesses increased by roughly 24% between 1997 and 2002, and businesses owned by Hawaiians and Pacific Islanders increased as much as 67% (Hoy, 2006). As these statistics reveal, a rapid growth in the minority- and women-owned business sectors prompted many buying firms to diversify their supply base. By including these underutilized business sectors in the supply chain network, the buying firm not only can increase its supply options, but also spur more innovative supply chain practices resulting from more diverse and multicultural supply bases. Another benefit of supplier diversity is the increased opportunity for federal government (e.g., Department of Defense) contracts by avoiding any penalty imposed on the lack of "good-faith" efforts of sourcing from minority- and women-owned business sectors (Min, 2009). Examples of best-practice firms that successfully leverage the supplier diversity program include Caterpillar, Wells Fargo, Exxon, and Dell.

Generally, a supplier diversity program refers to a key business strategy that aims to provide equal (or maximum) opportunity to qualified minority suppliers to compete and participate in the buying organization's procurement award process for goods and services. Herein, a *minority supplier* is defined as a supplier with at least 51% minority ownership or, in case of a corporation, at least 51% of the stock owned by minority persons, while its management and daily operations are controlled by one or more minorities. Minorities are those U.S. citizens who are at least 25% of one of the following ethnic groups—African-, Hispanic-, Asian/Indian-, Pacific Islander-, and Native-Americans—and those U.S. citizens who are disabled veterans or women, regardless of their race and ethnicity. The rationale for supplier diversity includes federal and state legislation (e.g., affirmative action) requiring the greater use of minority-owned suppliers, set-aside quotas for government appropriations, and the firm's "social-consciousness," which helps foster its positive public image among the potential minority consumers.

Supplier Relationship Management

According to a 2008 survey conducted by the APQC Consortium, supplier partnership and its integration with strategic sourcing initiatives improved supply chain efficiency. In addition, the Hackett Group, a U.S.-based business process consulting firm, found that the best-practice firms used 75% of their sourcing contracts as "long term," whereas the average firms used only 37% of their sourcing contracts as "short term" (http://www.scribd.com/doc/14171678/Deploying-Best-Practices-in-Supplier-Relationship-Management). This fact affirms the importance of a long-term supplier relationship in sourcing stability and subsequent supply chain success. Recognizing such importance, a growing number of firms have attempted to transform their adversarial relationship with their suppliers into a more cooperative partnership with them. A cooperative supplier partnership is an agreement between the buying firm and its supplier that involves a mutual commitment over an extended period of time, including the sharing of information, risks, and rewards (Ellram, 1990). As such, a cooperative supplier partnership differs from traditional supplier relationships in that it necessitates mutual trust, information/risk sharing, and joint problem solving, as summarized in Table 8.8 (Stuart, 1993).

Table 8.8. Traditional Supplier Relationship versus Cooperative Supplier Partnership

Traditional Supply Relationship	Cooperative Supply Partnership
Price emphasis for supplier selection	Multiple criteria including delivery and quality performances for supplier selection
Short-term purchase contracts	Long-term purchase contracts
Competitive bid evaluation	Intensive evaluation of supplier value added
Large supplier base	Few supplier base
Proprietary information	Information sharing
Power-driven problem solving	Mutual problem solving

By forming the cooperative supplier partnership, the buying firm can reap various managerial benefits, as listed next (Ellram, 1995a; Maloni and Benton, 2000):

- Reduced uncertainty for the buying firm in material cost, quality, and delivery schedules
- Cost savings thanks to reduced administrative costs, lower switching costs, and economies of scale in ordering, production, and transaction
- Increased supplier loyalty
- Risk sharing through joint investment, joint research and development, and joint product design
- More stable supply bases
- Better demand forecast and the subsequent reduction in inventory through information sharing
- Reduced time looking for new suppliers/gathering competitive bids
- Enhanced supply chain process integration

Despite these numerous benefit potentials, a supplier partnership can fall apart, unless it is properly managed. Among others, the most common causes of supplier partnership failures are poor communication, lack of top management support, lack of trust, lack of commitment to total quality management by a supplier, poor up-front planning, lack of distinctive supplier value-added/benefits, lack of strategic directions for the relationship, and lack of shared goals (Ellram, 1995b).

Intermediaries for Sourcing

As sourcing complexity increases with the globalization of business activities, a growing number of firms often utilize sourcing intermediaries rather than sourcing directly from suppliers. As opposed to direct sourcing, where the buying firm makes all the sourcing decisions (such as supplier selection and contract negotiation) by itself, "mediated" (indirect) sourcing delegates all the sourcing decisions to intermediaries, which will consolidate purchase orders from their clients and then leverage their economies of scale or greater bargaining power for more favorable contractual terms. From the supplier's standpoint, mediated sourcing through intermediaries may help the supplier better protect its proprietary information from potential leakage. From the buyer's perspective, mediated sourcing enhances the buyer's sourcing flexibility by increasing the access to many sourcing alternatives available from the intermediaries, while minimizing the risk of supply disruptions in volatile sourcing environments. For example, dramatic fluctuations in foreign currency exchange rates, a sudden increase in customs duty, or rapidly rising wages in the sourcing country may change the low-cost supply bases and consequently require the development of new sources of supply in a hurry. In this situation, sourcing intermediaries such as trading companies, which usually have extensive supply networks across different industries, will be in a better position to seek, monitor, and control other viable sourcing alternatives in a shorter period of time than the individual buying firm. Also, sourcing intermediaries, which deal with a wide variety of products, can adjust their purchase prices in some items to make up for lower profit in others, or can shift their businesses to product lines that are less affected by the higher customs duties and are therefore less vulnerable to pricing/duty fluctuations.

Sourcing intermediaries are broken down into two categories: agents and merchants (Scheuing, 1989, p. 345). Agents are simply go-betweens or matchmakers who help buyers find suitable suppliers, and vice versa, on a commission, which is a percentage of the amount transacted. On the other hand, merchants act for their accounts, buying and selling products at their own risk by taking physical possession of products. In other words, merchants acquire the title of products and live on the profit earned on the resale of those products. Some of these agents and merchants may be based overseas and are primarily dealing with overseas purchases. To elaborate, an overseas purchasing agent is an intermediary who assists the buyer in identifying and selecting foreign suppliers by gathering and screening information about them and their products. Also, the overseas purchasing agent who serves as a point of contact for foreign suppliers can assist the buyer in negotiating and smoothing out contract details. An export broker is an intermediary who specializes in bringing both foreign suppliers and domestic buyers together for a fee but does not take part in actual purchase transactions (http://www.termwiki.com/EN:export_broker).

Overseas-based merchants can be used when the purchasing manager has little contact with foreign suppliers and limited knowledge/experience of global sourcing. One of the overseas-based merchants is the export (sales) subsidiary. An export sales subsidiary essentially removes the export function from the parent company and places the function in a separate wholly owned subsidiary. The export sales subsidiary purchases goods from the parent company and then resells and exports the goods. The subsidiary performs much the same function as an export department of the multinational firm except that the subsidiary usually has better (and simpler) access to export financing, is able to add products from outside the parent company in order to diversify its product line, and is able to segregate costs and expenses more efficiently than an internal export department of the multinational firm (http://www.blackwellreference.com/public/tocnode?id=g9780631233176_chunk_g97806312349378_ss1-102). Another form of the overseas-based merchant is a state trading agency, which is either a government entity or a private enterprise that has been granted special or exclusive privileges (e.g., government grants and loans, loan guarantees, priority for obtaining foreign currency) by the government to export government surpluses or other goods such as agricultural products in a monopolistic manner (http://www.ers.usda.gov/publications/aer783/aer783b.pdf).

Regardless of the differences in sourcing intermediaries, the decision on mediated sourcing through intermediaries rests on the following checklist (Hickman and Hickman, 1992):

- Does the size/volume of the purchase warrant any extra effort and time?
- Does the purchasing manager need help in translation or documentation?
- Will source inspection be essential for quality assurance?
- Is direct sourcing too risky?
- Does the purchasing manager have captive sources of supply who may not want to do business with him or her directly?

Supply Risk Management

You cannot sell goods that you cannot make. You cannot make goods without acquiring the necessary parts and materials. You cannot acquire the necessary parts and materials if they cannot be delivered to you. That is to say, if there is any form of disruption in the supply chain process of sourcing, making, and delivering, the company cannot generate revenue because it cannot sell goods. Indeed, a FM global study of more than 600 financial executives around the world revealed that supply chain disruption caused by the presence of supply chain risk was the major source of a revenue decline (Bosman, 2005). Considering the significant impact of supply chain risk on the company's bottom line, supply chain risk has to be identified, monitored, and mitigated in a systematic fashion. However, inherent unpredictability associated with supply chain risk makes the task of managing risk extremely difficult. To ease such difficulty, supply chain professionals need to answer questions concerning where risk may originate from, when it may occur, and how it affects supply chain practices.

Supply chain risk originates in two forms: internal risk and external risk. *Internal risk* represents any negative events that appear in normal supply chain operations, including late deliveries, excessive inventory, material shortages, employee error, work site/traffic accidents, equipment failure, forecast error, and so forth. *External risk* is often a uncontrollable negative outcome that comes from outside the supply chain, such as natural disasters (e.g., earthquakes, hurricanes, tornadoes, flooding, drought), wars, political upheavals, terrorist attacks, disease outbreaks, banking crisis, and so forth (Waters, 2007). These risks can be further classified into four different types of risks—sourcing risk, operational risk, financial risk, and demand volatility risk—as shown in Figure 8.6. *Sourcing risk* is associated with the acquisition of needed raw materials, parts, components, and supplies. This risk occurs more frequently and widely than other supply chain risk, as discovered by an Accenture survey (Malone, 2006). *Operational risk* arises from the company's routine business activities that affect the firm's ability to produce goods and services. *Financial risk* is associated with the value and flow of money such as interest rates, inflation, payments, cash flows, investments, liquidity, and the like. *Demand volatility risk* is incurred from the fluctuation of customer demands over time and the subsequent difficulty in forecasting demands.

In this section, our discussions will focus on sourcing risk due to its influence on strategic sourcing decisions. Managing sourcing risk starts with the discovery of the sourcing risk, which will be followed by the recovery from sourcing risk and the redesign of a supply chain, as displayed by Figure 8.6. The discovery process involves the identification of the sources of sourcing risk via detailed supply chain mapping and then the assessment of risk. The following list represents warning signals for the presence of potential sourcing risks:

- The supplier has unstable labor-management relationships.
- The supplier lacks well-skilled, well-educated, and well-trained labor.
- The supplier has limited production capacity and know-how.
- The supplier's response to new orders or order changes is slow.
- The supplier is concentrated in a certain geographical region vulnerable to natural disasters such as earthquakes and tornadoes.
- The supplier relies heavily on too many middlemen.
- The supplier has poor communication and information technology infrastructure.
- The supplier has loose quality and environmental standards.
- The supplier is located in a country that is politically and economically instable.
- The supplier uses proprietary technology.
- The supplier is a single source of (critical) supply.

In addition to checking for the various warning signs discussed earlier, the purchasing manager can use the *supply chain vulnerability map* to categorize and then identify the potential weakest point/link of the supply chain, as illustrated by Figure 8.7. Once the sourcing risk is identified and its probability of occurrence is estimated at the discovery phase, a purchasing manager can devise the risk prevention or preemptive mitigation initiatives as follows:

- **Buffering practices with redundancy**—Unless there is a catastrophic disaster, a simple way to cope with sourcing risk is to stockpile safety stocks sufficient enough to prevent supply disruptions until the supply chain returns to normal (Handfield et al., 2008).
- **Supply chain collaboration**—By making a risk-sharing agreement with key suppliers, the buying firm can minimize the loss resulting from unexpected supply disruptions.
- **Information visibility**—If a firm has the greater access to real-time demand information by sharing information with its suppliers and is capable of identifying most vulnerable points/links of the supply chain, regardless of where they exist, it can better forecast what is coming ahead and thus better prepare for future sourcing risk. Also, information visibility can be enhanced by utilizing a product tracking system such as radio frequency identification (RFID) that can trace the location of the product on a real-time basis.
- **Volume and mix flexibility**—If a firm can adjust its volume of production and a range of different product lines in response to supply disruptions, it can better

mitigate the adverse impact of supply disruptions (Zhang et al., 2003; Tomlin, 2006).

- **Postponement**—If a firm can delay sourcing activities to a much later point in the supply chain after the recognition of customer needs and actual demands, it will mitigate the risk of holding wasteful extra inventory.
- **Supply chain redesign**—To redesign the entire supply chain to be more resilient to any supply disruptions, the firm should first identify the "weakest" (most vulnerable) links or any choke points in the supply chain and beef them up with alternative transportation routes or distribution channels.

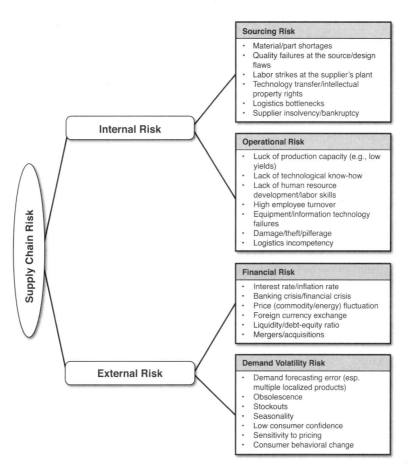

Figure 8.6. *Typology of supply chain risk*

Despite the availability of these risk prevention or preemptive risk mitigation initiatives, many firms are not ready to obviate supply risks and therefore suffer from sourcing failures and the subsequent revenue decline, as illustrated in Table 8.9 (Martha and Subbakrishna, 2002).

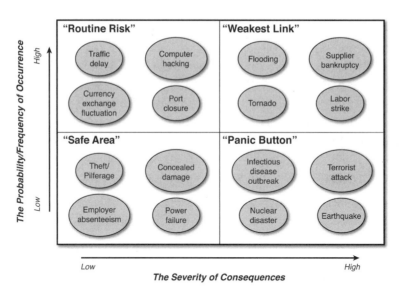

Figure 8.7. *A supply chain vulnerability map*

Table 8.9. Examples of Firms Prepared for Potential Supply Risk versus Firm Unprepared for Potential Risk

Risky Events	Firms Prepared for Supply Risk	Firms Unprepared for Supply Risk
Outbreaks of mad cow and foot-and-mouth diseases in Europe that forced the mass slaughter of cattle (Spring 2001)	**Naturalizer, Danier**, and **Justin Boot** relied on their safety stocks of cow leather to cope with shortages.	**Etienne Aigner** shifted purchases to other regions due to insufficient inventories of leather but faced stiff cost increases in leather.
Terrorist attack on the World Trade Centers that crippled the air transportation network (9/11/2001)	**Daimler-Chrysler** used alternate modes of transportation such as trucks and rails as part of its contingency plans.	**Ford** was forced to close five plants for several days due to auto part shortages.

Because full supply risk preparedness cannot be achieved overnight, many firms usually transform themselves to more risk-resilient organizations by following the four incremental phases of risk preparedness described next (Bowman, 2008):

- "Pre-compliant" Organization
  - Places disaster recovery plans on the back burner.
  - Plays catch-up when its supply chain is disrupted by the unforeseen events. For example, Land Rover was not prepared for the sudden supplier bankruptcy that cut off the supply of chassis frames in 2001. Similarly, when an unexpected tsunami hit Japan in 2011, Sony halted its production and evacuated six factories, Toyota closed three factories, and Nissan closed four factories due to part shortages caused by the tsunami.

- "Compliant" Organization
  - Views supply resiliency or security merely as the cost of doing business.
  - Complies with Customs-Trade Partnership Against Terrorism (C-TPAT) and thinks that imposed measures are enough.
- "Secure" Organization
  - Is proactive about working with suppliers and customers to head off supply disruptions.
- "Resilient" Organization
  - Views risk management as part of its business strategy and an opportunity to boost its competitiveness in the marketplace.
  - Builds some measure of redundancy into its supply chains and thus can detect and respond to potential disasters.

Competitive Bidding versus Negotiation

One of the final stages of sourcing process is to enter a purchase contract. Basically, there are two ways to enter the purchase contract—competitive bidding and negotiation—which are often mutually exclusive. Competitive bidding may be more favorable than negotiation if the following conditions hold true (Burt et al., 2010):

- The dollar value of purchase (e.g., more than $50,000) is large enough to justify the bidding process.
- Production specifications are explicitly clear to both the buying firm and the potential suppliers.
- There exists an adequate number of suppliers (e.g., three to eight) who are willing to participate in the bidding process.
- All the suppliers (bidders) must be (technically) qualified and actively want the contract.
- Suppliers have sufficient time to prepare and submit their bids before the given deadline.

If these conditions do not hold, the purchasing manager can finalize the purchase contract through negotiation. Generally, negotiation is a back-and-forth communication process of arriving at a common ground that is acceptable to both the buyer and the supplier with a feeling that they have won through bargaining on the essentials of a purchase contract, such as price, delivery term, and product specifications. Negotiation is more appropriate than competitive bidding in the following situations:

- It is nearly impossible for the buyer to estimate costs with a high degree of certainty.

- Price is not the only important consideration for the purchase contract.
- There exists a high level of supply risk.
- Early supplier involvement is needed for new product development.
- A buying firm anticipates the need to make changes in product specification.
- A buying firm is contracting for a portion of the supplier's production capacity.
- Tooling and setup costs represent a large percentage of the total purchase costs.
- A long period of time is required to produce the items to be purchased.
- A long-term contract is needed.
- A product to be purchased is patented.
- Supplier collusion is suspected.
- Only a single source of supply is considered.

An effective purchase negotiation strategy can be formulated by following these four steps:

1. Preparation
    a. Determine what the desired final outcomes are.
    b. Develop available options.
    c. Understand the authority level and role of each negotiator.
2. Opening
    a. Take the initiative.
    b. Listen to the other party more than talking to him or her. In other words, let the other party do most of the talking.
    c. Never accept the first offer.
3. Persuasion
    a. Consider the interests of the other party.
    b. Put the other party in your shoes.
    c. Do not divulge confidential information.
    d. Find "common ground."
    e. Begin with the easy issue(s) first.
    f. Go for smaller concessions initially by using facts and proofs.
4. Agreement
    a. Do not agree to something that is considered "unfair."
    b. Do not insist on last-minute additional changes.
    c. Do not let a deadline force you into a bad deal.
    d. Put agreements in writing after summing up the agreement.

Negotiation is situation specific; therefore, the same negotiation strategy that worked well in the past would not guarantee the same outcome for a different negotiation situation. Also, depending on the negotiation skills and experience of the negotiator, the outcome may differ from one situation to another. Figure 8.8 lists all the potential factors that may contribute to negotiation success by increasing the negotiator's bargaining power. The bargaining power of the negotiator may be strengthened depending on demand/supply dynamics in the marketplace. For instance, if a buyer is interested in buying more standardized commodities that are available from many suppliers, the buyer will have the bargaining edge (so-called "power of competition") over his or her supplier. On the other hand, if there is a shortage of parts that the buyer needs to buy in a monopolistic environment, the supplier has the upper hand over the buyer. Likewise, a negotiation site may influence the negotiator's bargaining strength. For example, if the buyer meets his supplier in his own office for a negotiation where he has direct access to the files that contain crucial information about the market and the product, the buyer may have an edge over his supplier. Also, being in his office may boost the "home court" advantage. Similar arguments can be made regarding the impact of negotiation time, communication channel (e.g., direct face-to-face versus phone negotiation), and team size/composition on the potential negotiation outcome. Although the real impact of some of these factors on the negotiation outcome is not scientifically proven and is inconclusive at best, as indicated by Min et al. (1995), a purchasing manager still needs to realize the potential influence of these factors on the negotiation effectiveness.

Also, because one's negotiation skill can be enhanced from either training/education or years of experience, it should be factored into the formation of a negotiating team representing the buying firm's interest. Likewise, the negotiator's personal attributes, such as natural aptitude, negotiation style, given role, and authority, may influence the bargaining power and the subsequent negotiation outcome. Such bargaining power will be termed the *power of expertise*. In particular, among these attributes, the negotiator's patience, knowledge of contract agreements, and honesty and politeness are known to enhance the effectiveness of the negotiation through the greater power of expertise (Min and Galle, 1993).

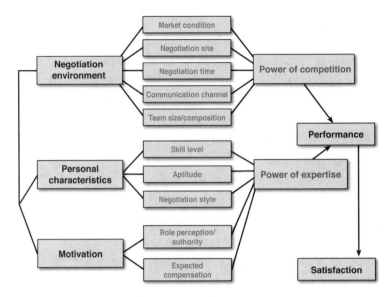

Figure 8.8. *Potential factors influencing the negotiation outcome*

Global Sourcing

In a broad sense, global sourcing refers to the practice of buying goods and services outside the geopolitical boundary of the country. Global sourcing first took off in the 1980s and has steadily increased over the last three decades as the supply base across the globe gradually expanded. The emergence of global sourcing also parallels with the evolution of business strategy over time, as shown in Figure 8.9. As such, the rationale and the focus of global sourcing should also change over time, as the priority of business strategy shifts. Regardless, global sourcing is often motivated by the following types of leverage that the buying firm tries to take advantage of:

- **Cost savings**—Wages and material costs are often much lower in developing countries than those in developed countries due to lower costs of living in the developing countries. Although developing countries will not always equate to low-cost countries (LCC), it is common that the buying firm can find more price bargains in developing countries such as China, India, Mexico, and Philippines than those in developed countries.
- **Higher quality**—Certain countries have a long tradition of producing specialized goods or materials better than other countries due to accumulated know-how and skills. For example, Brazil is known for its high-quality leather products, whereas Korea is known for its high-quality computer chips. Therefore, a buyer who seeks to buy such products may have a better chance of finding them in those

foreign countries than on his or her domestic soil.

- **Flexibility**—Even if the buying firm has a reliable group of domestic suppliers to choose from for future sourcing, broadening the supply base globally increases competition among the potential suppliers and thus gives the buying firm an opportunity to obtain the better deal. Also, in case of unexpected supply disruptions in the domestic region, the addition of foreign suppliers to the supply base provides protection from supply risks.
- **Declining trade barriers**—The Reciprocal Tax Agreements Act of 1934 began the U.S. trend toward lower tariffs and freer trade, a trend that continues to this day. Similarly, all across the globe, the formation of the World Trade Organization (WTO) and a series of free trade movements increased the incentive to source globally. For example, the U.S. has a relatively low duty rate and does not impose a value-added tax (VAT) on imported goods from other countries. Therefore, global sourcing is more popular among many U.S. firms (Assaf et al., 2006).
- **Compliance with local content regulations**—Local content regulations require multinational firms that do business in foreign countries to increase the content/percentage of locally produced goods for their purchases. To stay in business and maintain larger customer bases in foreign countries, these firms need to source from those host countries where they actively sell and promote their goods and services to local customers. Though not critical, a compliance with local content regulations may pressure the multinational firms to increase global sourcing (Monczka and Giunipero, 1990).

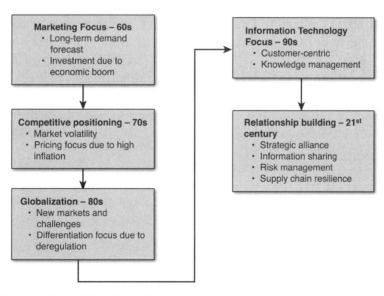

Figure 8.9. *The evolution of business-level strategy*

Before jumping onto the bandwagon of global sourcing, the purchasing manager has to understand its potential drawbacks that can easily undermine its benefits. These drawbacks include the following:

- **Logistical challenges**—Global sourcing often involves a lengthy transportation process due to long distances imported goods have to travel and multiple transfers of those goods from one mode of transportation to another (e.g., air to truck). In addition, the process of going through the customs office can further lengthen delivery time. In particular, goods that are time sensitive, fashionable, seasonal, and perishable may lose their value as a result of ill-timed delivery. This concern was echoed by the purchasing professionals who were surveyed by Min and Galle (1991). They indicated that transportation delay was the biggest obstacle to global sourcing.

- **Communication difficulty**—Although English is frequently spoken as the universal language for international commercial transactions, there are many non-English-speaking countries where those transactions will take place. For example, with an exception of Canada, the United Kingdom, and Hong Kong, the top sources of foreign supply for U.S. purchasing professionals include China, Japan, Mexico, Germany, Korea, and Taiwan (Min and Galle, 1991). Due to language barriers, negotiation with these foreign sources of supply may take much longer than domestic sourcing, and some of the contractual terms may be misinterpreted and misunderstood.

- **Hidden costs**—One of the common mistakes that many purchasing managers have made is their premise that cost of global sourcing is tantamount to the quoted price of the product plus delivery cost and customs duty. However, the actual landed cost of the imported product can be much higher than the anticipated cost of global sourcing due to other added costs, such as relationship costs, as shown in Figure 8.10. Part of this miscalculation stems from the difficulty in quantifying these costs and opaque business practices or government rules and regulations in foreign countries. For example, Min and Chen (2003) discovered that an import trade with China requires the payment of many hidden logistics-related fees, as follows:
    - Navigating fees
    - Trans-anchoring fee
    - Mooring/unmooring fee
    - Harbor fee
    - Groundage fee
    - Demurrage fee
    - Terminal handling charge
    - Freight/fuel surcharge

- Port charge
- Loading/unloading cost
- Insurance cost
- Additional cost due to required compliance with container security initiatives

In addition to these drawbacks, there are other challenges to deal with in global sourcing. These challenges include price instability, different quality/environmental standards, varying labor laws, difficulty in legal recourse and conflict resolution, complex payment terms, and so forth. Some of these issues will be discussed in Chapter 10, "Global Supply Chain Management."

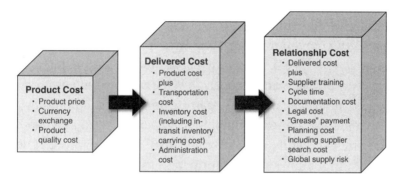

Figure 8.10. *Cost elements of global sourcing*

E-purchasing and Auctions

As advancement in electronic commerce (EC) continues to revolutionize today's business practices, it is apparent that electronic commerce has become an integral part of the business landscape. EC has generated a high level of interest because of its ability to expedite the purchase ordering process, simplify purchase payment, expand supplier bases, reduce paperwork, and eliminate order errors. Realizing such managerial benefits, a growing number of purchasing organizations are exploring the possibility of utilizing e-purchasing (Min and Galle, 2003). E-purchasing generally refers to paperless purchasing practices that utilize an inter-organizational information system intended to facilitate business-to business electronic communication, information exchange, and transaction support through a web of either public access or private value-added networks (VANs). This inter-organizational information system includes Electronic Data Interchange (EDI), direct link-up with suppliers, Internet, intranet, extranet, XML (Extensible Markup Language), electronic catalogue ordering, and online collaboration with groupware and email (Min and Galle, 2001). Among these, EDI, Internet, and

XML are the popular means of e-purchasing. EDI is intended to facilitate computer-to-computer exchange of data among multiple business trading partners in standard, machine-readable formats through a web of business-to-business communication networks (Min, 2000). For instance, Ford, Chrysler, Walmart, RJR Nabisco, and Black & Decker electronically generate purchasing documents such as requests for quote (RFQ), purchase requisitions, purchase orders, and invoices through EDI and have reported to save a tremendous amount of time and cost during the pre-ordering and ordering processes. Also, EDI can support purchasing bids. For example, the Union Pacific Railroad encouraged its bidders to use EDI to receive requests for bids and quotations and send invoices. In a broader perspective, EDI plays three important roles in supply chain management: electronic integration, information diffusion and sharing, and electronic marketplaces (Min, 2000).

Electronic integration involves integrating two or more independent organizational units or business processes by exploiting the capabilities of computers and communication technologies (Venkatraman and Zaheer, 1990). Increased integration of efforts across supply chains, in turn, enables EDI-participating firms to compress time to market. Also, electronic integration formed by EDI can help reduce manufacturing costs and enhance product quality by involving suppliers in the product design and development process. EDI allows more information to be communicated across the supply chain in the same amount of time through faster electronic transmission of data. Thus, EDI is likely to reduce information search cost associated with the acquisition of information about product offerings and market prices, thereby expediting information diffusion. EDI linkages also increase interdependence across the supply chain and subsequently require sustainable long-term relationships and risk sharing (Min, 2000).

EDI can serve as an intermediary connecting the buyer and the supplier in a vertical supply chain, thus creating an electronic marketplace where both buyers and suppliers can get access to market information in a more efficient and timely fashion and subsequently making comparison shopping much easier (e.g., Bakos, 1991). With abundant market information available from the electronic marketplace, a purchasing manager can compare a fairly large number of product offerings in a short period of time by using EDI.

XML is a technology that transmits data in flexible formats. Whereas EDI uses rigid, predefined data sets to transmit data, XML can be used for communication and electronic transmission using a wide range of different kinds of data and related data-processing instructions among the computers. Also, XML can facilitate computer-user interfaces through web browsers (Hugos, 2003). Although both EDI and XML provide more data security during electronic transmission than the Internet, they require specific data standards for communication. The Internet, however, does not require such standards and therefore can facilitate direct communication between the buyer and the supplier without any data transformation. Also, the Internet is ubiquitous, open, and inexpensive to use. Despite these merits, the Internet is not without its shortcomings. For instance, the Internet is more vulnerable to purchase fraud than EDI or XML (Min and Galle, 1999).

To summarize, the benefits of e-purchasing are as follows:

- Reduction in paperwork due to paperless purchasing process
- Improved accuracy due to less human error
- Reduced lead time (order cycle time) due to faster transmission of data
- Reduction in administrative costs due to less amount of documentation
- Enhanced communication and information sharing
- Potential tax breaks due to online sales tax reprieve

And here are its potential shortcomings:

- Initial startup investment in computer equipment and software
- Resistance within the organization to changes in traditional purchasing practices
- Backup system requirements in case of computer downtime and failures
- "Grey" (ambiguous) legality of electronic purchase contracts

As electronic commerce continues to evolve, e-purchasing will be a wave of the future. One of the recent evolutions of e-purchasing includes e-auctions. In particular, an e-auction is frequently applied to the reverse auction situation where many suppliers are vying to obtain the purchase contract from a single buyer. In a reverse e-auction, the buyer is not obligated to award the purchase contract until at least one of the participating suppliers bids at the "reserve" (qualification) price or lower. Generally, a reverse e-auction is "an online, real-time auction between a buying organization and two or more invited suppliers, where suppliers can submit multiple bids during the time period of the auction, and where some degree of visibility exists among suppliers regarding the actions of their competitors" (Carter et al., 2004).

The potential benefits of e-auction are as follows (http://wiki.answers.com/Q/Advantages_of_e-auction):

- Online tendering is a method of standardizing the procurement process.
- Preferred bidders are all contained within a single database.
- Bidders can be monitored in real time.
- Good control of bidders' submissions.
- Quick and easy comparison of bids.
- Confidence in the validity and integrity of contractual documentation.

- Reduction in paperwork, mailing cost, and photocopying.
- Ease of communication to multiple bidders.
- Audit trail for documentation.
- Secure bidding.
- The supplier is allowed to submit more than one bid.

Chapter Summary

The following are key lessons learned from this chapter:

- One of the first steps of sourcing is the recognition of the need for buying. This need cannot be realized without making a "make-or-buy" (or in-housing vs. outsourcing) decision. To make a sound make-or-buy decision, a purchasing manager should weigh the pros and cons of such a decision. If buying (or outsourcing) turns out to be more favorable to a company than making the product (in-housing), the company should trigger the outsourcing process.
- Outsourcing is a popular means of enhancing a company's flexibility and competitiveness because it allows the company to focus on what it does best (i.e., core competency) and thus better utilize its given resources. However, the scope of outsourcing may vary from one company to another depending on the company's range of business activities and level of outsourcing experiences.
- The effectiveness of sourcing can be gauged through the buying firm's ability to save purchasing costs. Cost-saving measures cannot be developed without knowing the sources of cost and then estimating the true cost of sourcing. Therefore, cost analysis is one of the most important responsibilities of the purchasing manager. In the sourcing context, the main purpose of cost analysis is to obtain the "lowest fair and reasonable" purchasing price from suppliers. Challenges of cost analysis are the misinterpretation of some costs (e.g., overhead expenses) due to the arbitrary breakdown of cost elements, a difficulty in estimating the present value of a product due to the incremental incurrence of costs over time, and a difficulty in incorporating the "learning curve," which affects the unit cost of production over time, into cost estimations. To overcome some of these challenges, alternative nontraditional costing methods such as activity-based costing and life cycle costing were introduced.
- Value analysis is one way of controlling costs by improving product/service quality and/or eliminating the unnecessary costs. In an effort to identify and eliminate the avoidable cost of doing business, value analysis often targets items or services (1) of high value or expense, (2) that are created through an overly complex process or outdated technology, (3) that are not standardized, (4) that

- are relatively new to the market, or (5) that are fashionable and have many optional features that customers may not necessarily want.
- Finding the "right" supplier(s) is one of the most important responsibilities of a purchasing manager. Once the potential candidates of suppliers are located through multiple information sources, including published industry sources, internal sources, professional sources, and international sources, the purchasing manager needs to figure out which supplier(s) is most appropriate for upcoming sourcing. The ideal profile of a right supplier may include a supplier (1) who is technically and financially sound; (2) who would openly share information about the product, its costs and quality, and service requirements with the buyer; (3) who is willing to participate in early supplier involvement in new product development/design; (4) who is flexible enough to adjust his or her company's capacity to changing demands, delivery schedules, and supply chain operations; and (5) who is oriented toward taking unnecessary costs out of the supply chain process.
- There are systematic ways to select a right supplier. These are the categorical method, weighted-point method, cost-ratio method, analytic hierarchy process, and multiple attribute utility theory.
- The development of a long-term partnership with the right suppliers is one of the greatest untapped frontiers in supply chain management, because it facilitates supply chain integration. Some high-profile firms such as General Electric, John Deere, Motorola, Chrysler, Toyota, Honda, and Hyundai began to leverage their supplier partnership as the competitive edge over their rivals. Building a supplier partnership may also entail early supplier development, supply base reduction, supplier certification, supplier co-location, and risk/gain sharing.
- The presence of risk is nothing new to sourcing, but its level of significance to the company's bottom line has been dramatically increased after a series of supply chain disruptions caused by 9/11 terrorist attacks, a Tsunami in Indonesia, the Katrina Hurricane in New Orleans, earthquakes, the subsequent nuclear disaster in Japan, and the like. With its increasing significance, supply chain risk (especially sourcing risk) has to be managed and controlled properly after careful planning. Supply chain risk management aims to safeguard the continuity of supply chain operations and maximize supply chain efficiency against unexpected supply chain disruptions. To make the supply chain more resilient against risk, the focal firm and its supply chain partners should identify supply chain risk and the most vulnerable points (including chokepoints) of the supply chain, develop recovery plans, and then redesign the existing supply chain by beefing up its weakest links.
- Once a particular sourcing need arises, purchasing contract details should be negotiated with the supplier. These details may include purchase price, delivery/freight terms, warranty terms and conditions, payment terms, liability/damage

control clauses, incentive arrangements, and escalation clauses. To obtain favorable deals through negotiation, the purchasing manager needs to enhance his or her bargaining power. This power is classified into power of competition and power of expertise. There are a multitude of factors influencing this power. One of those factors is the negotiator's aptitude. The desirable characteristics of an "effective" negotiator include (1) a habit of entering into negotiations with more demanding negotiation objectives than his or her counterparts, (2) a flexible mindset to deal with different negotiators using different negotiation styles (e.g., from "hardball" to "collaborative"), and (3) a persistence and patience even when the negotiation stalls.

- Global sourcing and e-purchasing represent the new wave of sourcing practices. Global sourcing, however, is fraught with risk and uncertainty. This uncertainty includes the difficulty in assessing hidden cost elements. To fully estimate the true cost of global sourcing, the purchasing manager should factor into both relationship and production costs.
- With the continued growth of e-commerce spurred by advances in information technology, e-purchasing has become a viable alternative to traditional, paper-based sourcing. E-purchasing relies on inter-organizational information systems such as EDI, XML, and the Internet.

Study Questions

The following are thought provoking questions that can be answered through class discussions and exercises:

1. What may be the biggest contribution of sourcing to supply chain management or the company's bottom line?
2. How would you justify your company's outsourcing decision?
3. What are the biggest obstacles for sustaining the productive outsourcing relationship with suppliers?
4. Explain the four stages of outsourcing.
5. How do you search for potential suppliers?
6. Do you have to use the cross-functional team to select the "right" supplier(s)?
7. How do you define the "right" supplier? What are the most important attributes of the "right" supplier?
8. How can you distinguish "problem" suppliers from "pre-qualified" suppliers?
9. Which cost components comprise the total cost of ownership?
10. Compare and contrast traditional costing versus activity-based costing.

11. How would you define value for your customers?
12. What benefits can value analysis bring to sourcing?
13. What are the most important prerequisites to the successful development of supplier partnerships?
14. Is supplier diversity detrimental to quality assurance at the source?
15. What are the pros and cons of direct sourcing? What are the pros and cons of mediated sourcing? How would you choose sourcing intermediaries?
16. Think of a casual restaurant franchise in your neighborhood that specializes in selling steaks. Consider its beef supply chain and then identify its supply chain links that are most vulnerable to the outbreak of mad cow and/or food-and-mouth disease by drawing a beef supply chain map.
17. How can you increase supply chain visibility to foresee the presence of potential supply risks?
18. Is competitive bidding and negotiation mutually exclusive? When is competitive bidding appropriate?
19. Explain the power of competition and describe the potential factors influencing the power of competition.
20. How can the purchasing manager improve his or her power of expertise?
21. What makes global sourcing differ from domestic sourcing?
22. Which historical events or business trends trigger a growth in global sourcing?
23. Can the worldwide financial crisis adversely affect global sourcing?
24. What are the potential drawbacks of "low-cost" country sourcing?
25. What can be considered the hidden cost elements of global sourcing?
26. What makes e-purchasing differ from traditional sourcing?
27. When is e-auction appropriate? What are the advantages of e-auction?

Case: Lucas Construction, Inc.

This case was originally written by Dr. Hokey Min based on an actual situation. It has been substantially modified to keep the identity of the company confidential.

A Need for Global Sourcing

Lucas Construction, Inc., is one of the largest construction and civil engineering companies in the world. Its headquarters are located in Atlanta, Georgia. As a leading construction company, it is involved in the construction of many logistics infrastructure such as railroads, highways, oil pipelines, bridges, shipyards, and airport terminals.

Under the new leadership of Steve Landrieu, the company started diversifying and expanding its business activities, including the construction of office buildings, warehouses, theaters, and schools throughout the world. Typically, Lucas Construction handles a project from its site planning and design to its construction and maintenance. Due to its do-it-all, one-stop service offering, Lucas Construction is responsible for sourcing all the construction materials such as concrete, plaster, tile, wire ropes and cables, structural steel beam, bricks, plumbing fixtures, stud, plywood, treated lumber, and so on. Its construction operation is a continuous, seven-days-a-week process and therefore its interruption can result in a serious delay of the project and substantial financial loss. That is to say, the timely acquisition of needed construction materials is a key to its business success. Lucas Construction's purchasing department is organized into four divisions:

- Commodity and operating materials
- Capital equipment (e.g., earth-moving equipment)
- Replacement parts and components for capital equipment
- MRO (maintenance, repair, and operating) supplies

In terms of purchase order activity, the commodity and operating materials division is the largest of the four.

Quality Failure at the Source

Lucas Construction recently won a bid for building a new 600,000 square-feet mega-warehouse for a fast-growing third-party logistics (3PL) company in a suburb of Atlanta in Fulton County. Based on the bill of materials (BOM) and blueprint drawings for the warehouse that were on file, Lucas Construction's purchasing department has a list of specific building materials required for the warehouse construction. One of the key building materials includes treated lumber. However, the purchasing manager, Sandy Wheeler, who is primarily responsible for buying treated lumber, noticed a substantial shortage of treated lumber due to booming construction activities along the Atlanta Beltline corridor. If treated lumber is available, its price is excessively high. So, Sandy believes that the local sourcing option is not a good idea. After consulting with her boss and colleagues, Sandy decided to buy treated lumber from a building material supplier called Pearl Trade Co., Ltd in Foshan, China. To avoid any construction delay, Sandy was in a hurry to order hundreds of lumber from Pearl, which promised on-time delivery of lumber and guaranteed its quality. When the ordered lumber finally arrived in the Port of Long Beach, Sandy was notified by the U.S. Customs Office that two-thirds of lumber shipped by Pearl were infested with insects and therefore the U.S. Customs Office would put a hold on its unloading. Sandy never thought that treated lumber could be attacked by insects. Sandy was further puzzled by the fact that the lumber sent by Pearl showed a clear mark indicating the IPPC (International Plant Protection Convention) certification. Knowing that the lumber should be sent to the job site within three weeks to avoid serious construction delays and subsequent penalties, Sandy was panicked and bewildered by this unexpected setback.

Urgent Plan B

Sandy picked up a phone in the middle of the night and called Pearl's salesman, who could barely speak English, and then threatened to take a legal action against Pearl unless replacement lumber could be shipped via air express in a matter of two weeks. Pearl's salesman sounded unfazed by Sandy's strong warning and claimed that insect infestation happened more than likely during long transit. Sensing the urgency, Sandy checked with the local supplier, Smokey Lumber, with whom she has done business in the past. Smokey was a reliable supplier, with a proven record of quality excellence, but quoted a price three times higher than Pearl's. This series of unexpected events has left Sandy speechless—she does not know what to say and do next. Now, her job may be on the line.

Discussion Questions

1. What is Sandy supposed to do now?
2. If you were Sandy, how would you select the overseas supplier? Do you think that the bidding should have been the right way to procure lumber in this sourcing situation? (Be as specific as possible and explain why bidding is appropriate for this sourcing decision.)
3. What kind of supply risk created Sandy's dilemma?
4. What kind of sourcing mistakes did Sandy make? (List at least three sourcing decision errors made by Sandy.)
5. How would you prevent the recurrence of this kind of sourcing failure?

Bibliography

APQC Consortium (2008), "Supplier Relationship Management: Collaboration for Win-Win Competitive Advantage," *Supply Chain Brain*, April 18, http://www.supplychainbrain.com/content/research-analysis/apqc/single-article-page/article/supplier-relationship-management-collaboration-for-win-win-advantage-1/, retrieved on June 15, 2011.

Assaf, M., Bonincontro, C., and Johnsen, S. (2006), *Global Sourcing & Purchasing Post 9/11: New Logistics Compliance Requirements and Best Practices*, Fort Lauderdale, FL: J. Ross Publishing.

Baily, P., Farmer, D., Jessop, D., and Jones, D. (2005), *Purchasing Principles and Management*, 9th edition, Harlow, England: Pearson Education.

Bakos, J.Y. (1991), "A Strategic Analysis of Electronic Marketplaces," *MIS Quarterly*, 15, 295–310.

Bosman, R. (2005), *The New Supply Chain Challenge: Risk Management in a Global Economy*, Unpublished White Paper, Johnston, RI: FM Global.

Burt, D., Petcavage, S., and Pinkerton, R. (2010), *Supply Management*, 8th edition, New York, NY: McGraw-Hill Irwin.

Business Week (2011), "Strength in Diversity," May 23, 2011, http://www.businessweek.com/adsections/2011/pdf/110523_SupplierDiversity.pdf, retrieved on January 15, 2012.

Carter, C.R., Kaufmann, L., Beall, S., Carter, P.L., Hendrick, T.E., and Petersen, K.J. (2004), "Reverse Auctions—Grounded Theory from the Buyer and Supplier Perspective," *Transportation Research Part E*, 40, 229–254.

Choi, T.Y. and Krause, D.R. (2006), "The Supply Base and Its Complexity: Implications for Transaction Costs, Risks, Responsiveness, and Innovation," *Journal of Operations Management*, 24(5), 637–652.

Cooper, R. and Kaplan, R.S. (1988), "Measure Costs Right: Make the Right Decisions," *Harvard Business Review*, 66(5), 96–103.

Cooper, R. and Slagmulder, R. (1977), *Target Costing and Value Engineering*, Portland, OR: Productivity Press.

Ellram, L.M. (1995a), "A Managerial Guideline for the Development and Implementation of Purchasing Partnerships," *International Journal of Purchasing and Materials Management*, 31(2), 9–16.

Ellram, L.M. (1995b), "Partnering Pitfalls and Success Factors," *International Journal of Purchasing and Materials Management*, 31(2), 35–44.

Ellram, L.M. (1990), "The Supplier Selection Decision in Strategic Partnerships," *Journal of Purchasing and Materials Management*, 26(4), 8–14.

ExactCost, Inc. (2011), "Traditional Costing vs. Activity-Based Methods," http://www.exactcost.com/Traditional_Costing_vs_Activity_Based_Methods, retrieved on June 5, 2011.

Forrest, E. (1996), *Activity-Based Management: A Comprehensive Implementation Guide*, New York, NY: McGraw-Hill.

Grieco, Jr., P.L., Gozzo, M.W., and Claunch, J.W. (1988), *Supplier Certification: Achieving Excellence*, Plantsville, CT: PT Publications, Inc.

Handfield, R.B., Blackhurst, J., Elkins, D., and Craighead, C.W. (2008), "A Framework for Reducing the Impact of Disruptions to the Supply Chain: Observations from Multiple Executives," in *Supply Chain Risk Management: Minimizing Disruptions in Global Sourcing*, edited by R.B. Handfield and K. McCormack, Boca Raton: Auerbach Publications, 29–49.

Hickman, T.K. and Hickman, Jr., W.M. (1992), *Global Purchasing: How to Buy Goods and Services in Foreign Markets*, Homewood, IL: Business One Irwin.

Hicks, D.T. (1999), *Activity-Based Costing: Making It Work for Small and Mid-Sized Companies*, 2nd edition, New York, NY: John Wiley & Sons, Inc.

Hoy, P. (2006), "Minority- and Women-Owned Businesses Skyrocket," May 1, *Inc.*, http://www.inc.com/news/articles/200605/census.html, retrieved on June 15, 2011.

Hugos, M. (2003), *Essentials of Supply Chain Management*, Hoboken, NJ: John Wiley & Sons, Inc.

Information Age (2006), "P&G to Outsource 80% of Back-office Functions," Information Age, February 9, 2006, http://www.information-age.com/industry/uk-industry/305321/p%26g-to-out-source-80-of-back-office-functions, retrieved on July 27, 2007.

Innes, J., Mitchell, F., and Yoshikawa, T. (1994), *Activity Costing for Engineers*, Somerset, England: Research Studies Press LTD.

Kaufman, J.J. (1990), *Value Engineering for the Practitioner*, Raleigh, NC: North Carolina State University.

Kraljic, P. (1983), "Purchasing Must Become Supply Management," *Harvard Business Review*, 61(5), 109–117.

Malone, R. (2006), "Growing Supply Chain Risks," www.Forbes.com, retrieved on May 5, 2010.

Maloni, M.J. and Benton, Jr., W.C. (2000), "Power Influences in the Supply Chain," *Journal of Business Logistics*, 21(1), 42–73.

Martha, J. and Subbakrishna, S. (2002), "Targeting a Just-In-Case Supply Chain for the Inevitable Next Disaster," *Supply Chain Management Review*, 6 (5), 18–23.

Mearig, T., Coffee, N., and Morgan, M. (1999), *Life Cycle Cost Analysis Handbook*, Juneau, Alaska: State of Alaska Department of Education & Early Development.

Min, H. (1994), "International Supplier Selection," *International Journal of Physical Distribution and Logistics Management*, 24(5), 24–33.

Min, H. (2000), "Electronic Data Interchange in Supply Chain Management," in *Encyclopedia of Manufacturing and Production Management*, edited by Paul Swamidass, Norwell, MA: Kluwer Academic Publishers, 177–183.

Min, H. (2009), "The Best-Practice Supplier Diversity Program at Caterpillar," *Supply Chain Management: An International Journal*, 14(3), 167–170.

Min, H. and Chen, G. (2003), "Challenges and Opportunities for Entering the Chinese Logistics Market." *Supply Chain Forum: An International Journal*, 4(2), 36–46.

Min, H. and Galle, W.P. (2001), "Electronic Commerce-based Purchasing: A Survey on the Perceptual Differences between Large and Small Organizations," *International Journal of Logistics: Research and Applications*, 4(1), 79–95.

Min, H. and Galle, W.P. (1999), "Electronic Commerce Usage in Business-to-Business Purchasing," *International Journal of Operations and Production Management*, 19(9), 909–921.

Min, H. and Galle, W.P. (2003), "E-Purchasing: Profiles of Adopters and Non-Adopters," *Industrial Marketing Management*, 32(3), 227–233.

Min, H. and Galle, W.P. (1993), "International Negotiation Strategy of U.S. Purchasing Professionals," *International Journal of Purchasing and Materials Management*, 29(3), 40–50.

Min, H. and Galle, W.P. (1991), "International Purchasing Strategies of Multinational U.S. Firms," *International Journal of Purchasing and Materials Management*, 27(3), 9-18.

Min, H., LaTour, M., and Jones, M. (1995), "Negotiation Outcome: The Impact of Initial Offer, Time, Gender, and Team Size," *Journal of Supply Chain Management*, 31(4), 19–24.

Min, H., LaTour, M., and Williams, A. (1994), "Positioning Against Foreign Supply Sources in an International Purchasing Environment." *Industrial Marketing Management*, 23(5), 371–382.

Monczka, R.M. and Giunipero, L.C. (1990), *Purchasing Internationally: Concepts and Principles*, Chelsea, MI: Book Crafters.

NetStride Internet Solutions (2010), *Hosting: In-House vs. Outsourcing*, Unpublished White Paper, New Town, PA: NetStride Internet Solutions.

Nydick, R.L. and Hill, R.P. (1992), "Using the Analytic Hierarchy Process to Structure the Supplier Selection Procedure," *International Journal of Purchasing and Materials Management*, 28(2), 31–36.

Ogden, J. (2003), "Supply Base Reduction within Supply Base Reduction," *Practix: Good Practices in Purchasing & Supply Management*, 6(1), 1–7.

Ostring, P. (2004), *Profit-Focused Supplier Management: How to Identify Risks and Recognize Opportunities*, New York, NY: AMACOM.

Pohlen, T.L. (1994), "Activity Based Costing for Warehouse Managers," *Warehousing Forum*, Vol. 9, No. 5, p. 1.

Sanders, N.R. and Locke, A. (2005), "Making Sense of Outsourcing," *Supply Chain Management Review*, 9(2), 38–45.

Scheuing, E. E. (1989), *Purchasing Management*, Englewood Cliffs, New Jersey: Prentice Hall, Inc.

Stuart, F.I. (1993), "Supplier Partnerships: Influencing Factors and Strategic Benefits," *International Journal of Purchasing and Materials Management*, 29(4), 22–18.

Sysque Quality (2011), "Value Analysis: Example," http://syque.com/quality_tools/toolbook/Value/example.htm, retrieved on June 8, 2011.

The Outsourcing Institute Membership Database (1998), *Survey of Current and Potential Outsourcing End-User*, http://www.outsourcing.com/content.asp?page=01b/articles/intelligence/oi_top_ten_survey.html, retrieved on September 9, 2009.

Timmerman, E. (1986), "An Approach to Vendor Performance Evaluation," *Journal of Purchasing and Materials Management*, 22(1), 2–8.

Tomlin, B. (2006), "On the Value of Mitigation and Contingency Strategies for Managing Supply Chain Disruption Risks," *Management Science*, 51(5), 639–657.

Venkatraman, N., and Zaheer, A. (1990), "Electronic Integration and Strategic Advantage: A Quasi-experimental Study in the Insurance Industry," *Information Systems Research*, 1(4), 377–393.

Verma, R. and Pullman, M. E. (1998), "An Analysis of the Supplier Selection Process," *Omega*, 26(6), 739–750.

Waters, D. (2007), *Supply Chain Risk Management: Vulnerability and Resilience in Logistics*, London, United Kingdom: Kogan Page Limited.

Zenz, G.J. (1994), *Purchasing and the Management of Materials*, 7th edition, New York, NY: John Wiley & Sons, Inc.

Zhang, Q., Vonderembse, M., and Lim, J. (2003), "Manufacturing Flexibility: Defining and Analyzing Relationships among Competence, Capability, and Customer Satisfaction," *Journal of Operations Management*, 21(2), 173–191.

9

Logistics Intermediaries

The good middleman does not lack goods.
—Sicilian Proverb

Learning Objectives

After reading this chapter, you should be able to:

- Understand the importance of logistics intermediaries to supply chain operations.
- Explain the specific roles of logistics intermediaries in the dynamically changing global marketplace.
- Understand the impact of the country's government policy and regulations on the role of logistics intermediaries.
- Comprehend the differences in service offerings among various types of logistics intermediaries.
- Identify a host of factors influencing the intermediary selection decision.
- Recognize the managerial benefits and shortcomings of hiring logistics intermediaries.
- Learn to exploit the emerging third-party logistics (3PL) industry across the globe and leverage their diversified services to enhance the firm's competitiveness in the global marketplace.
- Understand the evolution and future trends of 3PL industry.
- Learn to build long-term partnerships with logistics intermediaries including 3PLs.
- Learn to effectively handle potential role conflicts or contract disputes with logistics intermediaries.

The Role of Intermediaries

Some large firms have the resources, skills, and expertise to handle all kinds of supply chain activities, including logistics and purchasing, by themselves, whereas others do not. If the firm does not have such resources, skills, and expertise, it should no doubt outsource some of the supply chain activities by hiring intermediaries. Generally speaking, an *intermediary* is an individual or a business entity that facilitates business transactions between two or more trading partners as a mediator and is put in contact with those trading partners as a conduit for various supply chain activities. Examples of intermediaries are agents, brokers, wholesalers, distributors, third-party logistics providers (3PLs), and retailers, which help make a product or service available for the end customer's use or consumption. The use of these intermediaries is more common in international trade because they can help the multinational firm (MNF) navigate through uncharted territory in the foreign market and identify any potential business opportunities there with their connections and expertise in local sales and distribution. Sometimes the use of intermediaries is mandatory for business transactions under local jurisdictions (e.g., restriction on the number of expatriates to be hired). For example, some Middle Eastern countries such as Saudi Arabia require the foreign MNF to hire a local intermediary to do business in those countries.

Regardless of a local mandate or requirement, the use of intermediaries is essential for smooth logistics operations in the global setting. For instance, the international cargo-handling process at the seaport may require 100 different ancillary logistics activities that can be more effectively and efficiently performed by logistics intermediaries specializing in those activities (Wood et al., 2002). In general, logistics intermediaries are responsible for arranging and coordinating logistics activities throughout the supply chain for compensation. That is to say, logistics intermediaries allow their users to focus on their core competency and thus improve their competitive position in the market without committing any resources and efforts to logistics operations. Also, these intermediaries reduce direct contact lines between the customers and the producers and subsequently help avoid duplicated logistics efforts. Furthermore, because the intermediaries typically work with many different manufacturers and service providers, they can increase product assortments and service offerings to ultimate customers; thus, the use of intermediaries can improve customer services. From the customer service perspective, the users of intermediaries will be the eventual beneficiaries because the intermediaries can help their users to search and find customers by becoming vital links between the producer and the customer. Examples of frequently used logistics intermediaries include freight forwarders, overseas distributors, non-vessel-owning common carriers, export management companies, trading companies, export packers, customs brokers, overseas distributors, and third-party logistics providers. Details of these intermediaries will be discussed in the next section.

Types of Intermediaries

Reflecting the increasing complexity and diversity of logistics operations, there are many different types of logistics intermediaries. For example, some are asset-based intermediaries, whereas others are non-asset-based intermediaries. Some are domestically oriented, whereas others are internationally oriented. Some offer a narrow scope of services, such as freight forwarding or warehousing, whereas others offer more integrated services ranging from transportation to information technology development. Given the many different forms of intermediaries, the firm should carefully select the right intermediary that meets its specific needs and requirements. Before selecting the right intermediary, the firm should familiarize itself with various alternative intermediaries, as described next.

Freight Forwarder

The freight forwarder is one of the most frequently used forms of logistics intermediaries. Freight forwarders organize and arrange shipping activities encompassing the booking of cargo space, freight consolidation, documentation, insurance coverage, language translation, freight rate negotiation, and freight charge payment (Wood et al., 2002; Coyle et al., 2008). Some freight forwarders may focus on the arrangement of a particular mode of transportation, such as air or ocean carriers. For example, an international air freight forwarder will perform the following logistics services:

- Promote intermodal air-surface transportation.
- Solicit freight from shippers.
- Book air cargo space for shippers.
- Consolidate small shipments into a larger shipment to get freight-rate discounts.
- Provide pickup/delivery by surface transport means.
- Track shipments during transit.
- Utilize containers.

In addition, freight forwarders are responsible for ensuring supply chain security by stemming the flow of illegal exports and by helping to prevent weapons of mass destruction (WMD) and other sensitive goods and technologies from falling into the hands of proliferators and terrorists under the Export Administration Regulations, or EAR (U.S. Department of Commerce, 2010). In the United Kingdom, a freight forwarder is not licensed, whereas a U.S. freight forwarder involved with international ocean shipping is licensed by the Federal Maritime Commission (FTC). Also, U.S. freight forwarders that handle air freight will frequently be accredited with the International Air Transport Association (IATA) as a cargo agent (Wikipedia, 2010). The important benefits accrued from the use of a freight forwarder include cost savings resulting from favorable freight rates and time savings resulting from the professional handling of various ancillary services associated with global logistics operations.

Overseas Distributor

When selling products to an unfamiliar overseas market, the MNF can utilize an overseas distributor who purchases products from an original equipment manufacturer (OEM) as the middle man and then takes full responsibility for distributing and selling them to ultimate foreign customers. Thus, an overseas distributor can be considered a customer as well. The pros and cons of the overseas distributors are summarized here (Business Link, 2010):

- Advantages
  - An overseas distributor enables the OEM to access unfamiliar foreign markets while avoiding local logistics hassles and trade-related risks in the foreign market.
  - The overseas distributor usually handles the product shipment and the accompanying customs formalities and paperwork.
  - The OEM can leverage the overseas distributor's established reputation and contacts to enter a newly emerging foreign market.
  - The overseas distributor takes care of sales and marketing to promote products in the foreign market.
  - The overseas distributor often offers credit terms to potential customers.
  - The overseas distributor can carry inventories of the products and thus perform warehousing operations in the overseas market.
- Disadvantages
  - In return for taking supply chain risks and sales burdens in the foreign market, an overseas distributor often expects heavy discounts and generous credit terms from the OEM.
  - The OEM may lose control of local marketing and pricing.
  - An overseas distributor often demands a long period of exclusivity; therefore, the OEM can be locked up with the overseas distributor for a long time without many other alternative means of selling products overseas.

Non-Vessel-Operating Common Carrier (NVOCC)

A non-vessel-operating common carrier (NVOCC) is a modified form of a foreign freight forwarder that does not own or operate its own vessel; however, the NVOCC issues its own bills of lading or airway bills to provide a variety of (ocean) transportation services for point-to-point movement of goods. An NVOCC specializes in less-than-container load shipments and often utilizes containers rather than vehicles or vessels (Gourdin, 2001). It is sometimes called a shipment consolidator because it combines small shipments (partial container loads of goods) destined for the same location into a full container load. As of January of 2014, Top 100 NVOCCs in the U.S. handled 28% of the total U.S. imports (3PL News.com, 2014). However, NVOCC lost its clout due in part to its specific but small-scale operations, which were vulnerable for mergers and acquisitions in the shipping industry (Wood et al., 2002).

Shipping Agent

A shipping agent is a local licensed intermediary that arranges for the ship's arrival, berthing, and customs clearances, including imported goods inspection, insurance, loading/unloading, cargo claims settlement, cargo related document preparation/delivery, and the payment of all fees when the ship is in the port's dock on behalf of the ship's owner (Coyle et al., 2003; Maritimeknowhow, 2010). The duties of shipping agents may vary depending on their service specialties and categories such as port agents, liner agents, and owner agencies. Because of the diversities of duties, the shipping agent, within the framework of his or her responsibilities/competencies, often performs the tasks of other intermediaries such as chartering brokers (or cargo brokers) and booking agents (Maritimeknowhow, 2010).

Container Leasing Company

A container leasing company facilitates intermodal movement by relieving individual carriers of the financial burden and managerial responsibility associated with container equipment (Wood et al., 2002). Although a container provides an efficient and secure means of transportation, it can incur additional transportation costs because the container itself has to be shipped around. The transportation of empty containers for backhaul trips can be especially costly. Furthermore, the containers can get lost at sea in stormy weather. Considering such hassles, it is worth considering a container leasing company, which has the experience and expertise to reposition empty containers and recycle damaged containers. Some container leasing companies can offer ship finance services for container shipping and provide a lease/purchase option.

Customs (House) Broker

A customs (house) broker is an agent who clears shipments through the importing nation's customs, prepares and submits the documents necessary for customs clearance, estimates taxes and duties, pays the smallest applicable duty, and facilitates communication between the importer and the government. In the U.S., a customs broker is licensed by the U.S. Customs and Border Protection and regulated by the U.S. Treasury. Other ancillary services offered by the customs (house) broker include the following (Long, 2003):

- **Electronic documentation**—The customs broker allows the importer's carrier to pick up the cargo without any paper documents.
- **Immediate delivery**—Because the customs broker can clear customs before the shipment arrives, the importer can save lead time by employing the customs broker.
- **Door-to-door delivery service**—The customs broker can arrange for local delivery services.
- **Classification**—The customs broker can assist the importer in classifying products.

Export Packer

An export packer provides packaging (including moisture-resistant packaging) and protection services for all types of goods, including hazardous products. The export packer often works in conjunction with international freight forwarders for physically assembling export shipments and packing those for ocean-going containers (Wood and Johnson, 1996).

Export Management Company

An export management company (EMC) assists noncompeting firms in marketing their products overseas. EMC can be either local or foreign owned, and operates on either a commission or a fee basis with three to five exclusive contracts. An EMC can appoint sales representatives in importing countries, conduct market research, promote the goods and services of its clients, arrange for transportation and packing, prepare documentation and buy insurance for its clients, provide warranties and after-sales service, and extend the importer's credit (Wood et al., 2002; InvestorWords.com, 2010). An EMC functions like its client's external export department but primarily acts as an advisor to its clients.

Export Trading Company

The role of an export trading company (ETC) is very similar to that of an EMC in that both of them handle nearly all facets of export operations, including sales, marketing, promotion, documentation, transportation, warehousing, insurance, and communication. A key difference may be that an ETC takes the ownership of a cargo, whereas an EMC does not. It is also intriguing to note that an ETC can be owned and operated by the U.S. commercial banks under the Bank Export Services Act of 1982. Therefore, financing export operations through an ETC can be easier than through other intermediaries.

Third-party Logistics Service Provider (3PL)

One of the most dominant forms of logistics intermediaries is the third-party logistics service provider (3PL), whose industry grew rapidly in the late 1990s and the early 2000s. According to Armstrong and Associates (2009), more than three quarters (77%) of the Fortune 500 companies used 3PL services in 2009 and annual gross revenue of the 3PL industry reached $107.1 billion in 2009. Generally, 3PL refers to an intermediary that supplies, coordinates, and integrates multiple logistics functions, including transportation and warehousing across multiple links in the supply chain and acts as a "third-party" facilitator between seller/ manufacturer (the "first party") and the buyer/ customer (the "second party"; see Lieb and Maltz, 1995). Table 9.1 shows the number of different logistics functions that were frequently outsourced to 3PLs. The benefits that can be gained through the use of 3PLs include the following:

- Logistics cost reduction resulting from increased logistics efficiency provided by 3PL. Eyefortransport (2005) reported that 3PL users slashed their logistics cost by an average of more than 15%.
- Asset reduction and the subsequent improvement in cash flow due to the utilization of the 3PL's logistics assets such as warehouses, trucks, and airplanes.
- Order cycle time and cash-to-cash cycle time reduction thanks to streamlined logistics operations by 3PL.
- Inventory reduction resulting from reduced lead time.
- Overall customer service improvements.

The full realization of these benefits, however, may depend on the selection of the right 3PL or the right combination of multiple 3PLs among many available 3PLs all across the world. Some of the well-known, high profile 3PLs include the following: UPS Supply Chain Solutions, Ryder Integrated Logistics, FedEx Supply Chain Solutions, Schneider Logistics, TNT Logistics, Penske Logistics, DHL, Hub Group, and C.H. Robinson. Some of the factors influencing 3PL selection include years of service, financial stability, the range of service offerings, geographic coverage, pricing, gain-sharing provisions, past performances, and information technology support (Lieb, 2000).

Table 9.1. Logistics Activities Frequently Outsourced to 3PLs

Outsourced Logistics Activities	North America				Europe				Asia-Pacific			
	2006 (%)	2007 (%)	2008 (%)	2009 (%)	2006 (%)	2007 (%)	2008 (%)	2009 (%)	2006 (%)	2007 (%)	2008 (%)	2009 (%)
Domestic transportation	83	77	78	75	95	91	92	92	96	85	91	95
International transportation	83	68	69	70	95	87	89	91	95	89	89	91
Warehousing	74	71	70	71	76	68	73	72	77	73	75	65
Customs clearance and brokerage	71	65	66	73	59	58	57	61	83	78	81	78
Forwarding	55	51	48	61	54	51	44	57	66	60	70	82
Shipment consolidation	44	44	46	N/A	50	44	43	N/A	53	45	55	N/A
Reverse logistics	28	32	31	31	44	33	42	43	36	29	41	47
Cross-docking	36	36	37	40	40	35	43	42	30	26	35	42
Transportation planning and management	27	33	39	32	36	41	38	33	48	27	36	34
Product labeling, packaging, assembly, and kitting	26	31	29	33	45	33	42	42	33	34	37	40
Freight bill auditing and payment	55	51	54	53	22	18	20	24	18	14	21	26
Supply chain consultancy	21	18	21	21	16	11	15	19	16	11	14	25
Order entry, processing, and fulfillment	14	13	12	12	10	7	14	8	14	15	21	20
Fleet management	13	11	9	14	20	21	15	26	21	12	14	28
4PL/LLP services	12	13	11	10	13	11	13	12	6	10	14	17

Customer service	8	10	11	N/A	9	10	10	N/A	13	17	12	N/A
Information technology (IT) service	N/A	N/A	N/A	28	N/A	N/A	N/A	34	N/A	N/A	N/A	30

Data sources: Langley, Jr., J.C. (2009), 14th *Annual Third Party Logistics Study*, Atlanta, GA: Georgia Tech, Cap Gemini, Oracle, and Panalpina

Langley, Jr., J.C. (2008), 13th *Annual Third Party Logistics Study*, Atlanta, GA: Georgia Tech, Cap Gemini, Oracle, and DHL

Langley, Jr., J.C. (2007), 12th *Annual Third Party Logistics Study*, Atlanta, GA: Georgia Tech, Cap Gemini, SAP, and DHL

Langley, Jr., J.C. (2006), 11th *Annual Third Party Logistics Study*, Atlanta, GA: Georgia Tech, Cap Gemini, SAP, and DHL.

Fourth-party Logistics Service Provider (4PL)

A fourth-party logistics service provider (4PL) is often dubbed a lead logistics service provider that coordinates, manages, and integrates supply chain activities of multiple 3PLs hired by its client. As such, it focuses on synchronization of supply chain activities through the collaboration among multiple 3PLs as its supply chain partners. A 4PL is typically non-asset based and relies heavily on its intellectual capital and information technology (IT) capability to perform its functions. However, its role is constantly evolving because the industry is still in its infancy. Nowadays, some 3PLs and software developers claim to play the role of 4PL. Examples of 4PLs include consulting firms such as Accenture, Ernst Young, Deloitte, KMPG, AT Kearny, Theodore Wille Intertrade (TWI), Arup, Rollins, BMT Limited, MVA Consulting, the Boston Consulting Group, and PricewaterhouseCoopers. Unlike a typical 3PL, a 4PL intends to add value (e.g., profitability enhancement, operating cost reduction, capital reduction, customer service improvement) to its client's entire supply chain by maintaining the principle of neutrality (avoiding any conflicts of interest) for its clients.

Potential Challenges for Using Logistics Intermediaries

Although the use of logistics intermediaries can bring numerous managerial benefits, caution should be exercised before entering into a contract with an intermediary. More often than not, a firm that does not clearly understand its core competency, strategic mission, value propositions, and specific performance metrics will run into totally unexpected disasters resulting from the logistics outsourcing contract. For example, the office product retail giant Office Max abruptly terminated its logistics outsourcing contract with Ryder Integrated Logistics after less than two years of a seven-year contract and filed a $21-million lawsuit against the 3PL for a contract dispute stemming from failed agreed-upon services in June 1997 (PR Newswire, 1998). This lawsuit was later countered by Ryder with a $75-million lawsuit for wrongful termination of the contract. In the end, both companies agreed to an out-of-court settlement that offered a $5.1-million letter of credit to Ryder Integrated Logistics. However, this messy divorce between the 3PL user and provider is a reminder of what could happen to logistics outsourcing without careful planning. As this example illustrates, common mistakes among 3PL users are their failure to specify their performance expectations prior to contractual agreements with the 3PLs. In other words, they often fail to develop performance metrics and communicate those metrics to 3PLs under the premise that the 3PLs will instantly understand what their users want and expect. Prior to hiring an intermediary, the firm has some fundamental strategic questions to ask itself (e.g., Lynch, 2000):

- What logistics/supply chain problems should be solved?
- Will logistics outsourcing enable the company to concentrate better on its core competency?
- What results are expected out of logistics outsourcing?
- How does logistics outsourcing through the intermediary affect the company?
- Is the timing right?
- Does the company try to outsource logistics/supply chain activities it cannot manage efficiently or effectively?
- Can the company designate its manager, who can coordinate the outsourced function and become a key point of contact?
- Does the company have any escape clauses when there is a contract dispute?
- How should the intermediary's performance be monitored on a periodic basis?
- What are the potential risks associated with logistics outsourcing?

Once the intermediary is hired, the next most important challenge for the user is to sustain a productive outsourcing relationship. As an effective way to deal with this challenge, the user should consider the following guidelines—see Lynch (2000) and Macronimous (2010) for some tips for outsourcing relationships:

- Create a single point of contact between the user's and the provider's (intermediary's) key personnel to enhance communication regarding every aspect of outsourced operations. The types of communication for this purpose may include daily telephone conversation between the key personnel, periodic conference calls between senior management of both the user and the provider, and client visits to the intermediary's facilities.

- Harmonize the business goals and objectives of both the outsourcing service user and the provider to supply a meaningful mechanism for continuous improvement. This effort includes the establishment of quantifiable performance goals and objectives.
- Provide incentives and rewards while penalizing poor performance. Good performance should not be taken for granted, but should be recognized to motivate the intermediary. Also, make sure that recognition is properly directed to the unit or the individual who deserves the credit.
- Treat the intermediary as an extension of the user's own business. In other words, outsourced operations should be treated in the same manner as the user's in-house operations and then should be integrated with the user's own business operations. This effort may include the education and training of the intermediary's personnel.

These guidelines can be incorporated into the various steps needed for sustaining a productive outsourcing relationship, as shown in Figure 9.1.

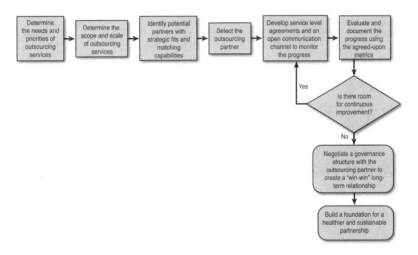

Figure 9.1. *Key steps for developing a successful outsourcing relationship*

3PL Market Trends

Parallel with the increased popularity of logistics outsourcing, the 3PL industry has been growing at a steady rate during a period of 1996 through 2013 (Armstrong, 2013). Although its growth temporarily slowed or slightly declined between 2009 and 2010 due to the economic downturn, 3PL has emerged as the most important and popular logistics intermediary in the past decade. As the 3PL industry has begun to mature, its competition recently intensified and consequently it was diversified with many different niche markets. In a broader sense, the 3PL market can be segmented into four different areas:

- Non-asset-based domestic transportation management
- Non-asset-based international transportation management
- Asset-based dedicated contract carriage
- Asset-based, valued-added warehousing/distribution service provider

3PLs that belong to the non-asset-based domestic transportation management category offer contractual transportation services such as valued-added shipments originating in and destined to the North American region in conjunction with freight brokerage. These 3PLs include C.H. Robinson, Hub Group, Landstar Logistics, Pacer, and Schneider Logistics. 3PLs that are categorized as non-asset-based international transportation managers primarily handle cross-border (intermodal) transportation in conjunction with freight forwarding. These 3PLs include DHL, EGL, Expeditors International, Kuehne & Nagle, and UTi. 3PLs that belong to asset-based dedicated contract carriers utilize their transportation assets such as tractors and drivers to offer long-term contract logistics services that usually last one to seven years. These 3PLs are Cardinal Logistics, J.B. Hunt Dedicated, Penske Logistics, Schneider Dedicated, Swift Dedicated, and Werner. 3PLs that are considered to be asset-based, value-added warehousing/distribution service providers utilize their warehouses/distribution centers to mainly provide long-term warehousing/distribution services. Examples of these 3PLs are: Americold, APL Logistics, Arnold Logistics, Cat Logistics, DSC, and Kenco (Armstrong and Associates, 2004).

In addition to these market segments, a couple of new market segments have emerged. These include financial service–based 3PLs such as Cass Information Systems, CTC (Commercial Traffic Corporation), and FleetBoston Financial Corporation, which focus on financial services ranging from freight payment and auditing to cost accounting/control. Another new segment was introduced by information-based 3PLs such as Transplace and Nistevo, which provide Internet-based, business-to-business, electronic logistics services without owning any assets (Coyle et al., 2008).

Regardless of the market segments and maturity of the 3PL industry, there is little doubt that the worldwide economic doldrum posed the greatest challenge for 3PLs. As a matter of fact, a 3PL market survey conducted by eyefortransport (2009) revealed that a majority (79% of the survey respondents) of 3PL users acknowledged the serious adverse impact of the economic slowdown on their 3PL usage. A similar pessimism prevailed among the majority (60%) of 3PLs, which expected either slow growth or the decline of the 3PL market in the near future. In addition to the economy, there are other challenges faced by 3PLs. As shown in Figure 9.2, here are some that require more attention from 3PLs:

- Supply chain forecasting due to increased supply chain risk and uncertainty
- Consumer/customer optimism with moving performance targets, which may increase the 3PL user's performance expectations
- Managing relationship that can be sustained between the 3PL user and the 3PL provider
- Fuel price volatility, which makes 3PL pricing unstable
- Capacity shortages and asset management challenges resulting from limited cash flows

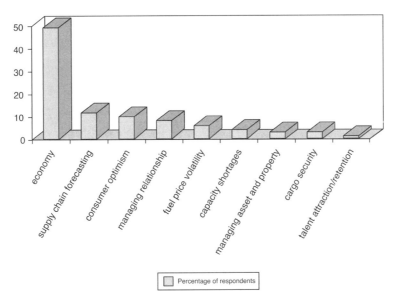

Figure 9.2. *Biggest challenges of 3PL services recognized by 3PL users (source: eyefortransport, 2009)*

Reflecting the intensified 3PL competition, it is also noted that the current 3PL market tilts toward the 3PL user's market. Therefore, 3PLs are losing their bargaining strengths for contract negotiation. This tendency also means that the 3PL users are less tolerant of their 3PLs' poor performance. As shown in Figure 9.3, poor performance is the biggest reason for 3PL contract nonrenewal. Although there are other reasons for contract nonrenewals, most of these reasons are tied to either service- or cost-related issues. In other words, today's 3PLs seem to receive greater pressure to improve their efficiency in terms of cost structures and service performance. Indeed, more than half (58%) of the surveyed 3PL users indicated that either cost or service is one of the main reasons why they switched their current 3PLs to another alternative (eyefortransport,

2009). Also, Min (2013) recently found that the two most important clauses in the 3PL contract were specifications of service standards and performance measures, including key performance metrics. To avoid disputes over the service performance, the 3PL industry should develop clearly defined performance metrics and understand which metrics are most important for improved 3PL services. According to a recent 3PL study conducted by Min (2013), the performance metric considered most important in setting the 3PL service standards is shipping accuracy. The next four most important metrics were on-time delivery, order accuracy, order fill rate, and frequency of customer complaints. On the other hand, asset turns, inventory turns, and cash-to-cash cycle time were considered relatively unimportant (see Table 9.2).

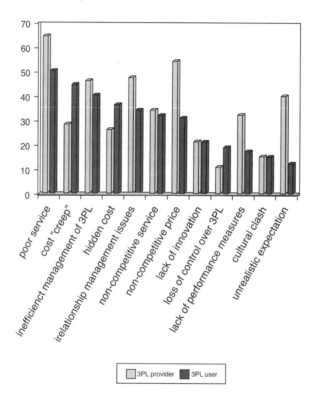

Figure 9.3. *Reasons for nonrenewal of 3PL contracts (source: eyefortransport, 2009)*

Table 9.2. Key Performance Metrics for 3PL Services (Source: Min, 2013)

Performance Metrics	Degree of Importance*	Rank
Shipping accuracy	4.73 (.486)	1
On-time delivery	4.70 (.502)	2
Order accuracy	4.52 (.603)	3

Order fill rate	4.33 (.695)	4
Frequency of customer complaints	4.25 (.775)	5
Order cycle time (lead time)	4.18 (.690)	6
Invoicing/billing accuracy	4.09 (.734)	7
Cost to serve/Cost of goods sold	4.07 (.843)	8
Responsiveness to inquiry	4.06 (.763)	9
Economic value added (EVA)	3.89 (.861)	10
Cash-to-cash cycle time	3.85 (.940)	11
Inventory turns	3.69 (.979)	12
Asset turns	3.61 (.979)	13

Note: Numbers in parentheses are standard deviations.

Scale: 5 = extremely important

4 = very important

3 = moderately important

2 = slightly unimportant

1 = not at all important

In light of increased cost and service pressures, the future of 3PLs lies in their ability to change their business model in such a way that their service offerings are more unique or innovative than their competitors, while enhancing their financial efficiency. For instance, J.B. Hunt improved its market position by streamlining its dedicated contract services and focusing on its niche market in full-truckload trucking services. Its success lies in its effort to retain its customers by improving its customer services through real-time shipment tracking systems. In particular, its deployment of the Terion Fleet View Trailer Tracking solutions greatly helped to utilize trailers and thus improved trailer revenue miles. In addition, its computerized yard management system allowed for more efficient utilization of yard space and subsequently reduced the frequency of yard checks and cargo theft loss. However, with the exception of J.B. Hunt, most of the asset-based 3PLs have struggled to enhance their financial efficiency as evidenced by the fact that asset-based 3PLs grossed their revenue three times less than their non-asset-based counterparts during the 2005 through 2007 (Min and Joo, 2009).

Overall, the recent trends of 3PLs can be summarized as follows:

- 3PLs' service offerings are getting broader and more comprehensive enough to include customized packaging, contract manufacturing, and final product assembly.

- The use of multiple 3PLs for a single company such as General Motors (GM), Ford Motors, Hewlett-Packard, and Walmart has become more common despite the broader service offerings of 3PLs. This trend means that the role of 4PLs and lead logistics providers will increase in the future.
- Many 3PLs have attempted to find their niches by focusing on a certain geographical region (e.g., North America, Europe, Asia-Pacific including China, Middle East), a certain industry (e.g., automotive, chemicals, consumer electronics), or a certain logistics function (e.g., reverse logistics, health-care logistics, freight bill auditing and payment) rather than trying to serve everyone.
- Although 3PL users tended to look for shorter-term contracts in uncertain economic times, 3PLs tended to seek longer-term contract arrangements for their stability. Although 3PLs are looking to establish a long-term partnership with their clients, Min (2013) noticed that a majority (73%) of 3PL users tended to have less than a three-year contract, as shown in Figure 9.4.
- While (especially large) 3PLs are more selective in choosing their clients (i.e., 3PL users), 3PL users have become more selective in choosing their 3PLs as well. Therefore, cultural fit and established rapport between 3PLs and their users will become even more important in the 3PL selection process (Lieb, 2000; Cooke, 2006; Lieb and Butner, 2007)
- Like other mature logistics industries such as railroads, the mergers/acquisitions of 3PLs will continue to take place in the more competitive 3PL market and to integrate 3PL services.

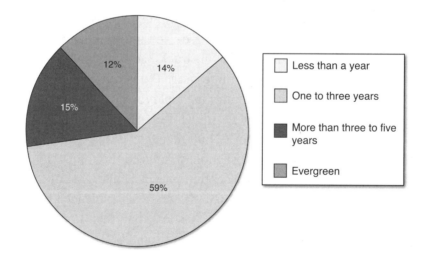

Figure 9.4. *The duration of the 3PL contract (source: Min, 2013)*

Chapter Summary

The following are key lessons learned from this chapter:

- With a strategic shift in today's companies to logistics outsourcing, the proper selection and use of logistics intermediaries are crucial for a company's business success. In particular, logistics intermediaries play a more significant role in global supply chain operations than in domestic operations due to the complexity and unfamiliarity associated with foreign distribution channels, business customs, and customer behaviors. Logistics intermediaries that are frequently used for essential parts of global supply chain operations include international (foreign) freight forwarders, overseas distributors, customs brokers, NVOCCs, export packers, export management companies, and export trading companies.

- Some logistics intermediaries are more frequently used in certain countries than other countries because of their history, scale of operations, and channel power. For instance, in Japan, an export trading company that is dubbed *sogo shosha* handles nearly every aspect of international trade, including cross-border transportation, banking, finance, insurance, warehousing, and sales/distribution in the foreign market. *Sogo shosha* such as Mitsubishi, Mitsui, Sumitomo, Marubeni, Nichimen, C. Itoh, Kanematsu-Gosho, and Nissho Iwai Corporation control about 10% of Japan's international trade by handling a range of 10,000 to 20,000 products, including food, clothing, automobiles, and appliances through a worldwide network (Simley, 2010). Similarly, the trading company plays a dominant role in facilitating international trade in South Korea.

- Third-party logistics providers (3PL) have become a dominant form of logistics intermediaries due to their quick adaptation to changing business environments that demand more "one-stop shopping" logistics services. In response to growing demand, many traditional trucking firms or warehousing companies have transformed themselves into 3PLs after broadening their logistics service offerings. However, the rapid growth of the 3PL industry has intensified competition and thus posed numerous challenges for 3PLs. Some of these challenges include increased cost and service pressures.

- To make logistics outsourcing through the intermediaries successful, companies should be aware of potential dangers associated with logistics outsourcing. Some of those dangers stem from a lack of strategic fit, incompatible business objectives, poor communication, and shaky business relationship between the logistics service user and the provider.

Study Questions

Following are thought provoking questions to answer through class discussions and exercises:

1. How would you justify logistics outsourcing (i.e., the use of logistics intermediaries) instead of in-housing logistics operations? Why are some companies reluctant to use a logistics intermediary such as a 3PL?
2. What are the most frequently outsourced logistics activities?
3. What are the different types of logistics intermediaries? Discuss each intermediary's role and explain the pros and cons of using each intermediary.
4. What are the most important decision factors for selecting the right intermediary (e.g., 3PL)?
5. What are the biggest challenges for using 3PLs and 4PLs?
6. What is the best way to develop a long-lasting, productive partnership with a 3PL?
7. Can some of the supply chain risks be transferred to 3PLs? If so, what are the elements of supply chain risks that can be transferred?
8. You are a supply chain executive working for the multinational U.S. firm that manufactures automotive parts and are currently pondering entering a new emerging market in India. Given your limited knowledge and experience with global supply chain operations, what rationale would you offer in support of the idea of using a 3PL? If you decide to use a 3PL, how would you select the right 3PL (or multiple 3PLs)? What specific attributes would you consider in selecting the 3PL(s)? Which specific clauses/terms should be included in the 3PL contract?

Case: Falcon Supply Chain Solutions

This case scenario was developed based on an actual situation, but was modified with fictitious data to disguise confidential information.

Global Presence

Falcon Supply Chain Solutions is a leading asset-based, third-party logistics service provider (3PL) headquartered in Cincinnati, Ohio, where more than 60% of the U.S. population can be reached within a one- or two-day driving distance. To serve customers in the West Coast and East Coast, it has recently established additional warehouses and terminals in Long Beach, California; Seattle, Washington; and Edison, New Jersey. With its latest information and communication technology (ICT), industry expertise, and growing customer bases across the world (especially the Asia-Pacific region), Falcon provides warehousing and distribution services, transportation management (including freight bill auditing/payment and freight forwarding services), customs/duty management, and value-added services (such as assembly, light manufacturing, kitting, sequencing, repairs, product return management, labeling, and packaging) primarily for office supply retail chains, including Office Depot, Office Max, and Staples. Because many of these customers' operations are based overseas due to the increased contract manufacturing and offshoring of their products, Falcon is often the sole U.S. presence for these customers. Falcon manages all receiving and shipping, responds to all customer inquiries, and handles the day-to-day inventory planning/warehouse operations on behalf of its customers. With a lean management philosophy to keep supply chain costs low, Falcon leverages the latest ICT, such as the Electronic Data Interchange (EDI), Extensible Markup Language (XML), and cloud computing to streamline its global supply chain operations and then improve communication with its clients and their supply chain (or trading) partners. Given the retail clients' preference for EDI or XML-capable 3PLs, Falcon's integrated EDI-XML system is one of the biggest selling points and core competencies.

Supply Chain Visibility through Information Technology

Exploiting its advanced ICT system, Falcon seeks to provide its customers with near "real-time" visibility to the shipment status of their orders and inventory levels throughout their global supply chains. Falcon, however, recently faced a tough choice of either increasing its customer support staff to handle incoming customer calls and e-mails regarding order, shipping, and inventory status at any time of day in different languages or implementing a new web-based solution that would provide globally based customers with the same information securely over the Internet. No matter their geographic location and time zone, Falcon's global customers would like to access inventory reports online, allowing them to view real-time inventory status, order status, shipment status, parcel-tracking information, and the like without revealing this kind of information to unauthorized individuals. In addition, they would like to transmit key documents (e.g., bills of lading, invoices, freight bills, receipts) electronically with the aid of Falcon's ICT system.

On some occasions, its customers asked Falcon to develop quick ICT solutions such as Enterprise Resource Planning (ERP) and radio frequency identification (RFID) devices in response to the sudden relocation of existing stores and opening of new store branches within a month or less. Such dynamic changes in the store location and establishment leave little or no time for Falcon to set up ICT solutions on-site. Regardless of its choice, adding ICT staff or installing a new ICT solution in a time of tight budgetary constraints can be very risky in that it not only defies Falcon's lean principle, but also limits its cash flow. More importantly, without clearly knowing the return on investment (ROI) of a new ICT solution, it would be difficult to gain the full support of the company's top management and board of directors. Another dilemma faced by Falcon is uncertainty and instability of its contract deals with its customers. In today's increasingly competitive 3PL market, a long-term 3PL contract exceeding a three-year commitment is rare and therefore many 3PLs may not recoup their new capital investment expenditures (including new ICT staff hiring and ICT infrastructure investment). On the other hand, Falcon is aware of the fact that its failure to meet the growing ICT needs of today's retail clients can lead to the loss of its 3PL accounts and a subsequent decline in market share and revenue. By offering online supply chain visibility and web-based EDI and XML services, Falcon may be able to deliver more streamlined logistics services to its customers while keeping its operational costs low and its productivity high. However, Falcon is wondering if this capability is sufficient enough to sustain its competitive advantage over its rivals.

Management Challenges

To deal with the aforementioned dilemmas, Wendy McDonald, Executive V.P. of the Supply Chain Planning for Falcon, has assembled her management taskforce group, including managers from the departments of sales and marketing, information technology (IT), and logistics. One of the intriguing options brought up by these managers is the outsourcing of its ICT solutions. Wendy grabs her pad of yellow sticky notes and then starts jotting down the potential pros and cons of ICT outsourcing after her taskforce group's most recent brainstorming session before she pondering it further. This is her list of discussion points:

Advantage of ICT Outsourcing	Disadvantages of ICT Outsourcing
Cost savings	Lack of control
Improved cash flow	Lack of flexibility
Quicker customer response	Potential loss of core competency
Immediate access to the latest technology	Difficulty in customizing services
Reduced risk of information technology obsolescence	Instability and lengthier learning curve for understanding customer needs

Because Wendy is unsure of the accuracy of this list and her managers continue to argue about the prospect of ICT outsourcing, she still gets confused and is undecided about Falcon's future course of action.

Discussion Questions

1. Which decision criteria have to be considered before making an ICT investment (in-housing) or outsourcing decision?
2. Which of the listed pros and cons of ICT outsourcing are inaccurate or misleading, if any?
3. What are the most important challenges (e.g., compatibility, security, affordability) of Falcon's global ICT operations? (List at least three potential challenges and explain how to cope with these challenges.)
4. What can be the right ICT solution that keeps Falcon's ICT overhead to a minimum and provides global access to Falcon's customers' order, shipment, and inventory data from any PC connected to the Web? (Be as specific as possible.)
5. How does Falcon's long-term relationship with its customers influence its ICT investment decision?
6. From a strategic perspective, how should Falcon protect its 3PL business from its competitors?

Bibliography

Armstrong, R. (2013), *Slow Dance - 2012 3PL Market Analysis and 2013 Predictions Report*, Unpublished Report, Stoughton, WI: Armstrong and Associates, Inc.

Armstrong and Associates (2009), *Trends in 3PL/Customer Relationships-2009*, Unpublished Report, Stoughton, WI: Armstrong and Associates, Inc.

Armstrong and Associates (2004), *Who's Who in Logistics: Armstrong's Guide to Global Supply Chain Management*, 12th edition, Stoughton, WI: Armstrong & Associates, Inc.

Business Link (2010), "Entering Overseas Market: Using an Overseas Distributor," http://www.businesslink.gov.uk/bdotg/action/detail?type=RESOURCES&itemId=1077658576, retrieved on August 6, 2010.

Cooke, J.A. (2006), "Sorry, You're Just, Not What We're Looking for," *DC Velocity*, December, 55–58.

Coyle, J.J., Bardi, E.J., and Langley, Jr., (2003), *The Management of Business Logistics: A Supply Chain Perspective*, 7th edition, Mason, OH: South-Western Thomson Learning.

Coyle, J.J., Langley, Jr., C.J., Gibson, B.J., Novack, R.A., and Bardi, E.J. (2008), *Supply Chain Management: A Logistics Perspective*, 8th edition, Mason, OH: South-Western Cengage Learning.

Eyefortransport (2005), *Outsourcing Logistics—The Latest Trend in Using 3PLs*, London, United Kingdom: FC Business Intelligence, Ltd.

Eyefortransport (2009), *The North American 3PL Market: A Brief Analysis of eyefortransport's recent survey*, Unpublished White Paper, United Kingdom: FC Business Intelligence, Ltd.

Gourdin, K.N. (2001), *Global Logistics Management: A Competitive Advantage for the New Millennium*, Oxford, United Kingdom: Blackwell Publishers Ltd.

Hergert, M. and Morris, D. (1988), "Trends in International Collaborative Agreements," in *Cooperative Strategies in International Business,* edited by F.J. Contractor and P. Lorange, Lexington, MA: Lexington Books.

InvestorWords.com (2010), "Export Management Company," http://www.investorwords.com/7281/export_management_company.html, retrieved on August 9, 2010.

Lieb, R.C. (2000), *Third Party Logistics: A Manager's Guide*, Houston, TX: JKL Publications.

Lieb, R.C. and Butner, K. (2007), "North-American Third-Party Logistics Industry in 2006: Provider CEO Perspective," *Transportation Journal*, 46(3), 40–52.

Lieb, R.C. and Maltz, A. (1995), "The Third Party Logistics Industry: Evolution, Drivers, and Prospects," Presented at the Annual Council of Logistics Management Conference, Chicago, IL.

Long, D. (2003), *International Logistics: Global Supply Chain Management*, Norwell, MA: Kluwer Academic Publishers.

Lynch, C.F. (2000), *Logistics Outsourcing – A Management Guide*, Oak Brook, IL: Council of Logistics Management.

Macronimous (2010), "Seven Tips for Successful Outsourcing Relationship," http://www.macronimous.com/resources/outsourcing_best_practices.htm, retrieved on August 14, 2010.

Maritimeknowhow.com (2010), "Persons and Businesses related to Shipping," http://www.maritimeknowhow.com/English/Know-How/Persons_and_Businesses_related_to_Shipping/shipping_agent.htm, retrieved on August 6, 2010.

Min, H. (2013), "Examining Logistics Outsourcing Practices in the United States: From the Perspectives of Third-party Logistics Service Users," *Logistics Research*, 6(4), 133–144.

Min, H. and Joo, S.J (2009), "Benchmarking Third-Party Logistics Providers Using Data Envelopment Analysis: An Update," *Benchmarking: An International Journal*, 16(5), 572–587.

PR Newswire (1998), "Office Max Sues Ryder Integrated Logistics in Federal Court," January 23, 1998, http://www.thefreelibrary.com/OfficeMax+Sues+Ryder+Integrated+Logistics+In+Federal+Court-a020184920, retrieved on October 11, 2008.

Simley, J. (2010), "Sogo Shosha," Reference for Business: Encyclopedia of Business, http://www.referenceforbusiness.com/encyclopedia/Sel-Str/Sogo-Shosha.html, retrieved on August 18, 2010.

3PLNews.com (2014), "3PL News NVO Update - Ranking of Top 100 NVOs in January 2014 by TEUs," February 20, 2014, http://www.3plnews.com/ocean-freight/3pl-news-nvo-update-ranking-of-top-100-nvos-in-january-2014-by-teus.html, retrieved on March 20, 2015.

U.S. Department of Commerce (2010), *Freight Forwarder Guidance*, Washington DC: The Bureau of Industry and Security, U.S. Department of Commerce, http://www.bis.doc.gov/complianceandenforcement/freightforwarderguidance.htm, retrieved on August 4, 2010.

Wikipedia (2010), "Freight Forwarder Roles in Different Countries," http://en.wikipedia.org/wiki/Freight_forwarder, retrieved on August 4, 2010.

Wood, D.F., Barone, A.P., Murphy, P.R., and Wardlow, D.L. (2002), *International Logistics*, 2nd edition, New York, NY: AMACOM.

Wood, D.F. and Johnson, J.C. (1996), *Contemporary Transportation*, 5th edition, Upper Saddle River, NJ: Prentice-Hall, Inc.

10

Global Supply Chain Management

"Globalization has changed us into a company that searches the world, not just to sell or source, but to find intellectual capital—the world's best talents and greatest ideas."
—Jack Welch, former CEO of General Electric

Learning Objectives

After reading this chapter, you should be able to:

- Understand the motivation behind the globalization of business activities.
- Understand the rationale and fundamental principles of free trade movements.
- Analyze the impact of free trade movements on global supply chain management.
- Comprehend the differences between domestic supply chain management and global supply chain management.
- Recognize the managerial benefits and potential challenges of global supply chain practices.
- Identify facilitators and inhibitors significantly influencing global supply chain practices.
- Be aware of potential risks involved in global supply chain management and learn to formulate viable business strategies to make the global supply chain more resilient.
- Understand the necessary changes and transformations required for the successful implementation of global supply chain strategies.
- Learn to prepare for import/export documentation and other global logistics-related paperwork.
- Learn to handle potential conflicts or contract disputes caused by language barriers, legal/ethical/cultural differences, varying business customs, and unequal quality standards.

The Impact of the Free Trade Movement on Global Supply Chain Management

Globalization continues to shape today's business landscape by presenting countless supply chain challenges and opportunities. Those challenges include the increased complexity and uncertainty created by varying national/organizational cultures, socio-economic conditions, national policies, currency fluctuations, communication protocols, quality standards, technical/ethical/legal practices, and business customs. For example, quality circles that worked well in Japan might not work at all in England due to differences in management styles and labor rules. Also, imposing high quality standards on the organizational setting in developing countries such as China can be a far more daunting task than doing so in developed countries such as Canada and Germany. On the other hand, globalization poses numerous business opportunities by expanding the multinational firm's customer and supplier bases. For example, many U.S. firms have actively sought to source their materials, parts, and components from overseas suppliers who have better access to cheap labor, stronger government support, and less-stringent environmental standards. Also, a company's increasing presence in the worldwide market will help it enhance its brand recognition and subsequently solidify its customer bases (Min, 2009). Reflecting this globalization trend, world merchandise exports grew by 3.5% in the years 2005 through 2012 (World Trade Organization, 2013). As a matter of fact, the volume of international trade has grown at an average rate of 6%, which doubled the 2.9% growth rate of world output from the years 1985 through 2008 (Loser, 2009). Indeed, between 1948 and 2010, the average international trade more than doubled (UNCTADSTAT, 2010).

The rapid growth of international trade has been fueled by increased opportunities for cost reduction, risk sharing, economies of scale, market expansion, new talent recruitment, global branding, and knowledge transfer. Generally, many nations engage in international trade for the opportunity to gain the following advantages:

- **Absolute advantage**—Countries will export whatever they do better than any other countries. A country has an absolute advantage in the production of a good relative to another country if it can produce the good at a *lower cost* or with *higher productivity*. The extent of an absolute advantage that the country has hinges on its level of industry productivity relative to other countries (Smith, 1776).
- **Comparative advantage**—A country will specialize in manufacturing those products for which it has the greatest comparative advantage and import those for which it has a comparative disadvantage. Countries trade whatever they do best. A country has a comparative advantage in the production of a good if it can

produce that good at a *lower opportunity cost* than other countries can (Richardo, 1817).
- **First mover advantage**—Because the cost of starting up a new venture is so great and the size of the market is so limited, those ventures that already exist in a certain country will hold a great advantage over any others when they enter emerging markets in foreign countries.

The preceding advantages cannot be fully exploited without free trade movement, which is a major catalyst for continuous growth in international trade. As such, free trade movement is inseparable from globalization of business activities and international trade. Free trade movement is a government policy that intends to minimize government interference in international business transactions (such as cross-border trade) by seeking to remove trade barriers, including tariffs, quota, and inequality. Although the root of free trade movement dates back to the nineteenth-century Victorian era, the idea of advocating the free flow of goods between countries was initiated by a conference in Geneva, Switzerland in 1947. This conference became a foundation for the General Agreement on Tariffs and Trade (GATT), which triggered a series of free trade movements across the world, such as the European Free Trade Association (EFTA) in 1958 and the Uruguay Round in 1993.

To elaborate, GATT is a treaty (a set of rules) created at the conclusion of World War II to quickly recover from the economic devastation caused by the war. GATT remained one of the focal features of international free trade agreements until it was replaced by the creation of the World Trade Organization (WTO) by former GATT members on January 1, 1995 (http://www.investopedia.com/terms/g/gatt.asp). GATT, however, is still in effect within the WTO framework. Its key objectives are to encourage nations to stop placing tariffs on foreign goods and to avoid preparing technical specifications (e.g., quotas and subsidies) that create international trade obstacles. GATT works under three principles (http://www.worldtradelaw.net/uragreements/gatt.pdf):

- **Reciprocity**—If one country lowers its tariffs against another's exports, it can expect the other country to do the same.
- **Nondiscrimination**—One country should not give one member (or group of members) preferential treatment over the other members of the group.
- **Transparency**—GATT encourages nations to avoid "ambiguous" non-tariff barriers, simplifies trade documentation procedures, and discourages government subsidies for exports.

GATT principles later became the ideological foundation for a number of regional free trade movements, including North American Free Trade Agreements (NAFTA) and European Union (EU). NAFTA was created in 1994 among the United States, Canada, and Mexico with the intent of gradually eliminating tariffs on goods that cross the three border countries in the form of the free trade association, which is illustrated in Figure 10.1. A free trade association is a fairly loose confederation of countries that have agreed to reduce trade barriers between them. Unlike NAFTA, EU is based on the framework of a common market, which goes a step further than a free trade association. A common market is an ultimate form of regional cooperation characterized by a reduction or even elimination of internal tariffs and common external tariffs and allows a free flow of labor and capital, as shown in Figure 10.2.

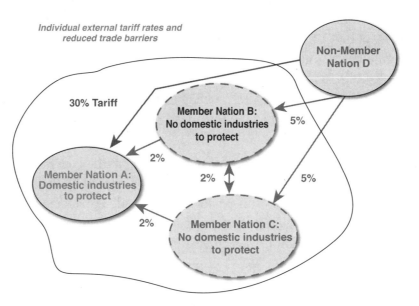

Figure 10.1. *Example of a free trade association*

The unprecedented growth of international trade fueled by free trade movements posed numerous supply chain challenges. These challenges include shipments of products to and from unfamiliar territories in foreign soils; acquisition of raw materials and components/parts from unknown sources of supply; lengthy lead time caused by a stretched global supply chain; selection of the optimal mode of transportation under different rules and regulations; selection of ports of entry and departure; utilization of foreign (free) trade zones and bonded warehouses; management of foreign labor and talents under varying labor rules; incompatible communication and information technology standards; establishment of uniform service and product quality standards across the multiple countries; development of global transportation networks; assurance of safe and secure cross-border transportation; assessment of potential supply chain risks in global environments; and resolution of international contract disputes and cross-cultural negotiations. Recognizing these challenges, supply chain strategies have to be altered in accordance with the unique needs and demands of globalization. Various concepts and ideas that may be useful for developing those strategies are discussed in detail in the following sections.

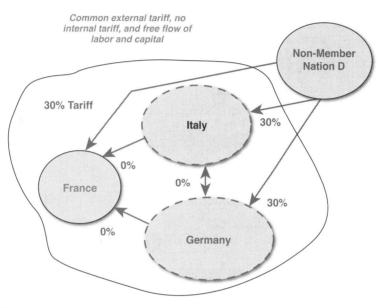

Figure 10.2. *Example of a common market*

Global Market Penetration Strategy of Multinational Firms

With widespread free trade movements all across the globe, most of the economies are increasingly open to foreign investors. For instance, in this open world economy, the U.S. invested approximately $3 trillion in foreign soils for the last five decades to capture the foreign market (http://en.wikipedia.org/wiki/Foreign_direct_investment). Foreign direct investment (FDI) is a viable business strategy for many multinational firms (MNFs), which would like to increase their customer bases, create global economies of scale, spread out their business risks, and establish their brands in the newly emerging global marketplace. Despite various benefits, the FDI decision can backfire without a careful formulation of foreign market penetration strategies. For example, more than one fourth of the licensed FDI inflows into Vietnam turned out to be failures partially because of poor infrastructure, lack of skilled labor, lax quality standards, and government controlled economy (Kokko et al., 2003). Winning the heart of foreign customers is somewhat analogous to the well-known scenario of popular *007* movies, as illustrated in Figure 10.3.

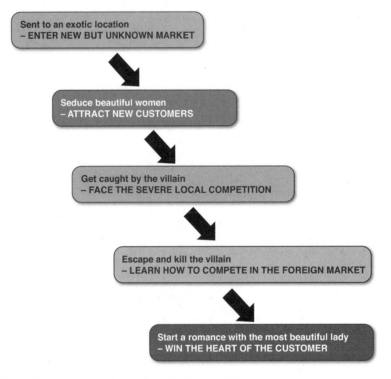

Figure 10.3. *The 007 analogy of winning the heart of foreign customers*

In general, foreign market entry strategies require a series of decision-making steps that will help MNFs bring about success in the unfamiliar global marketplace over a three- to five-year planning horizon (Root, 1994):

1. Assessing foreign markets and determining the suitability of products in those markets
2. Selecting foreign markets to target and developing/adapting the products suitable for those target markets
3. Setting business objectives and goals in those markets
4. Selecting the mode of proper market entry among the available options, such as exporting, joint venturing, establishing subsidiaries, and licensing

5. Designing specific marketing plans on pricing, sales and promotion, distribution, and post-sales follow-up
6. Controlling and monitoring market performance
7. Readjusting and revising market penetration strategies

Because the first two steps dictate the rest of the foreign market penetration strategies, a number of issues related to foreign market assessment/selection should be addressed. These issues may include the following:

- Who will be the key customer bases?
  - Who uses the product/service?
  - Who buys the product/service?
  - What are typical profiles of those who use or buy the product/service?
  - How are they willing to pay for the product/service?
  - How do they buy the product/service?
  - What is the customer's purchasing power?
- Who will be the key competitors in the target market?
  - What is the extent of competition in the target market?
  - How loyal are the current customers to the existing brands of the product/service?
  - How sensitive are the customers to price or promotion?
- What are the externalities?
  - What is the level of host government intervention in local business practices? (How free is the economy?)
  - How extensive are host government regulations?
  - What are typical customer sentiments toward foreign-brand products/services?
  - What are common traits of customer behaviors?
  - What is the current and future economic climate?
  - Is the host government supportive of foreign investments?

Once these issues are tackled, the MNF should carefully choose one of the specific market entry modes. Typical options of these modes are detailed in Table 10.1. Also, their pros and cons are summarized in Table 10.2. To elaborate, *exporting* is sending products/services out of a country of origin to sell them in other countries' markets. Exporting comes in two different forms: direct and indirect. Direct exporting can occur when products/services are shipped and sold to another country without the intervention of intermediaries. On the other hand, indirect exporting involves exporting products/services through various third parties, such as export management companies (EMC), which solicit and transact exporting activities on behalf of their clients (Capela, 2008). Indirect exporting makes sense for MNFs with limited global business experience and a risk-averse corporate culture.

As summarized in Tables 10.1 and 10.2, exporting regardless of its form is still riskier than other entry modes because it needs to overcome trade barriers and gives little chance for local adaptation of products/services. Unlike exporting, a *joint venture* allows the MNF to share the risks and costs with its partner. It also speeds up market penetration with the help of a local partner familiar with the local market conditions and distribution systems. A joint venture, however, limits the MNF's control over its foreign operations and increases the possibility of unwanted technological transfers to its foreign partner. *Licensing* is the extreme opposite of exporting because it does not require high capital investment and extensive marketing efforts in the targeted foreign market. Licensing, however, provides the MNF with little or no control over its manufacturing and marketing in the foreign market. Another entry mode similar to licensing is *franchising*, which gives foreign companies (franchisees) the right to use the original producer's brand names, logos, and operating techniques/know-how for a fee (royalty payment). An increasingly popular compromise between exporting and licensing is *contract manufacturing,* a viable entry mode when the MNF can enter an agreement with a local manufacturer to custom-produce its needed parts/components and finished products without making investments in its local manufacturing facilities and hiring/managing local workers. Thus, contract manufacturing is different from establishing a wholly owned subsidiary because the latter requires the full ownership of the foreign venture and the subsequent financial/operational responsibilities associated with foreign production, whereas the former does not.

Table 10.1. The Options of Foreign Market Entry Modes

Entry Mode	Exporting	Subsidiary	Joint Venture	Licensing
Location of Production	Domestic	Foreign	Foreign	Foreign
Foreign Firm's Involvement	None	Local sales force and intermediaries	Partner	Local marketing of patents, brand name, and trademarks by the licensee
Ownership	Domestic	Domestic	Shared ownership in agreed-on proportion	Joint ownership according to a contract
Capital Outlay	Low	High	Moderate	Low
Trade Barrier	High	Low	Low/none	None

Table 10.2. The Pros and Cons of Each Foreign Market Entry Mode

Entry Mode	Exporting	Subsidiary	Joint Venture	Licensing
Profit Potential	High	Excellent	Very good	3%–15%
Brand Recognition	Takes a long time to improve	Helps improve substantially	Helps improve substantially	Somewhat helps improve
Distribution Control	Yes	Yes	Partially/no	No
Price Control	Yes	Yes	Maybe/yes	No
Sales Promotion Control	Yes	Yes	Maybe/yes	No
Market Information	Limited	Abundant and detailed	Good amount	Fair amount
Expected Sales Volume	Low	Very high	High	High
Credit Risk	High	Moderate	Moderate	Very Low
Contractual Relationship	No (nonexistent)	Yes	Yes	Yes

Although the entry mode selection can have a profound impact on the effectiveness and efficiency of foreign market performance, it should be noted that the aforementioned foreign market entry mode can gradually change over time, depending on the MNF's experience/expertise in global supply chain operations, changes in the host country's socio-economic environment, and shifts in the host government's trade/economic policy.

Strategic Alliances among Multinational Firms

Considering the complexity and uncertainty involved in global supply chain operations, a growing number of MNFs seek potential business partnerships with foreign business ventures with the hopes of alleviating the business risks and financial burdens associated with globalized business activities. Such partnerships are often referred to as *global strategic alliances*. Put simply, global strategic alliances are international coalitions between two or more MNFs that pursue a set of agreed-upon goals by exploiting the partner's strengths and complementing the partner's given resources in the targeted global marketplace. The distinguished characteristics of global strategic alliances are as follows (Hergert and Morris, 1988):

- All partnering MNFs remain independent.
- All partnering MNFs share the benefits of allied relationships.

- All partnering MNFs contribute on a continuing basis in one or more key strategic areas, such as technology, products, distribution channels, intellectual properties, expertise, capital equipment, and funds.

The following are some motivating factors for entering into global strategic alliances:

- Gaining access to new customer bases in the targeted foreign market
- Utilizing the existing personnel or logistics network to reach new supplier bases
- Circumventing both trade and nontrade barriers to entering foreign markets posed by legal, regulatory, and political constraints
- Diversifying product lines that are tailored toward foreign customers
- Altering the technological base of global competition
- Accelerating the pace of research and development (R&D) with the help of foreign business partners
- Pooling resources in light of large outlays required in the expanded global marketplace
- Learning new technologies and skills and gaining know-how from foreign business partners
- Adding credibility to new products/services to be introduced in the foreign market through enhanced brand image
- Mitigating various risks associated with business practices and investments in unfamiliar foreign territories

Recognizing these benefit potentials of global strategic alliances, many MNFs have formed strategic alliances with their foreign partners. For example, in 2008, both Patheon and Solvias AG agreed to form a global strategic alliance to offer integrated development services to pharmaceutical and biotechnology companies. This alliance combined Patheon's experience and expertise in formulation development with Solvias AG's reputation of excellence in pre-formulation and solid state chemistry (Wheeler et al, 2008). Similarly, Trimeris entered into a strategic alliance with Roche in order to expedite the global development, approval, and commercialization of T-20 (Fuzeon) and T-1249. Roche brings the global clinical development capabilities needed to obtain marketing approvals in all major pharmaceutical markets to the collaboration. Additionally, Roche has a highly trained and experienced team of anti-HIV (Human Immunodeficiency Virus) pharmaceutical sales representatives and clinical support specialists. In return, Trimeris has granted exclusive marketing rights to Roche for the leading drug candidates in an important new class of anti-HIV agents. This alliance structure permitted Trimeris to retain a significant economic interest and a major strategic role in R&D, manufacturing, and commercialization of its fusion-inhibitor drugs.

More recently, diversified industrial manufacturer Eaton Corporation and Linde Hydraulics, a division of KION Group GmbH, entered into a global strategic alliance. This alliance strengthened product lines, distribution channels, and regional market coverage for both MNFs. Customers worldwide in construction, agriculture, vocational vehicles, civil engineering, processing, and other industrial markets can benefit from the expanded access to hydraulic system technologies and services through this alliance. Eaton can add Linde's comprehensive range of high-pressure piston pumps, motors and valves to its portfolio, while Linde can offer Eaton's medium-pressure piston products. In North, Central, and South America and in Asia Pacific, Eaton's Hydraulics business can supply Linde's Hydraulics products through Eaton's comprehensive sales, distribution, and services networks. Linde Hydraulics can continue to operate in these regions by providing application support and technical services. In Europe, the Middle East, and Africa, sales organizations and distributors for both MNFs continued to focus on their core businesses with the mutually expanded product lines (Eaton Corporation, 2010).

Despite these illustrated successes of global strategic alliances, a rushed decision to form a strategic alliance could lead to disastrous failure. As a matter of fact, the failure rate of strategic alliances was estimated to be as high as 70% (Kalmbach and Roussel, 1999). For instance, a once-promising global strategic alliance between the Korean auto manufacturer Daewoo and its U.S. partner General Motors (GM) turned sour after both parties experienced constant clashes of their corporate cultures and conflicting business philosophies (rapid business expansion versus conservative investment with a focus on cost-saving opportunities). Likewise, a global strategic alliance between Anamatic, UK's semiconductor manufacturer, and its Japanese counterpart turned out to be a failure due to their reluctance to share technological know-how and resources as a result of a lack of mutual trust between them (Elmuti and Kathawala, 2001). Generally speaking, there are three common causes for the failures of global strategic alliances, as displayed in Figure 10.4. These causes are national cultural differences between partnering MNFs, organizational incompatibility (e.g., differences in corporate culture between partnering MNFs, including differences in their corporate missions and strategic objectives as well as ambiguity in alliance relationships and roles), and technical incompatibility (e.g., different levels of advancement in technology, differing communication/information technology standards, differing data formats, and differing technological infrastructures).

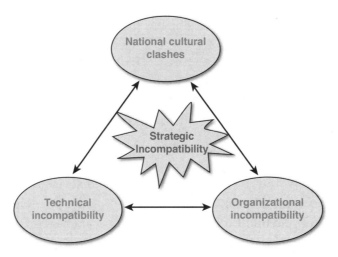

Figure 10.4. *The triads of global strategic alliance failures*

Global Outsourcing Trends

To focus on core competency and enhance competitive advantages in the rapidly changing global marketplace, a growing number of MNFs have sought for global outsourcing opportunities. These opportunities allow the MNF to better utilize its limited resources for what it can do best rather than diverting its efforts toward business activities that others in foreign countries can perform more efficiently. Generally, global outsourcing is an act of moving a firm's internal business functions and decision responsibilities to outside manufacturers and service providers in foreign countries. In other words, it is a way of increasing the firm's flexibility by subcontracting out noncore business tasks to foreign outsiders. For example, Ely Lilly, a major drug maker in the U.S., once slashed its $1.1 billion R&D costs to $800 million by subcontracting its chemistry lab experiments to Chem-Explorer Company in China. Likewise, a Penske trucking firm subcontracted out its back-office operations involving the driver log recordkeeping, billing, scheduling, invoicing, and fuel purchases to both Indian and Mexican suppliers and saved $15 million direct labor costs. As these examples illustrate, global outsourcing can bring a number of managerial benefits, including the following:

- Reduce business risks by passing them on to subcontractors
- Increase operational efficiencies with lower total cost through reduction of investments in noncritical assets
- Increase speed to market by working with a subcontractor with the expertise and capacity to bring new products and services to market quicker

- Enhance sales and promotional opportunity for a firm entering a new foreign market with the help of an experienced subcontractor who has already established itself in that market
- Allow the firm to focus on core competencies and build those focused world-class skills that can directly add value to foreign customers

According to the Outsourcing Institute Membership survey (1998), there are a number of reasons for global outsourcing. Among these, cost reduction and focus on the company's core competency are the two most frequently cited reasons. In addition to these two, there may be other reasons, such as the improvement of company focus, access to world-class capabilities and resources, and the free-up of resources for other business activities.

The importance of cost-saving opportunities to global outsourcing explains why many of the outsourcing destinations are low-cost countries such as China and India. As a matter of fact, the Global Outsourcing Survey conducted by A.T. Kearney (2009) reveals that China, India, and Malaysia represent the three most popular destinations for outsourcing. These countries gradually replaced some Eastern European countries such as Poland, the Czech Republic, and Hungary as outsourcing destinations for the last decade. In selecting the outsourcing destination, three factors usually came into play (A.T. Kearney, 2009):

- **Financial attractiveness**—Measured by compensation costs (e.g., wage), infrastructure costs (e.g., road conditions, electricity, water supply, telecommunication support), and taxes and regulatory costs
- **People score**—Measured by people's availability (e.g., working age populations), labor skills, educational levels, language competency, expertise in information and communication technology, work ethics, and attrition levels
- **Economic and political environments**—Assessed by economic freedom, infrastructure quality (e.g., quality of telecommunication and Internet networks), cultural compatibility, and intellectual property right protection

Despite numerous merits, global outsourcing should be planned with some caution because many global outsourcing efforts have failed. Indeed, *ABA Banking Journal* (2004) reported that more than half of global outsourcing initiatives failed in the past. One of the most common reasons for outsourcing failures is the firm's unclear expectations about its subcontractor's performance and a lack of adaptation of outsourced tasks to constantly changing business environments. Therefore, prior to entering into the outsourcing contract, the MNF should build a firm business relationship with its potential outsourcing partner (subcontractor) and then establish clear communication lines with the partner through a single point of contact.

Hidden Inhibitors Affecting Global Supply Chain Operations

Blindsided by the enormous benefit potentials of global business activities, MNFs may overlook a myriad of hidden risks involved in global supply chain operations. These risks often stem from the increased vulnerability of stretched supply chains across the globe, increased government interventions and surveillance in global supply chain activities due to growing concerns about terrorism, the global scale of the economic slowdown, communication difficulty, a lack of international quality standards, unforeseen natural disasters, volatile foreign currency fluctuations, rising terrorism and sea piracy, non-tariff barriers, a web of bureaucracy in customs procedures, multiple layers of distribution channels, indirect local taxation (e.g., value-added tax, withholding tax), and mounting paperwork. Because the ignorance of these risks can increase the cost of doing business globally, MNFs that want to engage in global supply chain activities should develop risk mitigation strategies. These strategies will not only make the global supply chain more resilient, but also help the MNF control its business expenditures associated with global supply chain operations. With this mind, the following subsections identify specific risk elements associated with global supply chain operations and discuss potential remedies for each one of those risk elements.

Effects of Foreign Exchange Fluctuations on Global Supply Chain Operations

International transactions trigger many different types of cash flows between trading partners (e.g., buyers and sellers, parent companies and their subsidiaries, licensors and licensees, manufacturers and distributors). These cash flows can be in the form of cash, dividends, royalties, and fees. Regardless of their form, unless international business transactions are made in U.S. dollars, the actual value of cash flow can be heavily affected by foreign exchange risks. Because this value will directly influence the MNF's profitability, the MNF should estimate foreign exchange risks and then hedge against such risks. Generally, these risks can be classified into three types: transaction exposure risk, translation exposure risk, and economic exposure risk. A *transaction exposure risk* arises when there is a change in foreign currency exchange rate between the time the contract agreement is made and the time it is paid in a foreign currency. A *translation exposure risk* occurs when subsidiary financial statements are consolidated at the parent company level for the company-wide financial reports. An *economic exposure risk* emanates from unexpected exchange rate fluctuations at the firm-wide level in the long run as opposed to the individual transaction level. Thus, this risk can affect the dollar value of the MNF's assets and liabilities (Ball et al., 2010). Figure 10.5 illustrates how the exchange rate works with the presence of economic exposure risk and how it impacts international trades. For example, let's suppose that the U.S. dollar depreciates relative to the Euro and thus its exchange rate changes from 0.7358 Euros per U.S. dollar to 0.7269 Euros per U.S. dollar. Therefore, the cost of goods to the U.S. buyer increases by 1.1%.

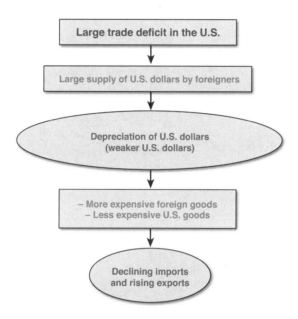

Figure 10.5. *An illustration of the way exchange rates work*

Given the profound impact of foreign currency rate fluctuations on the price of imported/exported goods, it is important for the MNF to predict the future exchange rate. However, the rate change is extremely difficult to predict due to a multitude of volatile factors influencing the exchange rate, as shown in Figure 10.6. In particular, individual factors such as the speculative behavior of the currency buyers and sellers in the FOREX (foreign exchange) market are hard to predict. Therefore, the forecast of the rate change alone cannot help the MNF prepare for the risk of foreign currency exchange fluctuations. To cope with the unpredictability and volatility of foreign currency exchange fluctuations, the MNF should carefully select the most appropriate option among the five different alternatives shown in Figure 10.7.

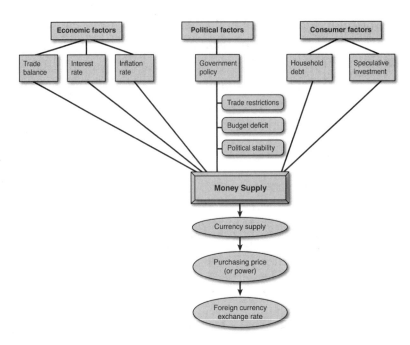

Figure 10.6. *Factors influencing the foreign currency exchange rate*

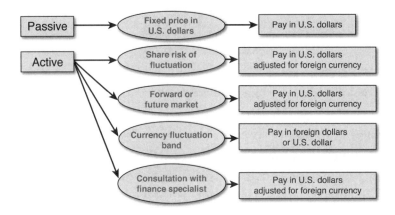

Figure 10.7. *Various options for foreign currency exchange fluctuations*

Effects of Terrorism on Global Logistics

The events following the terrorist attacks on the World Trade Center and Pentagon on September 11th of 2001 dramatically altered people's mindset regarding supply chain security. In particular, the fact that the primary means of terrorist attacks happened to be airplanes, which represent the mode of transportation essential to global logistics operations, only increased scrutiny for global logistics activities involving the use of transportation carriers filled with explosive fuel such as aircraft, oil tankers, freight trains, and tractor trailers. Such scrutiny makes global logistics operations far more challenging and costly than ever before. One of those challenges was posed by the increased security rules and regulations limiting global intermodal movements. For example, after the 9/11 events, the U.S. federal government reorganized the FAA (Federal Aviation Administration) and moved it under the newly created Department of Homeland Security (DHS). As a result, the FAA's name was changed to the Transportation Security Administration (TSA). This change means that transportation and supply chain security are inseparable concepts. Soon after its inception, the TSA introduced tighter air cargo security rules that forced all air cargo carriers to accept cargo only from shippers and freight forwarders with security programs and mandated criminal history background checks for workers handling cargo. Also, thorough screening is required for pilots and people traveling on all-cargo aircraft. These kinds of tougher air cargo security rules increase the carrier's compliance costs and lengthen its lead time, which has already been extended for cross-border shipments. To cope with these kinds of challenges, both the carriers and the shippers involved in cross-border shipments should understand the ramifications of influential security rules and then develop best-case management practices that can help them overcome the logistics inefficiencies created by a variety of supply chain security initiatives and rules/regulations.

Some of those initiatives and rules include the following:

- Resolution A. 924(2) of the International Maritime Organization (IMO), which calls for a review of existing international legal and technical measures to prevent and suppress terrorist acts against ships at sea and in port, as well as to improve security aboard and ashore
- International Ship and Port Security (ISPS) Code, created by the Safety of Life At Sea (SOLAS) convention
- Customs-Trade Partnership against Terrorism (C-TPAT)
- Container Security Initiatives
- Bio-Terrorism Act of 2002
- Security Trade in the APEC Region (STAR), enacted by the Association of the Southeast Asian Nations (ASEAN)
- Operation Safe Commerce (OSC)
- Free and Secure Trade (SAFE), involving the Canada Border Services Agency, Citizenship and Immigration Canada, and the U.S. Bureau of Customs and Border Protection (CBP)

Among these, the ISPS Code, C-TPAT, Container Security Initiatives, and SAFE are considered to be landmark security initiatives that continue to make profound impact on global logistics operations. As such, those initiatives deserve further discussion.

To elaborate, the ISPS is aimed at enhancing maritime security on board ships and at ship/port interface areas. It contains detailed maritime security measures–related requirements for government, port authorities, and shipping companies in a mandatory section (Part A), together with a series of guidelines about how to meet these requirements in a second, nonmandatory section (Part B). These requirements include approval of ship and port facility security plans; issuance of International Ship Security Certificates (ISSC) after verification; carrying out and approval of Port Facility Security Assessment; determination of port facilities that need to designate a Port Facility Security Officer; and an exercise of control and compliance measures such as port state control procedures (Branch, 2007). In addition, the ISPS code was designed to establish a framework for international cooperation among governments, their agencies, the shipping industry, and port authorities to detect security threats and take preventive measures against security incidents affecting maritime transportation by exchanging security-related information among them (Bragdon, 2008). Other related regulatory disciplines that complement the ISPS code are Contracting Government aka Flag States (e.g., U.S. Maritime Transportation Act of 2002) and Port State Control (e.g., Panama Maritime Authority).

Unlike the ISPS Code, which focuses on the maritime security, C-TPAT covers broader supply chain security issues because it intends to address the security procedures of all components of the global supply chain: production, transportation, storage, importation, and distribution (Assaf et al., 2006). C-TPAT is a voluntary government/business initiative that improves global supply chain security through cooperative relationships between the U.S. government and business entities, including manufacturers, freight carriers, and brokers that get involved in importing goods to the U.S. Though C-TPAT membership is not mandatory, its members can enjoy numerous benefits. For instance, C-TPAT members are less subject to customs inspections and will be routed to the front of customs lines when they do need to be inspected; thus, their customs procedures will be expedited. They can also take advantage of C-TPAT training for themselves and their employees, to learn more about ways to tighten the security of their global supply chain (http://www.wisegeek.com/what-is-c-tpat.htm).

To further beef up global supply chain security, the U.S. Customs launched the Container Security Initiative (CSI) in January 2002. Basically, CSI is intended to prevent containerized cargo destined to the U.S. from being exploited by terrorists by properly sealing and maintaining the shipping containers. The core elements of CSI are as follows:

- Use information technology (IT) to identify high-risk containers.
- Allow U.S. Customs inspectors to pre-screen high-risk cargo containers at the major foreign seaports (e.g., top 20 mega ports in the world) before they are shipped to the U.S. ports.
- Use detection technology to quickly pre-screen high-risk containers.
- Use smarter, tamper-proof containers.

The main benefits of CSI are the increased port security by detecting high-risk containers early, smooth flow of merchandise through seaports, and timely movement of merchandise and speedy customs procedures (http://www.globalsecurity.org/security/ops/csi.htm).

The FAST program is a joint Canada-U.S. supply chain security initiative that supports moving preapproved eligible goods across the Canada-U.S. border quickly and verifying trade compliance away from the border. Thus, this program increases speed for customs procedures and reduces the cost of compliance. These are its main benefits (Cook, 2008):

- Reduction in information requirements for customs clearance
- Elimination of the need for importers to transmit data for each transaction
- Dedicated lanes for fast clearance
- Reduction in the rate of border examinations
- Streamlined accounting and payment processes for all goods imported by approved importers (Canada only)

Impacts of Maritime Piracy on Global Shipping

Despite the wider media coverage of recent piracy attacks on U.S.-owned vessels in the Gulf of Aden, many people still have a romanticized image of lovable pirates swigging rum and plying the high seas in search of gold and treasure, as often portrayed by popular motion pictures. However, the reality is that modern-day pirates have been wreaking havoc on the global shipping industry for decades, which in turn has begun to disrupt global supply chain operations. As a matter of fact, the International Maritime Bureau (2007) reported a total of 1,552 worldwide piracy attacks between 2003 and 2007, and shows no sign of abatement. Actually, piracy incidents rose in 2008 by 11% from 2007 with an estimated $30 million loss due to ransom payment (O'Reilly, 2009). During the first six months of 2009, worldwide piracy attacks doubled. During this period, pirates boarded 78 vessels, fired on 75 vessels, hijacked 31, and ended up taking hostages of 561 crew members, among whom 19 were injured, six were killed, and eight are still missing. The greatest increase came from the Gulf of Aden and off the coast of Somalia, where no less than 130 attacks occurred. Ransom money paid in recent years in the Gulf of Aden alone was estimated to be as high as $100 million (Akiva, 2009).

Generally, maritime piracy refers to

(a) Any illegal acts of boarding a vessel for the purpose of committing a crime as defined by either an international treaty or a domestic law. These illegal acts are directed against:
   - Another ship on the high seas, or people or property on board such a ship
   - A ship, people, or property in a place outside the jurisdiction of any state
(b) Any act of voluntary participation in the operation of a ship with the intent to make it a pirate ship
(c) Any act of inciting or of intentionally facilitating an act described in subparagraph (a) or (b). (See Dillon, 2005, for legal definition of maritime piracy)

Considering the fact that as much as 90% of international commerce and trade involves ocean shipping, maritime piracy can have a devastating impact on global supply chain operations (Toth, 2009). Here are some examples of the piracy's impact (Min, 2011):

- Companies operating shipping vessels in the East Mediterranean and Black Sea region have an unpleasant choice to make: Use the Suez Canal and Gulf of Aden and subject the ship and its crew to the threat and dangers of piracy, or instead send the ship around the safer but far longer route through the South Atlantic Ocean and then around the Cape of Good Hope. The shorter, traditional route includes much higher expenses in the form of hijacking risk, kidnapping and ransom, and cargo insurance premiums. The far longer Cape route entails much higher fuel costs and other operational costs, in addition to transit times that may range from 17 to 26 days longer. If progressively greater numbers of vessels are forced to take the longer, safer Cape route, total shipping costs will balloon dramatically. If only one third of cargo that normally traverses the Suez Canal route was instead shipped via the Cape of Good Hope, the total additional shipping costs were estimated to be US$7.5 billion annually (De Monie, 2009).
- For ships traversing the traditional Suez Canal and Gulf of Aden route, insurance costs as a result of piracy risk increase as much as US$20,000–$25,000 per transit, and hiring a security guard for the trip might cost the carrier a maximum of up to $60,000 (De Monie, 2009).
- The Nigerian navy estimated that up to 200,000 barrels of crude oil—valued at around $200 billion annually—were lost daily as a result of piracy in 2013 (Starr, 2014).

- When the Japanese tanker *Takayama* was attacked on April 21, 2009, oil prices immediately spiked (Korteweg, 2009).
- Piracy in the gulf of Aden and the Red Sea, which ships must traverse to reach the Suez Canal, has already increased the maritime insurance risk premium for simply crossing the gulf from $500 per voyage in 2007 up to $20,000 per voyage in 2008. As such, the piracy threats could increase shipping insurance and transportation costs by $400 million annually, thereby increasing the consumer prices of goods and services worldwide (Sell, 2008).
- The cost of keeping global trade routes open could result in a growing "piracy tax" that would be felt by a wider range of businesses and consumers, already battered by the worldwide recession (Thomson Financial News, 2009).

Impacts of Natural Disasters on Global Supply Chain Operations

According to an executive survey about supply chain risks conducted by AMR Research, natural disasters was listed as one of the three leading risk factors causing supply chain disruptions (Hillman and Keltz, 2007). This survey finding was further supported by the fact that the early twenty-first century had been marred by an unprecedented level of natural disasters. Some well-known examples of these disasters include the 2004 Tsunami that hit ten countries and killed approximately 230,000 people in Southeast Asia and Africa; Hurricane Katrina in 2005, which created massive flooding in the Gulf Coast area (especially New Orleans); the 2008 earthquake that hit Sichuan, China; and the 2010 volcanic eruption at the Eyjafjalla Glacier in Southern Iceland. All of these disasters created devastating supply chain problems due to massive destruction of key manufacturing plants and logistics infrastructure in the affected regions and countries. For example, the earthquake in Sichuan almost paralyzed Dongfang Electric Corp., a power generator maker, whose main plant in the Sichuan province did not return to 80% of normal production for six months. Similarly, Intel, whose central processing chips serve as the "brains" for about 80% of the world's personal computers, shut a chip packaging factory in the Chinese city of Chengdu after the earthquake in Sichuan, which was centered about 100 km away from the epicenter of the quake. Considering the magnitude of adverse impacts of natural disasters on global supply chain operations, MNFs that engage in global supply chain operations should respond to the following questions:

- What risks (vulnerabilities) does the MNF face at each stage of the global supply chain?
  - How predictable are they?
  - What types of forecasts/intelligence need to be used to detect those risks?

- Is the supply chain in a high-risk area (e.g., earthquake epicenter, cyclone/hurricane path, flood zone, tornado alley)?
- What are the most critical assets to protect in the supply chain?
- What are the major sources of such risks at each stage of the global supply chain?
  - What would be the extent of the impact of those risks on the supply chain's performance?
  - Can the company transfer the risks? (Is it insured?)
- How can the company design or redesign its global supply chain processes in such a way that supply chain risks can be minimized?
  - How would customers respond to supply chain disruptions (e.g., shortages of materials/products)?
  - What are the repercussions of customers' responses?
  - How would competitors respond?
  - Could the brand and the company survive after an unexpected disaster?
  - Can MNFs competing in the global marketplace collaborate with each other and commit to knowledge/risk sharing in times of a natural disaster?

Managing International Distribution Channels

A distribution channel represents "Place" in the Four P's of the marketing mix. In other words, the distribution channel creates place utility by having products where the customer wants them and when the customer wants them. Given the long distance that products must travel in international trade, the selection of a particular distribution channel has a long-lasting impact on the MNF's success in the foreign market. Therefore, a proper distribution channel has to be carefully planned, developed, and managed in global supply chain operations. In simple terms, a *distribution channel* is a way of getting a company's products into the marketplace either directly or indirectly via intermediaries. The distribution channel can be classified into three types:

- **Direct distribution channel**—A company sells its product directly to customers without the involvement of middlemen. Examples include catalog sales through mail order, online sales via the Internet, door-to-door sales through selling agents, and sales at factory outlet malls. This channel is less costly because it avoids middleman markups on products, but it cannot create economies of scale due to limited sales volume to each customer.
- **Indirect distribution channel**—Utilizes various intermediaries such as distributors, wholesalers, brokers, dealers, export trading companies, third-party logistics providers, and retailers. This channel allows the manufacturer to

focus on producing goods, although it loses control over sales and distribution. Also, this channel can create the undesirable bullwhip effect and lengthen the company's market response time.

- **Hybrid distribution channel**—Uses multiple channels to sell the company's products to wider customer bases and thus increases potential revenue streams. For example, Starbucks sells its coffee at its franchised stores and grocery stores while also selling its coffee via direct mail. Likewise, Walmart sells its products at its worldwide retail stores while also selling its products on the Internet. This channel, however, can create channel conflicts between multiple sales outlets.

In addition to these three, there may be a grey market in international settings where products are sold outside the normal (authorized or originally intended) distribution channel when the price of a product in one country is significantly higher than one in another country. Unlike a black market, a grey market is still considered legal. For example, a foreign retailer may be able to acquire genuine Nike footwear in Malaysia for half the unit price of the same footwear in South Africa; the retailer can then import this merchandise and sell it for a profit in the local South African market, thus undercutting local retailers. Regardless, this practice is still legal in South Africa (Buys, 2009).

Among several different choices of the international distribution channel, the MNF should select the most appropriate channel based on the following decision criteria:

- **Marketing strategy**—If the MNF needs mass customer appeal and rapid market penetration, the MNF may choose an indirect channel that utilizes the existing distribution in the target market.
- **Type of product**—The sales of perishable products or nondurable goods with limited product life spans require faster customer responses and thus a direct distribution channel may make sense.
- **Market location**—Many foreign countries have their own unique distribution practices and therefore may limit the availability of a certain distribution channel. For example, in Japan, where its distribution practices have been traditionally conducted by wholesalers, the MNF may have no choice but to rely on an indirect distribution channel (Min, 1996).
- **Buying habits of customers**—If a majority of local customers in a target country tend to use the Internet to purchase products, a direct distribution channel may be an effective means of penetrating that market. On the other hand, if local customers prefer to shop at brick-and-mortar retail structures, an indirect distribution channel may be a more appealing option for that target market.
- **Years of experience in foreign supply chain operations**—Unless the MNF has gained considerable sales and marketing experience in the target foreign market, an indirect distribution channel may be more suitable for the MNF that attempts to enter a new market.

Foreign Trade Zones and Free Trade Zones

Despite worldwide free trade movements, the presence of a tariff alone can easily discourage an MNF from entering a foreign market. Because tariffs will not completely go away anytime soon, many governments that are eager to facilitate cross-border trade have established specially designated areas where customs duties can be postponed or exempted under certain conditions. Those areas include free trade zones (often dubbed "export processing zones" or "foreign trade zones" in the U.S. or "export distribution centers" in Canada) and bonded warehouses. These duty-free areas can be critical links in international logistics flows and the subsequent global supply chain. Therefore, to enhance global supply chain efficiency, the MNF should consider exploiting these areas more actively. In general, a free trade zone is a specially designated area within a country where normal trade barriers such as tariffs and quotas are removed and the bureaucratic necessities are reduced in order to attract new businesses, jobs, and foreign investments (EconomyWatch, 2010). The concept equivalent to a free trade zone is a foreign trade zone (FTZ) in the U.S. An FTZ is considered to be a duty-free area and is therefore not governed by the usual customs and tariff controls. In other words, merchandise permitted in the FTZ may be stored, sold, exhibited, labeled, repacked, assembled, distributed, and mixed with other merchandise without paying customs duty until the merchandise is released from the zone. Deferred customs duties in the zone can contribute significantly to the profitability of MNFs that get involved in global supply chain activities, because it can lower the supply chain costs associated with cross-border trade. In addition, FTZ status improves the cash flow of the MNFs utilizing it by waiving customs duties, taxes, and restrictions on goods until they actually enter either the U.S. or the foreign market (Min and Lambert, 2010). There are a number of other benefits that FTZ can bring (see Carver, 1999; Grant, 2004; Min and Lambert, 2010):

- "Inverted tariff" relief allows any FTZ importer or manufacturer to pay the duty rate applicable to either the imported components or the finished good itself—whichever is lower.
- Increased flexibility and expedited customs clearance through an FTZ facilitates just-in-time delivery.
- Tougher customs security requirements and federal criminal sanctions imposed by an FTZ can work as deterrents against product pilferage, which may lead to lower insurance costs and fewer incidents of cargo loss.
- FTZ users can avoid quota restrictions because they are allowed to store most merchandise until a quota is opened.

- Improved quality inspection at the FTZ site reduces the risk of quality failures for the FTZ manufacturers because only the products that meet the buyer's specifications will be imported through an FTZ.
- With an FTZ in place, imported goods are shipped "in bond" directly to the FTZ firm's local warehousing or manufacturing site, without going through time-consuming customs inspections and paperwork procedures required by the U.S. customs offices. Therefore, the use of an FTZ may reduce lead time and smooth out the product flow in the global supply chain.

Despite these numerous managerial benefits, the FTZ is one of the least utilized options among MNFs due in part to the misconception that the FTZ is merely intended for duty waivers and exemptions for goods exported from the FTZ. In particular, the FTZ is known to enhance competitive advantages by improving global supply chain efficiencies for high-value products such as automobiles, consumer electronics, and pharmacies (Min and Lambert, 2010).

Import and Export Documentation

One of the biggest hassles of global supply chain operations may be the preparation of various documents needed for imports and exports. These documents include shipping documents, administrative documents, commercial documents, sales documents, payment documents, security documents, and insurance documents, which can add up to well over 200 different documents to complete international transactions. For example, a single international shipment alone can generate 46 separate documents, with 360 copies consuming 46 man hours to process. In some cases, it requires as many as 158 documents with 790 copies (Mapp, 1978). Especially in many developing countries, the cost of necessary documentation could be one of the most expensive elements of international trade. For example, Indian exporters typically spent $350 for export documentation preparation, $120 for customs clearance, $150 for port and terminal handling, and $200 for inland transportation (World Bank Doing Business Study, www.doingbusiness.org). Unfortunately, documentation is an integral part of global supply chain operations and therefore cannot be done away with because it is essential for collecting necessary data and information regarding international trade, which has different shipping origins and destinations, different customs procedures, different government rules, different jurisdictions, and different currencies. These are some examples of common documents often required by international trade:

- **Application form for import/export registration**—Most countries require MNFs that engage in imports or exports to register with a government authority such as the Ministry of Commerce or the Registry of Companies. This form reveals the name and address of the MNF and its auditors, its key officers, and the business activities of the MNF. It will be the basis of a registration number unique to the MNF that applied for import/export activities. This registration number will be used in every international transaction as part of trade documentation.

- **Commercial invoice**—A commercial invoice serves as a bill for the goods from the importer to the exporter, while serving as evidence of a transaction. It contains information about terms of sale, a complete description of the product, country of origin, shipping destinations (i.e., name, address, and telephone number of a consignee), and the number of units, unit value, and total value of a shipment (Capela, 2008).

- **Consular invoice**—A consular invoice can be obtained from the consulate of the importing country at the point of shipment. It ensures that the exporter's trade papers are in order and the goods being shipped do not violate any laws or trade restrictions. It is also used as the basis for assessing *ad valorem* or specific import duties for goods (http://www.answers.com/topic/consular-invoice). Although some less developed countries may require this specific invoice, many importing countries have phased out this invoice.

- **Bill of lading (B/L)**—B/L functions as a contract between the owner of the goods (usually the exporter) and their carrier, a receipt from the carrier for the goods shipped, and a certificate of ownership (Ball et al., 2010). As such, B/L can be used as the evidence of contract of carriage and shipment.

- **Certificate of origin (CO)**—A certificate of origin traditionally states from what country the shipped goods originate. However, "originate" in a certificate of origin does not mean the country the goods are shipped from, but the country where the goods are actually made (http://en.wikipedia.org/wiki/Certificate_of_origin). It is often used for duty and import control purposes.

- **Inspection certificate**—An inspection certificate (or inspection report or report of findings) may be required by some importers when shipping certain items such as industrial equipment, meat products, and perishable merchandise. It verifies that the imported item meets required (quality) specifications, was in good condition, and was in the correct quantity prior to shipment through independent testing organizations (http://www.businessdictionary.com/definition/certificate-of-inspection.html).

In addition to these documents, other supporting documents commonly required for international trade include an insurance certificate, packing list, export license, quotations and pro forma invoice, shipper's letter of inspections, dock and warehouse receipts, and shipper's export declaration.

Incoterms and International Payments

Unlike domestic freight terms, international freight terms are compounded with the inclusion of terms of international sales/purchase under different jurisdictions and changing responsibilities associated with intermodal transfers. To ease this complexity and avoid confusion stemming from the misinterpretation of trade terms, the International Chamber of Commerce in Paris, France introduced the so-called Incoterms (International commercial terms) in 1936. After a series of revisions and recent simplifications, the most recent Incoterms came into effect on January 1st of 2011, with 11 different terms/codes that can be grouped into four broad categories: departure, main carriage unpaid, main carriage paid, and arrival. Notice that the five "D" acronyms (i.e., DAF, DES, DEQ, DDU, and DDP) used in the past were reduced to only three (DAP, DAT, and DDP) due to increased point-to-point sales and containerization. Specifically, the new term DAP (Delivered at Place) means that the seller pays all the costs of transporting goods to the final destination, including export fees, carriage, insurance, and destination port charges. The buyer is responsible for paying only the import duty/taxes/customs-processing costs. The buyer is also responsible for unloading the goods from the vehicle at the final destination (International Chamber of Commerce, 2010). Another new term, DAT (Delivered at Terminal), means that the seller covers all the costs of transportation and assumes all the risks until after the goods are unloaded at the terminal, which encompasses a quay, warehouse, container yard or road, or rail or air cargo terminal. The buyer covers the cost of transporting the goods from the terminal or port to the final destination and pays the import duty/taxes/customs-processing costs (Roos, 2011). Herein, the main carriage deals with the carriage from a terminal in the country of origin to a terminal in the country of destination (i.e., foreign country). Therefore, it does not include "pre-carriage," consisting of point-to-point carriage from the shipper's premises to the first carrier's terminal, and "on-carriage," consisting of a point-to-point carriage from the carrier's terminal to the consignee's (e.g., buyer's) premises. Basically, Incoterms comprise a set of uniform rules for the interpretation of international commercial (freight) terms defining the costs, risks, and obligations of buyers and sellers. In other words, Incoterms determine how the costs and risks of international shipments are allocated to each party involved in international trade. More specifically, Incoterms address the following questions:

- Which tasks will be performed by the exporter?
- Which tasks will be performed by the importer?
- Which activities will be paid by the exporter?
- Which activities will be paid by the importer?

- When will the transfer of responsibility (not necessarily title) between the exporter and the importer take place?

As such, parties involved in international trade should pay as much attention to the Incoterms as the price of the item when negotiating an international sales/purchase contract. A detailed breakdown of the Incoterms of 2010 is provided in Table 10.3 (Skymart, 2010). The selection of a particular Incoterms is usually based on the following factors:

- The type of product sold (cheap or expensive?). Some specific industries prefer using some specific terms of trade rather than others.
- The method of shipment (safe or unsafe?). Goods shipped by barge or ocean carriers may be sold under different Incoterms than containerized goods.
- The country of shipping origin and destination. For example, Poland, Romania, and India will not allow the exporter to purchase insurance abroad (e.g., a CIF term).
- The ability of either of the parties to perform the tasks involved in the shipment.
- The amount of trust placed by either of the parties toward the other.

Regardless of the specific terms, however, Incoterms explains what needs to be done in the following cases (http://www.iccwbo.org/incoterms/):

- Transfer of property rights in the goods
- Relief from obligations and exemptions from liability in the case of unexpected or unforeseeable events
- Consequences of various breaches of a contract, except those relating to the passing of risks and costs when the buyer is in breach of his or her obligations to accept goods or to nominate the carrier under a freight term

In addition, it should be noted that a seller (shipper) is obliged to pack and mark the goods in such a manner as is required for the transport and the extent of danger during transit. Regardless of the terms, the seller is often responsible for checking quality and conformity with the contract of sale, whereas the buyer (consignee) must take delivery of the goods when they have been placed at his or her disposal (Stapleton and Saulnier, 1999). Once Incoterms are settled, the next terms of sale/purchase that need to be agreed on by both the importer and the exporter are payment terms. These payment terms will dictate who will bear the financial and currency risks extant in international transactions. Here are the most common terms of payment:

- **Cash in advance**—Cash in advance is clearly the most desirable method of payment from the exporter's standpoint, because the exporter gets paid prior to shipment. On the other hand, it creates a cash flow problem for the importer. To make matters worse, the importer will not have a chance to receive and then inspect the ordered merchandise before payment and is therefore likely to refuse this kind of payment terms unless he or she made a small order.

- **Open account**—This can be seen as the total opposite of cash in advance in that the exporter sends the merchandise and the bill and then expects to get paid by the importer under the agreed-upon terms at a later date without the involvement of banks. Therefore, this payment term is very risky for the exporter, whose capital is tied up with the transaction, and legal recourses in case of payment disputes are very limited. Given a high credit risk for the exporter, this option may be feasible only when the importer has built a solid credit history and formed a long-term partnership with the exporter. This option may make sense for an international transaction between a parent company and its foreign subsidiary.

- **Sight draft (bill of exchange)**—A sight draft is similar in many respects to the personal check used in our daily transactions in that it is considered to be a "documentary collection" (not real money), which is more favorable to the importer (buyer). With this term, the exporter entrusts the collection of payment to the exporter's bank, which sends the necessary documents, such as a bill of lading, to the importer's bank along with payment instructions. Then, the payment received from the importer is remitted to the exporter through the bank involved in the collection in exchange for the documents (Capela, 2008).

- **Letter of credit (L/C)**—A letter of credit is issued by the importer's bank guaranteeing that the exporter will get paid, as long as the exporter presents the stipulated documents to his or her (advising) bank before or on the specified due date. With the involvement of both the importer and the exporter's banks, this option protects the interest of both parties and can be used for high-value international transactions (Monczka and Giunipero, 1990). Figure 10.8 also shows how the L/C process works in practice.

Table 10.3. Incoterms 2010 and Payment Responsibility

	EXW (Ex Works)	FCA (Free Carrier)	FAS (Free Alongside Ship)	FOB (Free on Board)	CFR (Cost and Freight)	CIF (Cost, Insurance & Freight)	CPT (Carriage Paid To)	CIP (Carriage & Insurance Paid to)	DAP (Delivered at Place) = Previously DAF (Delivered at Frontier)	DAP (Delivered at Place) = Previously DES (Delivered Ex Ship)	DAT (Delivered at Terminal) = Previously DEQ (Delivered Ex Quay)	Eliminated DDU (Delivered Duty Unpaid)	DDP (Delivered Duty Paid)
	Who pays?	Who pays?	Who pays?	Who pays?	Who pays?	Who pays?	Who pays?	Who pays?	Who pays?	Who pays?	Who pays?	Who pays?	Who pays?
Warehouse storage at point of origin	Seller	Seller	Seller	Seller	Seller	Seller	Seller	Seller	Seller	Seller	Seller	Seller	Seller
Warehouse labor at point of origin	Seller	Seller	Seller	Seller	Seller	Seller	Seller	Seller	Seller	Seller	Seller	Seller	Seller
Export packing	Seller	Seller	Seller	Seller	Seller	Seller	Seller	Seller	Seller	Seller	Seller	Seller	Seller
Loading at point of origin	Buyer	Seller	Seller	Seller	Seller	Seller	Seller	Seller	Seller	Seller	Seller	Seller	Seller
Inland freight	Buyer	Buyer	Seller	Seller	Seller	Seller	Seller	Seller	Seller	Seller	Seller	Seller	Seller
Port receiving charges	Buyer	Buyer	Seller	Seller	Seller	Seller	Seller	Seller	Seller	Seller	Seller	Seller	Seller
Forwarder's fee	Buyer	Buyer	Seller	Seller	Seller	Seller	Seller	Seller	Seller	Seller	Seller	Seller	Seller
Loading on ocean carrier	Buyer	Buyer	Buyer	Seller	Seller	Seller	Seller	Seller	Seller	Seller	Seller	Seller	Seller
Ocean/air freight charges	Buyer	Buyer	Buyer	Buyer	Seller	Seller	Seller	Seller	Seller	Seller	Seller	Seller	Seller
Charges at foreign port/airport	Buyer	Buyer	Buyer	Buyer	Buyer	Buyer	Buyer	Buyer	Seller	Buyer	Seller	Seller	Seller
Customs, duties & taxes abroad	Buyer	Buyer	Buyer	Buyer	Buyer	Buyer	Buyer	Buyer	Buyer	Buyer	Buyer	Buyer	Seller
Delivery charges to final destination	Buyer	Buyer	Buyer	Buyer	Buyer	Buyer	Buyer	Buyer	Buyer	Buyer	Buyer	Seller	Seller

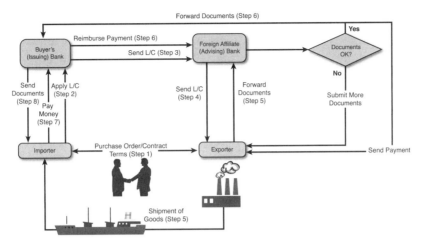

Figure 10.8. *The process cycle of the letter of credit*

Countertrade

Put simply, *countertrade* is a form of international transactions between two countries that involve exchanging goods or services for other goods or services rather than for hard currency. As such, countertrade is not affected by foreign currency rate fluctuations while reducing trade imbalances through reciprocal dealings. Countertrade is relatively common in trade with or between cash-poor countries with debt crises, such as Russia and the Philippines. Russia once exchanged its oil with another country's automobiles. Similarly, India and Iraq agreed on Iraq's oil for India's wheat and rice deals in 2000. By some estimates, countertrade accounts for 20%–25% of the world trade volume (Campbell, 2004). Typical reasons for countertrade include the following (Elderkin and Norquist, 1987):

- Expand or maintain foreign markets
- Increase sales in foreign markets by developing a new foreign market
- Sidestep liquidity problems
- Repatriate blocked funds
- Clean up bad debt situations
- Obtain new technology from a foreign country
- Keep from losing markets to competitors
- Gain foreign contracts for future sales
- Find lower-cost purchasing sources

Chapter 10: Global Supply Chain Management 373

Countertrade, however, can limit trade negotiation flexibility and increase transaction costs (e.g., additional customs duties, brokerage fees) resulting from bilateral trade agreements. Countertrade has more than several variants, including these:

- **Counter-purchase**—Counter-purchase is the most common form of countertrade, where the exporter pledges to buy goods or services from the importer that are equal in value to a predetermined percentage of the sales arrangement (contract). Counter-purchase is thought to be a better type of countertrade because it rewards the exporter with immediate payment for the goods delivered and gives the exporter more time to find a suitable product to fill the purchase arrangement (Procknow, 2010). For example, PepsiCo agreed to sell its cola concentrate to Russia, where it will be bottled and sold. In exchange, PepsiCo gained exclusive rights to export Russian vodka for sale in other countries (Ball et al., 2010).
- **Barter**—Barter is the simplest and most ancient form of countertrade, involving the direct exchanges of goods or services for other goods or services without using money. One of the largest barter deals to date involved Occidental Petroleum Corporation's agreement to ship sulfuric acid to the former Soviet Union for ammonia urea and potash under a two-year deal worth 18 billion euros (http://en.wikipedia.org/wiki/Counter_trade).
- **Switch trading**—Switch trading may occur when the product received through countertrade has little or no market value for the exporter or cannot be converted to cash instantly. To make use of the imported product of little or no market value in the exporter's country, the exporter brings in a third party who will dispose and sell this product in his or her own country or other countries for cash.
- **Buy-back (compensation trading)**—In a buy-back trade, the importer who buys the exporter's technology, equipment, and manufacturing plants pays back the exporter by sending a share of the output produced by such technology, equipment, and plants.
- **Offset**—Offset is a common form of countertrade in the defense industry. In offset, the importer requires the exporter to purchase a portion of the imported product's materials, parts, components, or subassemblies in the importer's country. Offset often aims to reduce trade deficits. For example, McDonnell Douglas once sold its MD-82 jetliner to China with an agreement that it would subcontract the manufacture of its components (e.g., landing gear doors) to the Shanghai Aviation International Corp (Alex and Bowers, 1988).

Transfer Pricing

Transfer pricing refers to the setting of price levels at which the parent firm "sells" its output (e.g., goods, services, labor, or intellectual property) to its subsidiaries in other countries to avoid high tariffs and exploit tax differentials, as illustrated in Table 10.4—the rationale being that a parent firm in a high-tax country can reduce the market value of its profits by selling its output to its subsidiaries in low-tax countries, while inflating the price for its subsidiaries in low-tax countries. The multinational firm can exploit transfer pricing due to bilateral tax treaties, which attempt to tax all multinational corporate income (or profit) once and only once. For example, let's suppose that a U.S. firm, which has the manufacturing capacity, has its foreign subsidiary in Switzerland, which has no manufacturing capacity, but has a foreign sales and marketing function. This U.S. firm with its corporate tax rate of 40% in the U.S. territory can reduce its worldwide taxes by using a transfer pricing strategy for its controlled sale to its foreign subsidiary in Switzerland where the applicable foreign tax rate is a mere 18%. Table 10.5 illustrates the effects of transfer pricing on both the U.S. firm and its foreign subsidiary, when a transfer price of $700,000 is used to shift the worldwide combined profit of $200,000 entirely to the Switzerland subsidiary. In this scenario, this firm pays $36,000 corporate tax (i.e., $200,000 × 0.18), instead of $80,000 (or $200,000 × 0.40). As such, it saves $44,000, unless it brings back its foreign profit to the U.S.

Table 10.4. A Sample of Corporate Tax Differentials across the World (source: Campbell and Kirchfield, 2013)

Country	2011	2012	2013
U.S.	40.0%	40.0%	40.0%
Japan	40.7	38.0	38.0
Brazil	34.0	34.0	34.0
Canada	28.0	26.0	26.0
Germany	29.4	29.5	29.6
Ireland	12.5	12.5	12.5
Russia	20.0	20.0	20.0
Switzerland	18.3	18.1	18.0
United Kingdom	26.0	24.0	23.0
Isle of Man	0.0	0.0	0.0

Table 10.5. The Effects of the Transfer Price between the Intercompany Transaction

Transaction	U.S. Firms	Switzerland Subsidiary
Production cost	$700,000	Zero
Controlled (inventory) sales		Cost of goods sold = $700,000
Additional sales and promotional expense		Local sales and marketing expense = $100,000
Sales revenue	$700,000	$1 million
Net profit	Zero profit	$200,000

Also notice that if a subsidiary, in a country where a high-percentage tariff is levied, imports the item from its foreign parent (e.g., U.S.) at a very low nominal price, only a small import duty is assessed.

Cross-Cultural Negotiations

One of the finishing touches of international trade is contract negotiation between the importer and the exporter. Due to the different nationalities of these two parties, who may have come from different cultures, negotiation between them can be more delicate, time consuming, and challenging than a typical contract negotiation between people of the same nationality and culture. For example, the negotiation process usually takes longer and is more complex in some Far East Asian countries such as China, Japan, and Korea because Far East Asians prefer to nurture a close personal relationship before getting down to business. If an American businessman/businesswoman who is unaware of such tendencies begins using the first name of his or her foreign business partner and quoting prices at the first encounter, his or her behavior will be construed as boorish and thus is likely to undermine the actual business dealings. As such, international (or cross-cultural) negotiation needs a high level of skill, tact, diplomacy, and patience. In general, cross-cultural negotiation is a back-and-forth communication process in which at least two partners of different nationalities with different needs, thoughts, behaviors, and viewpoints must reach an agreement on matters of mutual interest. In addition to potential cultural clashes, an international negotiator should be aware of a number of complicating factors that can influence the ultimate negotiation outcome. Some of those factors include ideological differences, foreign laws and regulations, foreign government bureaucracy, business protocols (e.g., dress code, etiquette), political and economic environments, communication styles (e.g., language nuances, body gestures) and decision-making styles (e.g., top-down, bottom-round). Considering the complexity and subtlety of cross-cultural negotiation, a negotiator should do a lot of painstaking groundwork before coming to the negotiation table with the foreign partner. This groundwork may require the following steps:

- **Acculturation**—Acculturation is the process in which one group learns and adapts to the culture, behaviors, beliefs, traits, and values of another group (Kottack, 2005). As such, acculturation is the first step in reducing cultural gaps between negotiators from different cultures and subsequently helps to minimize differences in a negotiation stance. Although some negotiators may wonder why they should adapt to others' cultures rather than let others do the adapting, reluctance toward acculturation seldom contributes to the success in cross-cultural negotiations. In fact, Min and Galle (1993) discovered that a failure to compromise cultural differences was one of the top three reasons for ineffective international negotiation. Kublin (1995) also observed that culture played an important role in international negotiation. Tables 10.6 and 10.7 summarize several distinctive cultures and their traits that cause cultural differences (see, e.g., Simintiras and Thomas, 1998; Hofstede, 2011).
- **Competency**—Once a negotiator has a chance to enhance his or her cultural IQ through acculturation, the next step to take is to increase his or her competency in international negotiation in a particular country. This process includes the development of sufficient knowledge about the host country's economy, politics, laws, and rules, along with the specific profiles (decision authority, role, personality, experience, negotiation style) of a foreign negotiator. Basically, this process involves fact finding and learning.
- **Strategy**—At the final preparation stage, a negotiator should formulate a specific negotiation strategy tailored for a particular foreign negotiator in a particular country. This strategy should be aligned with a specific negotiation agenda (e.g., price, quality, freight terms, service, claims settlement), pace of negotiation, negotiation team composition, communication media, potential use of agents/translators, extent of the need for socialization (personal relationship), initial offer and expected final outcome, and follow-up.

Table 10.6. Cultural Types with Respect to Communication Styles and Their Distinguished Traits

	High Context Culture	**Low Context Culture**
Definition	Cultures in which context is at least as important as what is actually said.	Most of the information is contained explicitly in the words.
Communication	Intention and unspoken meaning ("subtle").	Direct verbal language.
Performance	Collectivism (loyalty).	Individualism (efficiency).
Relationship	"Feminine" (interpersonal relationship, reserved, less time driven), or "relationship driven."	"Masculine" (aggressiveness, assertiveness), or "task driven."

	High Context Culture	**Low Context Culture**
Priorities	Polychronic (multiple tasks simultaneously)	Monochronic (one task at a time), punctuality
Examples	Most of Asia, Latin America, Middle East, Spain, France, and Africa.	U.S., United Kingdom, Canada, Scandinavian countries

Table 10.7 Cultural Types with Respect to Status and Influence Differences and Their Distinguished Traits

	High Power Distance Culture	**Low Power Distance Culture**
Definition	Employees seek no decision-making role.	Individual employees will seek a role in decision-making and will question decisions and orders in which they had no input.
Responsibility	Little personal initiatives from employees who need direction and discipline.	More responsibility for employees. (Micromanagement can be an issue.)
Examples	Russia.	North America.

Chapter Summary

The following are key lessons learned from this chapter:

- Today's world is characterized as a single, unified global marketplace where buy and sell activities take place every day on a global scale. In this setting, supply chain activities associated with international trade have gotten more complex and volatile. Therefore, supply chain professionals should understand the dynamics of global supply chains in constantly changing scenes of international trade. Because the international trade scenes are often shaped by worldwide governing bodies such as WTO, supply chain professionals should understand their changing trade policies and their managerial implications.

- Free trade movements are the catalysts for cross-border trade because they gradually remove protective trade barriers such as tariffs. For example, the North American Free Trade Agreement (NAFTA) among U.S., Canada, and Mexico significantly increased cross-border trade among those three countries, which contributed to the economic growth of those three nations. However, its positive impact can be mitigated unless logistics efficiency for such trade can be sustained or continuously improved.

- In the presence of complexity, risk, and uncertainty in global supply chain operations, many MNFs considered forming long-term strategic alliances with foreign business partners. To make these alliances successful, MNFs should look into the strategic fit of their potential partners with respect to organizational compatibility, technical compatibility, and cultural harmony.

- A supply chain is as strong as its weakest link. In a stretched global supply chain, the MNFs are more vulnerable to unforeseen risks, which disrupt global supply chain operations and thus offset the various benefits of global business activities. These risks stem from foreign currency fluctuations, terrorist threats, natural disasters, maritime piracy, and financial interdependency of global supply chain partners. To cope with these risks, MNFs should enhance their supply chain resilience.
- Though rarely used, a foreign trade zone (FTZ) can be exploited to land, store, and process imported products without customs hassles and duty payments. It also helps to improve supply chain security because the imported products are well protected in the area under the government's supervision and control.
- The true landed costs of imported merchandise should include hidden costs incurred during transit. These hidden costs include insurance premiums and logistics costs associated with loading, unloading, customs clearance, modal transfers, and material handling. To minimize these hidden costs, supply chain professionals should clarify their roles and responsibilities for global shipping operations. Because these roles and responsibilities vary depending on the specific freight terms agreed upon as part of the contractual terms, supply chain professionals should understand the ramifications of the 11 different Incoterms from which they can choose.
- Cross-cultural negotiation differs from domestic negotiation in that the former takes more time to prepare for and settle the deal than the latter because the former requires more groundwork than the latter. As such, a hasty approach to cross-cultural negotiation can lead to an undesirable negotiation outcome. In particular, supply chain professionals must understand the cultural idiosyncrasies of international negotiation through acculturation.

Study Questions

The following are important thought starters for class discussions and exercises:

1. What are the key differences between domestic supply chain management and global supply chain management?
2. How do free trade movements affect global supply chain operations? How does NAFTA impact cross-border shipping activities between U.S. and Canada or U.S. and Mexico?
3. What are the different types of free trade movements? Give an example of how each type of free trade movement works.
4. What motivates the MNF to form a strategic alliance with a foreign business partner? How do you choose the right supply chain partner in a global setting?

5. What are the elements of risks in global supply chain management?
6. How do you mitigate the risks involved in global supply chain management?
7. Imagine that you want to import a textile product (e.g., bath towel) from Pakistan and ship it to the U.S. In this imaginary situation, draw the global supply chain map and then identify and assess the extent of supply chain risks.
8. Explain how the currency fluctuation risk enters into your global outsourcing decision and then how you would deal with such risk in international transactions.
9. How does the presence of risk and uncertainty in global supply chain operations affect the MNF's just-in-time production and logistics practices?
10. Discuss the major intents of various supply chain security initiatives and explain how those affect the global supply chain operations.
11. What makes your global supply chain more resilient?
12. Which documents are necessary for typical international transactions? How can you simplify the documentation process in international transactions?
13. What role do Incoterms play in global logistics? What are the 11 distinctive options of Incoterms? What are the specific risks and responsibilities of the importer for each Incoterm? How do you determine the proper Incoterm in international transactions?
14. What is the most favorable term of payment for the exporter? What is the most favorable term of payment for the importer?
15. In what cases does countertrade make sense? Classify the various types of countertrade with their pros and cons.
16. What are the biggest challenges of cross-cultural negotiation?
17. How do you improve your cultural IQ? How does your cultural IQ influence the outcome of international negotiation?
18. Compare and contrast high-context and low-context culture.
19. Imagine that you are meeting with a Chinese supplier in Shanghai to purchase a set of patio tables for a national retail chain in your country. How would you prepare for a contract negotiation with the Chinese supplier? Which terms of purchases should be included in this contract negotiation?

Case: Aurora Jewelers

This case was written based on fictitious but real world-like scenarios.

Impeccable Reputation

Aurora Jewelers (AJ) is a well-established fine jewelry retail chain in the U.S. that was founded by Aurora LaTour in 1925. It is headquartered in Trenton, New Jersey where Aurora was born as a daughter of French immigrants. It now sells and distributes luxury time pieces, silverware, clocks, and jewelry watches. Its popular product lines include bridal jewelry, finished jewelry, crystal jewelry, mountings, findings, diamonds, pearls, gemstones, holiday gifts, and wedding bands/rings that have become fashion symbols with affluent professionals. Thanks to its commitment to real-time sales/customization and just-in-time (JIT) delivery philosophy, it has received a lot of accolades from jewelry critics and professional organizations. For instance, it is one of only 25 U.S. retailers internationally certified by the Responsible Jewelry Council (RJC), an international not-for-profit, standard-setting and certification organization composed of more than 500 member companies.

Aurora has also been featured as one of the leading independent jewelers in media outlets such as *The New York Times* and the *Town and Country* magazine. The core of its business success is predicated on fresh designs and flexible styling that offer its customers endless possibilities with the so-called 3C (Choose/Change/Create) interactive option. This option allows the customer to choose the gemstone shape, size, color, quality, and additional features for further customization. In addition, Aurora has made every effort to meet the newly launched RJC Chain-of-Custody (CoC) Standard for the precious metal (e.g., gold and platinum group metals) supply chain. The CoC Standard aims to increase the awareness of corporate social responsibility (CSR) for sourcing, producing, processing, and trading jewelry materials through the jewelry supply chain. The RJC Code of Practices articulate the standards established for human rights, labor standards, environmental impacts, and business ethics from mine to retail. The RJC can also recognize comparable standards from other initiatives under the CoC Standard, and has already done so for gold refiners through due diligence audits and the artisanal mining sector.

Global Sourcing

A vast majority of jewelry materials that Aurora uses are imported from other countries, including South Africa, Brazil, China, India, Egypt, and Peru. These are typically transported to its distribution center in Newark, New Jersey in cases/crates by air express carriers, including FedEx and UPS. From its distribution center in Newark, AJ ships directly to its 300 retail stores in the U.S. and nine stores in Mexico via truck or air carrier. Customers who order online will receive their orders at their doorstep. The average order cycle time for online order varies from one location to another, ranging from three days to four weeks after receipt of customer order, compared to the industry average of two weeks. Regardless of order cycle time, the biggest logistics challenge facing Aurora is a limited choice of transportation carriers due to the liability of shipping high-value, precious items. Many carriers refuse to ship such expensive items unless they are allowed to limit their liability to $500 per crate. Some carriers such as FedEx often ask shippers to pay an additional charge for an item with a declared value exceeding $100. The typical air-bill or the Service Guide indicates that the highest declared value allowed is $500 per crate, except for items of extraordinary value such as artwork, collector's items, and antiques. As Aurora's customer and supplier bases have been expanded all across the world over the years with 20,000 different items, its shipping and insurance costs are beginning to eat up its profit margin substantially.

Case in point, AJ has been doing business in Mexico since 1990. It started with two stores in Monterrey and has expanded the business to a current total of nine retail stores in Mexico's Golden Triangle. The Golden Triangle is the region bound by Mexico City, Guadalajara, and Monterrey. This area contains over half of Mexico's population. The logistics of jewelry in Mexico reached a critical stage during the period of 2012 to 2013. Considerable growth in the Mexican economy, coupled with the increased importation of U.S. goods in the wake of NAFTA provisions, caused an explosion in the demand for AJ jewelry products. Even though, AJ increased the number of retail outlets in Mexico, it still continues to logistically support these stores from its Newark distribution center, which is in a designated foreign trade zone (FTZ). With the increased flow of goods into Mexico from the United States, as shown in Exhibit 10.1, the congestion at the border crossing points has created longer cycle times and increased stockouts, which conflicts with Aurora's JIT delivery principle. Also, U.S. foreign direct investment (FDI) in Mexico (stock) was $91.4 billion in 2011 (latest data available), an 8.4% increase from 2010, according to the U.S. Trade Representative. U.S. FDI in Mexico is primarily concentrated in the manufacturing sector, nonbank holding companies, and finance/insurance sectors.

Exhibit 10.1. U.S. and Mexico Trade Volume

Total Merchandise Traded	2011	2012	2013	Percent Change (2011–2012)	Percent Change (2012–2013)
U.S. to Mexico	$198.3 billion	$216.3 billion	$226.1 billion	9.1%	4.5%
Mexico to U.S.	$262.9 billion	$277.7 billion	$280.5 billion	5.6%	1.0%

Data Source: U.S. Census Bureau (2014), *Trade in Goods with Mexico*

The growth in the Mexican economy was not matched by an equal growth in its logistics infrastructure. The border crossing points became very congested, with two to three days being the normal customs clearance time during peak periods. The Mexican highway system consists mostly of dirt roads (about 55%) and secondary roads (about 25%). The Mexican trucking industry has been dominated by many small, regional carriers. Few U.S. trucking firms operate in Mexico, because the Mexican law has prohibited foreign ownership of Mexican trucking firms in the past. Some U.S. trucking firms have developed strategic alliances with Mexican carriers, and this has helped. Now, NAFTA permits 100% ownership by foreign investors of Mexican trucking firms that haul foreign commerce.

Supply Chain Headache

Given the growing demand for AJ's jewelry in Mexico, AJ's top management has made a strategic commitment to remain in the Mexican market. However, top management must make it a high priority to solve the long and inconsistent cycle time for jewelry going to the stores in Mexico. Also, in 2009, Mexico devalued the peso against the U.S. dollar. In December 2013, the exchange rate reached 13 pesos per dollar, an unprecedented loss in the value of Mexican currency. This devaluation had a very negative impact on the demand for U.S. goods (more so high-value jewelry items) in Mexico. As a matter of fact, Mexico has been anti-global. Overly dependent on oil exports, the country was hit hard by a series of recessions, chronic inflation, and high unemployment. Recently, Mexico's annual inflation rate rose to a two-year high of 5.70% in December, 2013, led by the mounting cost of electricity, gasoline, and food, cementing expectations that the country's central bank will refrain from cutting interest rates this year. A group of Mexican economists has forecasted their inflation in 2014 to be at 4.56%. However, considering the volatility of the Mexican economy with a peak inflation rate of 51.97% in 2005, you never know when the Mexican inflation rate will be stabilized.

Ever since Carlos Salinas became president in 1988, Mexico has adopted a new course by opening up its economy, privatizing hundreds of nonstrategic public enterprises, and deregulating industry. It also moved aggressively to become a global player and continues to be attractive for low-value-added manufacturing and labor-intensive assembly operations (the 2013 average rate of pay is $12,000 to $13,000 a year). Notice that Mexico's minimum wage is $4.92/hour in Zone A, including Mexico City. Mexico's hourly wages are about a fifth lower than China's, a huge turnaround from just 10 years ago when they were nearly three times higher, according to new research by Bank of America Merrill Lynch. Average hourly wages are now 19.6 percent lower in Mexico than China, whereas in 2003 they were 188 percent more costly, according to the Bank of America study. Mexico can maintain that competitive advantage for at least five years, thanks to a growing labor market that puts downward pressure on wages, Capistran Consulting said. Despite such a bright future, it will be some time before the consumer product industry fully recovers and even longer before a broad "middle-class" marketplace appears in Mexico. Another concern is that half of retail sales in Mexico were made through informal markets (e.g., street vendors). Considering this dilemma, AJ's top management should determine whether or not it is in their best interest to change the current precious metal supply chain system supporting the Mexican market. Also, they are wondering if the current shipping policy that Aurora has been using for its Mexican customers should change. The policy is as follows:

- Aurora offers free shipping on all orders over $449 within Mexico's Golden Triangle. If, for any reason, the customer decides to return or exchange a purchased item that is defective or was damaged during transit, the customer has 60 days to return it free of charge.
- All orders received after 11:00 a.m. EDT (Eastern Daylight Time) will be processed the next business day.
- There will be an extra $29.95 two-day express shipping charge on all emergency orders. There is no Saturday or Sunday delivery.
- In order to ensure proper delivery, Aurora requires an authorized signature on deliveries of orders over $500. Also, for customer protection, orders of $1,000 and above must ship to the billing address only.

Management Tasks at Hand

You are summoned to report to the office of AJ's CEO (Linda Herzberg) to explain the reason for global supply chain problems associated with the Mexican market entry. Linda Herzberg has also made it clear she wants you to come up with immediate solutions to these problems. Linda's four managers (i.e., production manager, human resource manager, sales/marketing manager, and logistics manager) have been assembled and informed that "quick and sensible solutions" are needed as soon as possible. As one of the AJ's management team, you should address the questions posed in the following section.

Discussion Questions

1. Do you support AJ's global sourcing strategy? If so, why? Do you think that it is a good idea for AJ to continue to source its jewelry materials from remotely located countries such as Egypt and South Africa?
2. What are the most important logistics challenges of AJ's global operations? (List at least three potential challenges.) Does it make sense for AJ to use trucks for both domestic and Mexican customers?

3. To cope with increasing logistics challenges in Mexico, do you think that the current logistics strategy (of using Newark as the distribution hub) should be changed? If so, then where and how?
4. What kind of shipping policy is more appropriate than the current one for Mexican customers?
5. What are the risks of doing business in Mexico? (Be as specific as possible.)
6. What would you suggest to AJ to tap into the growing market in Mexico? Should AJ focus on its current customer bases (geographically or demographically)?
7. Do you think that AJ should continue to rely on its "brick-and-mortar" (retail store) sales in Mexico? Otherwise, how should it expand its sales and distribution channels in Mexico? Do you think that AJ should consider selling some of its product lines online in Mexico? If so, what are the added supply chain challenges?
8. From the global supply chain perspective, how should AJ protect its business interest from the unexpected competition from "informal" markets? Also, is it possible for AJ to penetrate into the aforementioned informal markets? If so, how?

Bibliography

ABA Banking Journal (2004), "Recipe for Offshore Outsourcing Failures: Ignore Organization, People Issues," *ABA Banking Journal*, September 1, http://goliath.ecnext.com/coms2/gi_0199-1391240/Recipe-for-offshore-outsourcing-failure.html, retrieved on July 19, 2010.

Akiva (2009), "Maritime Piracy Doubled in First Half of 2009," International Maritime Bureau (IMB) Report, July 16, 2009, http://www.maritimeterrorism.com/2009/07/16imb-report-maritime-piracy-doubled-in-first-half-of-2009.html, retrieved on November, 22, 2009.

Alex, C.G. and Bowers, B. (1988), "The American Way to Countertrade," *BarterNews #17*, http://www.barternews.com/american_way.htm, retrieved on July 24, 2010.

Assaf, M., Bonincontro, C., and Johnsen, S. (2006), *Global Sourcing and Purchasing Post 9/11: New Logistics Compliance Requirements and Best Practices*, Fort Lauderdale, FL: J. Ross.

Ball, D.A., Geringer, J.M., Minor, M.S., and McNett, J.M. (2010), *International Business: The Challenge of Global Competition*, 12th edition, New York, NY: McGraw-Hill/Irwin.

Bragdon, C.R. (2008), *Transportation Security*, Burlington, MA: Butterworth-Heinemann.

Branch, A. (2007), *Elements of Shipping*, 8th edition, London, Great Britain: Routledge.

Buys, B. (2009), "Grey Markets: Advantages and Pitfalls," *Articlesbase*, January 15, http://www.articlesbase.com/online-business-articles/grey-market-advantages-and-pitfalls-722876.html, retrieved on July 22, 2010.

Campbell, R.H. (2004), "Countertrade," *The Free Dictionary by Farlex*, http://financial-dictionary.thefreedictionary.com/Countertrade, retrieved on July 24, 2010.

Capela, J.J. (2008), *Import/Export for Dummies*, Hoboken, NJ: Wiley Publishing, Inc.

Carver, (1999), *Understanding Foreign Trade Zones*, Los Angeles, CA: Collier.

Cook, T.A. (2008), *Managing Global Supply Chains: Compliance, Security, Dealing with Terrorism*, New York, NY: Auerbach Publications.

De Monie, G. (2009), "Economic Consequences of Piracy and Armed Robbery on Shipping," *European Commission Seminar*, http://ec.europa.eu/transport/maritime/events/doc/2009_01_21_piracy/critical_demonie.pdf, retrieved on October 22, 2009.

Dillon, D. (2005), "Maritime Piracy: Defining the Problem," *SAIS Review*, 15(1), 155–165.

Eaton Corporation (2010), "Eaton and Linde Hydraulics Forge Global Strategic Alliance to Advance Hydraulic Technologies for Mobile and Industrial Equipment," *Business Wire*, June 30, 2010, http://www.tradingmarkets.com/news/press-release/etn_eaton-and-linde-hydraulics-forge-global-strategic-alliance-to-advance-hydraulic-technologies-for-mob-1013859.html, retrieved on July 18, 2010.

EconomyWatch (2010), "International Free Trade Zone," http://www.economywatch.com/international-trade/free-trade-zone.html, retrieved on July 22, 2010.

Elderkin, K.W. and Norquist, W.E. (1987), *Creative Countertrade: A Guide to Doing Business Worldwide*, Cambridge, MA: Ballinger Publishing Company.

Elmuti, D. and Kathawala, Y. (2001), "An Overview of Strategic Alliances," *Management Decision*, 39(3), 205–217.

Grant, B. (2004), "Discover How Your Company Can Save Big $$$ by Qualifying as a Foreign Trade Zone," *CLM Logistics Comment*, 38, 14–15.

Hergert, M. and Morris, D. (1988), "Trends in International Collaborative Agreements," *Cooperative Strategies in International Business* edited by F.J. Contractor and P. Lorange, Lexington, MA: Lexington Books.

Hillman, M. and Keltz, H. (2007), *Managing Risk in the Supply Chain-A Quantitative Study*, Unpublished AMR Research Report, Boston, MA: AMR Research, Inc.

Hofstede, G. (2011), "Dimensionalizing Cultures: The Hofstede Model in Context," *Online Readings in Psychology and Culture*, 2(1), 8–33.

International Chamber of Commerce (2010), *Incoterms 2010: ICC Rules for the Use of Domestic and International Trade Terms*, Paris, France: ICC.

International Maritime Bureau (2007), *Piracy and Armed Robbery Against Ships*, Annual Report, London, United Kingdom: ICC International Maritime Bureau.

Kalmbach, C. Jr. and Roussel, R. (1999), "Dispelling the Myths of Alliances," http://www.2c.com.html, November 16, 1999, 1–8.

Kearney, A.T. (2009), *The Shifting Geography of Offshoring*, Unpublished White Paper, http://www.slideshare.net/fred.zimny/atk-global-outsourcing-survey, New York, NY: A.T. Kearney.

Kokko, A., Kotoglou, K., and Krohwinkel-Karlsson, A. (2003), "Characteristics of Failed FDI Projects in Vietnam," http://findarticles.com/p/articles/mi_6790/is_3_12/ai_n28172516/pg_2/?tag=content;col1, retrieved on July 15, 2010.

Korteweg, R. (2009), "Turbulent Waters in a Maritime Black Hole," http://www.hcss.nl/en/column/672/Turbulent-Waters-in-a-Maritime-Black-Hole-.html, retrieved on October 22, 2009.

Kottack, C.P. (2005), *Mirror for Humanity: A Concise Introduction to Cultural Anthropology*, 5th edition, New York, NY: McGraw-Hill.

Kublin, M. (1995), *International Negotiating: A Primer for American Business Professionals*, Binghamton, NY: International Business Press.

Loser, C. (2009), *Cross-Border Trade and Investment among Emerging Economies: Lessons from Differing Experiences in African, Asia, and Latin America*, Discussion Draft, Emerging Market Forum, Washington DC: The Centennial Group.

Mapp, W.D. (1978), "Documentary Problems of Intermodal Transportation," *Journal of World Trade Law*, 12, 511.

Min, H. (1996), "Distribution Channels in Japan: Challenges and Opportunities for the Japanese Market Entry," *International Journal of Physical Distribution and Logistics Management*, 26(10), 22–35.

Min, H. (2009), "Editorial on Global Operations," *International Journal of Services and Operations Management*, 5(6), 737–739.

Min, H. (2011), "Modern Maritime Piracy in Supply Chain Risk Management," *International Journal of Logistics Systems and Management*, 10(1), 122–138.

Min, H. and Galle, W. (1993), "International Negotiation Strategy of U.S. Purchasing Professionals," *International Journal of Purchasing and Materials Management*, 29(3), 40–50.

Min, H. and Lambert, T. E. (2010), "The Utilization of Foreign Trade Zones in the Global Supply Chain: An Exploratory Study," *International Journal of Services and Operations Management*, 6(2), 110–125.

Monczka, R.M. and Giunipero, L.C. (1990), *Purchasing Internationally: Concepts and Principles*, Chelsea, Michigan: Book Crafters.

O'Reilly, J. (2009), "Fighting Piracy on the High Seas," *Inbound Logistics*, http://www.inbound-logistics.com/articles/global/global0309.shtml, retrieved on October 28, 2009.

Outsourcing Institute (1998), "Top Ten Reasons Companies Outsource," http://www.outsourcing.com/content.asp?page=01b/articles/intelligence/oi_top_ten_survey.html, retrieved on July 19, 2010.

Porter, M.E. (1985), *Competitive Advantage: Creating and Sustaining Superior Performance*, New York, NY: The Free Press.

Ricardo, D. (1817), *On the Principles of Political Economy and Taxation*, London, England: John Murray.

Procknow, G. (2010), "Understanding the Different Types of Countertrade," http://www.helium.com/items/1873491-countertrade-barter-counter-purchase-buy-back, retrieved on July 24, 2010.

Roos, P. (2011), *Incoterminology 2010—Manual for Practical Use of Incoterms*, Rotterdam, Netherlands: NT Publishers.

Root, F.R. (1994), *Entry Strategies for International Markets*, revised and extended 2nd edition, New York, NY: Lexington Books.

Sell, J. (2008), "Piracy May Make Shipping Costs Soar," November 21, http://nl.newsbank.com/nl-search/we/Archives, retrieved on October 22, 2009.

Simintiras, A C. and Thomas, A H. (1998), "Cross-cultural Sales Negotiations: A Literature Review and Research Propositions, " *International Marketing Review*, 15(1), 10–28. Skymart (2010), *Incoterms 2010*, http://www.skymartworldwide.com/incoterms-2009.html, retrieved on April 15, 2011.

Smith, A. (1776), *The Wealth of Nation*, London, United Kingdom: W. Strahan and T. Cadell.

Stapleton, D.M. and Saulnier, V. (1999), "Defining Dyadic Cost and Risk in International Trade: A Review of INCOTERMS 2000 with Strategic Implications," *Journal of Transportation Management*, 11(2), 25–43.

Starr, S. (2014), "Maritime Piracy on the Rise in West Africa," *Combatting Terrorism Center Sentinel*, April 28, 2014, https://www.ctc.usma.edu/posts/maritime-piracy-on-the-rise-in-west-africa, retrieved on May 25, 2014.

Thomson Financial News (2009), "Shippers, Insurers Fear Somali Piracy May Escalate," June 29, http://www.reuters.com/article/latestCrisis/idUSLT153861, retrieved on October 22, 2009.

Toth, D. (2009), "Maritime Piracy and the Devastating Impact on Global Shipping," http://www.examiner.com/x-18100-Pittsburgh-Foreign-Policy-Examinery2009m7d31-Maritime-piracy-and-the-devastating-impact-on-global-shipping, retrieved on October 22, 2009.

Wheeler, W., Evans, E., and Treadwell, J. (2008), "Patheon and Solvias AG Announce Global Strategic Alliance to Provide Integrated Development Services," *News Release*, http://www.patheon.com/InvestorRelations/NewsReleases/tabid/69/ctl/NewsRelease/mid/510/releaseid/223/Default.aspx, retrieved on July 18, 2010.

World Trade Organization (2013), *International Trade Statistics 2013*, Unpublished Report, Geneva, Switzerland: WTO.

11

Legally, Ethically, and Socially Responsible Supply Chain Practices

"Corporate social responsibility is a hard-edged business decision. Not because it is a nice thing to do or because people are forcing us to do it... because it is good for our business."
—Niall Fitzgerald, former CEO of Unilever

Learning Objectives

After reading this chapter, you should be able to:

- Understand what constitutes a triple bottom line.
- Understand how the increasing adaptation of a triple bottom line reshapes the company's business philosophy and supply chain practices.
- Learn about the fundamental framework of the United States law and its impact on supply chain practices.
- Develop ways to resolve contract disputes wisely without damaging business partner relationships.
- Comprehend the nuances of business ethics and their potential impacts on supply chain practices.
- Learn about green supply chain management (GSCM) and understand how it can transform supply chain practices.
- Learn to improve the company's environmental performance and comply with environmental standards/regulations.

Triple Bottom Line

Warren Buffet was once quoted as saying, "It takes 20 years to build reputation and five minutes to ruin it." The company's reputation can be easily marred once it ignores its legal, ethical, and social responsibilities, as illustrated by corporate scandals involving high-profile companies such as Enron, Siemens, HealthSouth, Halliburton, K-Mart, Bristol-Myers Squibb, Nortel, and Tyco International. One of the biggest mistakes made by these companies was their misconception that short-term financial success matters most, without realizing the increasing importance of the "triple bottom line" to their ultimate corporate success. As a matter of fact, the *Economist* magazine reported that companies with their eye on that triple bottom line tend to outperform their less fastidious rivals on the stock market. Generally, *triple bottom line* refers to the three different pillars of the corporate bottom line (Economist, 2009):

- **Profit**—As a financial performance measure
- **People**—As a measure of the company's dedication to socially and ethically responsible business practices
- **Planet**—As a measure of the company's environmental commitment and performance

The concept of the triple bottom line has become a buzzword these days due to growing concerns among conscientious customers about environmental degradation and preservation of social well-being. Unlike the traditional business paradigm, which focuses on the economic prosperity of the corporation and its shareholders, the triple bottom line increases the growing awareness of sustainability by capturing an expanded spectrum of economic, social, and ecological values. That is to say, the triple bottom line is built upon the idea that a sustainable corporation is one that creates profit for its shareholders while protecting the environment and improving the lives of those (e.g., suppliers, workers, customers) with whom it interacts (Savitz and Weber, 2006). For example, in an effort to fulfill the company's environmental mission, FedEx Kinko increased its use of recycled materials and its renewable energy at its multiple store locations. Similarly, Walmart made great strides in reducing its carbon footprint by recycling its electronic product waste and plastic shrink wrap while using more fuel-efficient vehicles. Starbucks also made a conscious decision to pay a premium to coffee growers in developing countries who use traditional farming methods that can better preserve tropical forests. As these examples illustrate, a growing number of firms are increasingly aware of the corporate value of sustainability. In response to this changing value proposition aligned with the triple bottom line, the next sections will elaborate on a number of managerial issues related to the triple bottom line and their implications. In addition, the legal aspect of strategic supply chain decisions and activities will be discussed in detail.

Types of Laws

To maintain order and liberty in human society, law exists in every country. The law helps protect us from any harm or ill will, but can limit certain human acts or behaviors by imposing various forms of liabilities and bodies of rules. Good or bad, fair or unfair, every citizen should abide by these rules. One cannot excuse him- or herself from legal liabilities by not knowing the law or claiming special social status. Therefore, in order to protect yourself from others who intend to harm you and preserve your rights and privileges, you must understand the legal consequences of your everyday decisions. In general, law is a system of rules or guidelines prescribed by the supreme power in a State, commanding what is right and forbidding what is wrong (Bigelow, 1905). Because it shapes everyday human behavior, including important business decisions, supply chain managers should familiarize themselves with it, especially contract law, tort law, administrative law, and international law, as these may have a profound impact on supply chain activities. Prior to elaborating on these laws, let's start discussing the basic framework of the legal system in the United States, which is depicted in Figure 11.1 (Scheuing, 1989, p. 48).

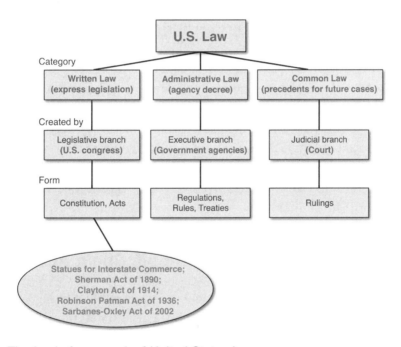

Figure 11.1. *The basic framework of United States law*

In a broader sense, U.S. law is composed of three categories, including laws used during the British Colonial period. These are written law, administrative law, and common (unwritten) law. To elaborate, *written law* represents a set of rules and conditions of social order that have been enacted in express legislations by duly elected parliament bodies (e.g., the U.S. Congress). Examples of written law include the U.S. constitution, the Sherman Act of 1890, the Clayton Act of 1914, the Robinson Patman Act of 1936, the Sarbanes-Oxley Act of 2002, and the Federal Rules of Civil

Procedures. *Administrative law* refers to the body (collection) of rules, regulations, orders, and decisions created by administrative agencies (e.g., the U.S. Department of Labor, the U.S. Department of Agriculture, the Federal Communication Commission, and the Securities Exchange Commission, the Federal Aviation Administration) of the government (The Free Dictionary by Farlex, http://www.thefreedictionary.com/administrative+law). Administrative law is considered a branch of public law and was expanded greatly in the twentieth century due to the increased need for regulating social, political, and economic activities. As a body of law, administrative law deals with the decision-making of administrative units of government (e.g., tribunals, boards, or commissions) that are part of a national regulatory scheme in such areas as police law, international trade, manufacturing, environment, taxation, broadcasting, immigration, and transportation (Koch, 2010). For example, a set of environmental regulations created by the Environmental Protection Agency (EPA), workplace safety rules created by the Occupational Safety and Health Administration (OSHA), and consumer protection rules created by the Consumer Product Safety Commission (CPSA) belong to the category of administrative law. Though eliminated in 1996, the rules created by the Interstate Commerce Commission (ICC) that aimed to regulate the freight rate charged to shippers by the interstate railroad are also considered part of the administrative law. *Common law* is a body of law evolved from judges' past rulings based on precedents and customs and is often called "case law," which originated and was developed in England during the Middle Ages based on court decisions and implicit doctrines derived from similar tribunals in the past rather than written statutory law. Because common law places greater weight on court decisions made by judge at their discretion, it is more complex and less transparent than statutory law.

Laws Applicable and Relevant to Supply Chain Activities

Supply chain activities such as freight shipment are often subject to legal terms and conditions defining the rights and duties of companies that initiated those activities for their trading partners and customers. Thus, companies that engage in supply chain activities should clearly understand the laws that govern those activities and plan those activities under legal restrictions. Some of these laws may be specifically designed to govern particular supply chain activities, such as transportation, warehousing, and purchasing, while others such as labor/employment law and tax law may be applicable to the entire spectrum of supply chain activities. Law with supply chain implications includes employment law (e.g., Equal Employment Opportunity, Fair Labor Standard Act, Child Labor Protection, Unemployment Compensation), antitrust law (e.g., Sherman Antitrust Act, Consumer Protection), intellectual property law (e.g., copyright, trade rights, patents, industrial design rights, trade secrets), environmental law (e.g., Clean Air Act, Clean Water Act, Resource Conservation and Recovery Act, Energy Policy Act), and commercial/business law (e.g., privacy law, corporate law, law of agency, contract law).

To elaborate, *employment law* is the large body of statues, judicial decisions, and regulations administrated by the Department of Labor that encompass all areas of the employer/employee relationship. Some of those areas include employment discrimination, wages, unemployment compensation, pensions, workplace safety and health, child labor, and workers' compensation (http://www.hg.org/employ.html). *Antitrust laws* are state and federal laws that were created to prevent noncompetitive behaviors (namely monopolies). Antitrust laws apply to both businesses and individuals. The motivation behind this type of law is that monopolies can stagnate the markets and prevent others from engaging in fair competition. Its roots date back to the Sherman Antitrust Act of 1890, which was passed by Congress to remove limits on competitive trade. The Sherman Antitrust Act affected all interstate business transactions (http://www.wisegeek.com/what-are-antitrust-laws.htm). *Intellectual property law* is the body of law that deals with protecting the rights of those who create original work. It covers everything from inventions, literary and artistic work, and company identification marks (e.g., symbols, names, images). The main purpose of intellectual property laws is to encourage new technologies, artistic expression, and inventions, while promoting economic growth (http://www.alllaw.com/topics/intellectual_property/). This law can be broken down into three subcategories:

- **Copyrights**—Copyright is a form of exclusive rights granted to the author or creator of an original work, including the right to copy, distribute, and adapt the work based on the U.S. Constitution. Copyright does not protect ideas and discoveries, only their tangible medium of expression (both published and unpublished works). In most jurisdictions, copyright arises upon fixation and does not need to be registered. Copyright owners have the exclusive statutory right to exercise control over copying and other exploitation of their work for a specific period of time, after which the work is said to enter the public domain (http://en.wikipedia.org/wiki/Copyright).
- **Patents**—A patent is a set of exclusive rights granted by a government to the inventor or their assignee for a limited period of time (e.g., 17 years for patent applications filed before June 8, 1995 and 20 years for patent applications filed on or after June 8, 1995) in exchange for public disclosure of an invention (http://en.wikipedia.org/wiki/Patent). There are three different types of patents: *utility patents* protect inventions with a specific function, including things like chemicals, machines, and technology; *design patents* protect the unique way a manufactured object appears; and *plant patents* protect plant varieties, including newly invented strains of asexually reproducing plants (http://www.alllaw.com/topics/intellectual_property/). Because a patent holder can sue any manufacturer,

seller, and/or user of a patented invention for infringement, a supply chain professional should consider including a patent indemnity clause in every contract (Scheuing, 1989).

- **Trademarks**—A trademark is a distinctive sign or indicator (e.g., name, word, phrase, logo, symbol, design, image, or a combination of these elements) used by an individual, a business organization, or other legal entities to identify that the products or services to consumers with which the trademark appears originate from a unique source, and to distinguish these products or services from those of other entities (http://en.wikipedia.org/wiki/Trademarks). The trademark owner can prevent others from using the protected item in the specific category unless this protection has been declared invalid by a U.S. court due to its generic usage (Scheuing, 1989). Unlike patents, trademark protection lasts for an indefinite period of time. For example, Owens-Corning obtained its trademark for the color pink in fiberglass to prevent any other companies from ever making pink fiberglass. In another example, Vermont Teddy Bear once filed lawsuit against the Walt Disney "Pooh-Grams" for the infringement of a trademark law after registering its name, logo, and marketing channel (personalized "Bear-Grams") as its trademark.

Environmental law is a body of state and federal statutes, treaties, and regulations intended to protect natural environments (e.g., wildlife, forest, mineral deposits, land and scenic beauty), prevent pollution, save endangered species, conserve water and energy, and mitigate the detrimental effects of human activities on natural environments (http://www.legal-explanations.com/definitions/environmental-law.htm). It also provides a basis for measuring and apportioning liability in cases of its violations and the failure to comply with its provisions (http://www.businessdictionary.com/definition/environmental-law.html). Its key components are pollution control and remediation, resource conservation and management, and planning land use and (transportation and energy) infrastructure (http://en.wikipedia.org/wiki/Environmental_impact_assessment). *Commercial law* (business law) refers to the body of law that governs how business parties enter into contracts with each other, execute them, and remedy problems that arise in the process (Scheuing, 1989). Commercial law covers the topics of agency, contract, bailments, labor relations, carriers, sales, product liability, partnerships, corporations, unfair competition, secured transactions, property, commercial paper, consumer credits, insurance, and bankruptcy. In the U.S., a large part of commercial law has been codified in the Uniform Commercial Code (UCC). All 50 states, the District of Columbia, and the Virgin Islands have adopted the UCC as law, although the state of Louisiana has selectively adopted Articles 1 (general provisions), 3 (commercial paper, negotiable instruments), 4 (bank deposits and collections), 5 (letters of credit), 7 (warehouse receipts, bills of lading and other documents of title), 8 (investment security), and 9 (secured transactions, sales of accounts, and chattel paper) (http://www.referenceforbusiness.com/encyclopedia/Clo-Con/Commercial-Law.html, http://uniformcommercialcode.uslegal.com/states-adopting-the-ucc/louisiana/).

Disputes and Claim Resolutions

Most contracts are bilateral and therefore indicate the mutual obligation to fulfill what was agreed upon in the contract. Nevertheless, a contract can be disputed by either one of the parties for performance failure, a partial fulfillment of the contract, a different interpretation of contractual terms, ambiguous and contradictory contractual terms, and unexpected/uncontrollable events preventing the successful fulfillment of contractual terms. Typically, there are four different ways to resolve the contract disputes. These are negotiation, mediation by the third-party facilitator, arbitration by the third-party decision maker, and litigation. Because issues related to negotiation were discussed in Chapter 8, "Sourcing," this section will elaborate on the last three methods of contract dispute resolution, as follows:

- **Mediation**—Mediation is a non-adversarial communication process in which a third-party, neutral facilitator (one, two, or a panel) assists in resolving a dispute between two or more other parties by focusing on the real issues, reframing the issues, and finding the common ground. Mediation is often used to resolve long-running, deep-rooted conflicts that can be rarely handled without the involvement of a third-party facilitator (Honeyman and Yawanarajah, 2003). Mediation is an alternative option when negotiation has failed. Generally, mediation is less expensive and more flexible than litigation. Also, its agreement remains confidential. However, the mediation agreement is fully enforceable in a court of law (Moore, 1996; Beer, 1997).
- **Arbitration**—Arbitration is a process by which the disputing parties submit their case to the impartial third-party intermediary (or panel) for their dispute settlement, which is usually binding. In binding arbitration, the losing party must comply with the terms of a settlement. Although arbitration is not as formal as court adjudication, it creates adversarial contests that typically lead to "win-lose" outcomes (http://www.colorado.edu/conflict/peace/treatment/arbitrat.htm). Regardless, because arbitration serves as a forum to resolve differences outside of the judicial system, it can provide quicker and easier solutions than litigation, which may drag on for years. Arbitration is also less expensive than litigation because the arbitrator fee is not usually as high as the attorney's fee. In addition, arbitration preserves privacy and therefore will not subject both disputing parties to any revelation of trade secrets and proprietary information during the dispute resolution process. On the other hand, arbitration does not allow the losing party to make an appeal for the final decision made by the third-party arbitrator, who

still can be biased and unjust. (http://alternative-dispute-resolution.lawyers.com/arbitration/Arbitration-and-Mediation.html). Arbitration is often used to resolve international commercial contracts and labor-management contracts.

- **Litigation**—Litigation is the last resort for contract disputes because it is time consuming and expensive, and its outcome is uncertain—and more importantly, it often alienates good business partners due to the adversarial nature of the process. Despite these drawbacks, when nonjudicial remedies are exhausted, one disputing party has the constitutional right to bring a lawsuit against another disputing party. Unfortunately, litigation is the most common form of dispute resolution. To bring the case in a court of law, one has to prove that there exists a factual "right-duty" relationship in the contract, a breach of the contract (conduct of violation of the contractual terms), and economic loss (including damage and injury) resulting from the breach of the contract. In a typical case of a contract dispute, the litigation starts with the filing of "pleading" (complaints) with the judicial court. A party who brings the lawsuit is called a "plaintiff," and the party who is being sued is called a "defendant." After pleading, the early stage of litigation involves pretrial discovery of facts and evidence relevant to the legal case, which may support or contradict each disputing party's position. If the case proceeds to trial, each party enters facts and evidence into the record and presents witnesses (including expert witnesses). In this process, the plaintiff usually has the burden of proof in making his or her case. During any phase of the litigation, the settlement can occur if both parties agreed on the terms. Unlike arbitration, an appeal can be made within an allotted time after the final judgment is made (see, e.g., for Kramer, 1998; Kerley et al., 2001).

Supply Chain Ethics

Generally, *ethics* is defined as a body of moral principles of value relating to human conduct with respect to the rightness and wrongness of actions and the goodness and badness of the motives and ends of such actions (http://dictionary.reference.com/browse/ethics). In that context, *supply chain ethics* refers to the well-founded rules and standards for right conduct and good practices that are generally acceptable to supply chain professionals. Put simply, supply chain ethics prescribes what supply chain professionals ought to do in acquiring, transforming, delivering, and promoting goods and services. Examples of unethical supply chain behaviors include the following:

- Taking a bribe or kickback to secure a sales/purchase contract
- Accepting samples without payments
- Accepting a supplier's quote that's obviously in error
- Using child labor or forced labor in producing goods

- Using animals for product testing
- Distributing counterfeit goods and engaging in brand piracy
- Distributing unsafe products
- Deceptive or dishonest advertising
- Price fixing
- "Low-ball" pricing
- "Bait-and-switch" selling
- Misusing of trade discounts for personal purchases
- Ignoring health, safety, and environmental standards

The following subsections detail some of these ethical issues.

Bribery versus Gift (Gratuity)

Bribery is the corrupted act of offering, giving, receiving, or soliciting something of value for the purpose of unduly influencing the action of an official in the discharge of his or her public or legal duties (Garner, 2006). Once an agent is guilty of taking a bribe without the consent of the principal by soliciting, accepting, or agreeing to accept any benefit from another person upon an agreement or understanding that such benefit will influence his or her conduct in relation to his or her principal's affairs, he or she will be punished for committing a class B misdemeanor. Both bribe takers and makers, along with their intermediaries, can be penalized for involvement in a bribery scheme. Despite the seriousness of the bribery, some guidelines for the bribery may be vague and the subsequent interpretation of the bribery may differ from one company to another or one country to another. Therefore, supply chain professionals should take extra care when they suspect their potential involvement in "questionable payment" practices.

Although many believe that questionable payments are not of great concern among the U.S. managers due to their perceived higher ethical standards, questionable payments in the U.S. is on the rise. During the period of 1999 to 2013, the U.S. Department of Justice took anti-bribery enforcement actions against 194 U.S. multinational firms and executives for alleged violations of the Foreign Corrupt Practice Act (*Bloomberg Businessweek*, 2014). Also, Fletcher (2009) reported that even U.S. federal government made $98 billion in questionable payments to contractors and individuals in 2008, which showed a sharp increase from the estimated $78 billion payout made in 2007. Basically, questionable payments are business transactions related payments that can raise questions concerning potential violations of moral principles. These may include illegal political contributions, excessive sales commissions, secret kickbacks for favorable business concessions, and hedges against disadvantageous treatment. However, so-called "grease payments" are often permissible as legitimate payments in many foreign countries. Examples of grease payments include gifts that can short-circuit bureaucratic red tape for processing documents, stamping visas, and scheduling product inspections; tips (gratuities) for services rendered, and fees for the use of local middlemen/agents (e.g., Baksheesh in the Middle Eastern countries). Also, gifts in the spirit of the holiday season (e.g., Christmas, Chinese New Year) along with nominal gifts (typically less than $25) that would not influence future business contracts/transactions can be considered "legitimate."

Fraud

In a supply chain setting, *fraud* is considered any act, deed, or statement made by either a buyer or a supplier *before* the business contract is signed or completed that intends to deceive the other party through the perversion of truth and/or the false representation of facts. Fraud can be proven by showing one of these five elements (http://legal-dictionary.thefreedictionary.com/fraud):

- A false statement of a material fact
- Knowledge on the part of the perpetrator that the statement is untrue
- Intent on the part of the perpetrator to deceive the alleged victim
- Justifiable reliance by the alleged victim on the statement
- Injury to the alleged victim as a result

However, an alleged perpetrator (e.g., a supplier) is not liable for fraud if evidence proves the following:

- The perpetrator or his/her sales representative made a false statement *after* the contract was signed.
- The perpetrator actually did not know that the quality of the merchandise was not as claimed in the sale contract but merely expressed an opinion that he believed the quality to be as represented.

Fraud can take a variety of forms, as detailed next:

- **Bait and switch**—In retail sales, a bait and switch is a form of fraud in which the fraudster lures in customers by advertising a product or service at an unprofitably low price and then reveals to potential customers that the advertised good is not available but that a substitute good is. Likewise, if the contractor quotes an exceptionally low price for his/her services and later charges the customer a much higher price (e.g., more than 10% above the quoted price), such practice is considered a fraudulent act that violates consumer protection laws in most of the U.S.

- **Confidence trick**—A "shell game" that swindles the wager, who places a bet on which moving shell contains the ball. In this scam, the wager has little chance of winning the money.
- **False advertising**—The most common form of this kind of fraud is a new credit card deal that offer low interest rates initially and then raises the rates greatly afterward.
- **Identity theft**—Identity theft can take the following forms:
  - *Financial identity theft* involves using another's name and social security number to obtain goods and services.
  - *Criminal identity theft* involves posing as another individual when a perpetrator is apprehended for a crime.
  - *Identity cloning* involves using another's personal information to assume his or her identity in daily life. For instance, Frank Abagnale, Jr., the impostor who wrote bad checks and falsely represented himself as qualified professional, such as an airline pilot, a doctor, and an attorney.
  - *Business/commercial identity theft* involves using another's business name to obtain credit.

According to the common law rescission, once the fraudulent act is proven, the business contract can be nullified and therefore the fraud victim (e.g., a buyer) has the right to return goods that he or she purchased and is seeking a refund for.

Conflicts of Interest

A *conflict of interest* generally refers to a situation where a person in a position of trust (e.g., a public official or a business executive) has a private or personal interest that clashes with his or her official obligations/responsibilities and thus undermines the impartiality of his or her official actions and decisions. For example, the company's board member cannot vote on or influence the company's supplier selection decision involving one of the candidate suppliers with which he or she is affiliated financially. A conflict of interest is ethically problematic in that it often does not protect the best interest of the organization that a decision or policy maker represents. In other words, a conflict of interest impairs the independent, unbiased judgment of public officials and business executives and subsequently hurts the organization's effectiveness and efficiency.

Confidentiality

Confidentiality is an ethical principle that forms the cornerstone of information security in today's organizations (http://en.wikipedia.org/wiki/Confidentiality). Confidentiality refers to a nondisclosure agreement between two parties that restricts the access of others to certain private information (e.g., a patient's medical history). Such an agreement is often used when a company has a proprietary supply chain process or a new product that it wants another company to evaluate as a precursor to a comprehensive licensing agreement. Also, it can be used when one party wants to evaluate another's existing commercial product for a new and different application. As such, a confidentiality agreement often precedes licensing agreements or the acquisition of intellectual property rights. Generally, a confidentiality agreement performs the following three functions (Radack, 1994):

- Confidentiality protects sensitive technical or commercial information from disclosure to others. One or more participants in the confidentiality agreement may promise not to disclose technical information received from the other party. If the information is revealed to another individual or company, the injured party has a cause to claim a breach of contract and can seek injunctive and monetary damages.
- The use of confidentiality agreements can prevent the forfeiture of valuable patent rights. Under U.S. law and in many other countries, the public disclosure of an invention can be deemed as a forfeiture of patent rights for that invention. A properly drafted confidentiality agreement can avoid the undesired—and often unintentional—forfeiture of valuable patent rights.
- Confidentiality agreements define exactly what information can and cannot be revealed. This is usually accomplished by specifically classifying the type of information as confidential or proprietary. The type of information that can be included under the umbrella of confidentiality is virtually unlimited. Any information that flows between the parties can be considered confidential—the company's data, know-how, prototypes, engineering drawings, computer software, test results, tools, systems, and specifications. This list is certainly not exhaustive but illustrates the breadth of information that can be deemed confidential. The definition of this term is subject to negotiation. As one would imagine, the company or individual disclosing the confidential information (the "discloser") would like the definition to be as all-inclusive as possible; on the other hand, the company receiving the confidential information (the "recipient") would like to see as narrowly focused a definition as possible.

Reciprocity

Reciprocity occurs when buyers exert pressure on their suppliers to buy products or services from them. Although it is still legal for buyers and their suppliers to buy from and sell to each other, it is considered unethical and potentially illegal to force the trading partners to exchange favors (Scheuing, 1989). In other words, cross-dealings between the buyer and the supplier themselves are not a violation of the anti-trust act (i.e., Sherman Act), but may constitute a problem if the chosen supplier based on reciprocal trading is not the best source (Burt et al., 2003). Reciprocity also hampers the best possible purchase and sales because it limits buyers and suppliers' purchasing/sales options and thus could lead to the sub-optimization of purchasing/sales practices and then potentially increase supply chain costs.

Despite the potential harm to fair business practices, reciprocity is one of the trading principles of the World Trade Organization (WTO). In international trade, under a reciprocity principle, one country can lower its tariff rates and trade barriers (e.g., quotas) in return for similar concessions from other countries with comparable economies. The European Union (EU), for example, practices this kind of reciprocal deals among the member nations (Britannica.com, 2011). Considering the two contrasting sides of reciprocity in international trade, reciprocity can be subdivided into two categories: negative reciprocity and positive reciprocity. *Negative reciprocity* occurs when an action that has an adverse effect upon someone else is reciprocated with an action that has approximately equal negative effect upon another. If the reaction is not approximately equal in negative value, or worse, and the reaction has a much greater negative effect upon the first person, then the reaction will likely be judged unfair. Negative reciprocity fairness requires that negative actions be reciprocated in kind—a "quid pro quo" type of response. For example, the U.S. countervailing duty allows the U.S. government to raise its trade barriers in response to a foreign government subsidy on an exported product—the retaliatory tariff must be set equal in value to the value of the foreign export subsidy. On the other hand, *positive reciprocity* occurs when an action that has a favorable effect upon someone else is reciprocated with an action that has approximately equal positive effect upon another (Suranovic, 2010).

Concealed Damage

Product damage in transit is a mundane risk that can occur during every shipment. Despite their frequency, some product damages might not be noticed by the consignee at the time of delivery through ordinary visual inspection due to protective packages. Regardless, once these concealed damages are not notated on the delivery receipt within a specified time period, ranging from 24 hours to 15 days after delivery, the consignee is implicitly endorsing that delivered products are "free and clear" and thus is waiving the right to obtain a replacement for free of charge. Although it is unethical for a shipper or a carrier to knowingly conceal product damages, a consignee needs to be aware of the potential risk of concealed damage. Concealed damage may occur due to vehicle accidents, poor packaging, poor shipment handling, improper loading/unloading, inadequate climate control, and comingling of shipments with contaminated substance (e.g., chemicals).

Concealed damage is difficult to prove and settle because neither the shipper nor the carrier will take the blame unless the consignee can establish the fact that products were properly packed in a container, yet damaged during transit. According to so-called "15-Day" rules, the consignee has only 15 days after the delivery of a shipment to report any concealed damage (Tyworth et al., 1987). Also, the National Motor Freight Classification, ITEM 300135, states, "When damage to contents of a shipping container is discovered by the recipient (consignee) which could not have been determined at time of delivery, it must be reported by the recipient to the delivering carrier upon discovery and a request for inspection by the carrier's representative should be made. Notice of loss or damage and a request for inspection may be given by telephone or in person, but in either event must be confirmed in writing by mail." The best way to deal with concealed damage is to open up the package/container immediately after its delivery and then thoroughly inspect all sides of the product content, while writing the specifics of what is found on the carrier's copy of the delivery receipt.

Foreign Corrupt Practice Act

The Foreign Corrupt Practice Act (FCPA) of 1977 is an amendment to the 1933 Securities and Exchange Act that prohibits the promise of any "questionable" payments or anything of value to corrupt foreign government officials by U.S. citizens. Examples of questionable payments include illegal foreign political contributions, secret kickbacks for favorable business concessions, and money paid to foreign agents with inadequate documentation of purposes and value. FCPA, however, excludes illusory payments made to foreigners who are not government officials and thus are not acting in an official capacity for the foreign government (including the state-owned enterprise) or "grease" payments for a foreign official from its scope. *Grease* (or *facilitating) payments* are payments made to expedite or secure the performance of a routine government action. Routine government actions include obtaining permits or licenses, acquiring telephone services, processing official papers (e.g., stamping visas), clearing goods through customs, scheduling inspections, loading and unloading cargo, and receiving police protection (Stackhouse, 1993).

The main rationale for the enactment of FCPA is to stop U.S. firms from initiating or perpetuating the corruption of foreign governments and thus upgrade the public images of U.S. firms overseas. Also, FCPA may shield U.S. firms from the extortionate demands of foreign companies and governments for bribery. In addition to the anti-bribery provision, the accounting and record provision of FCPA requires U.S. firms to devise and maintain an accounting system that tightly controls and accurately records all dispositions of company assets. In other words, this provision intends to prohibit the establishment of "slush funds." U.S. firms that violate the anti-bribery provision are subject to severe penalty. To elaborate, U.S. firms can be fined up to $2 million while U.S. individuals (including officers and directors of companies that have willfully violated the FCPA) can be fined up to $100,000 and imprisoned for up to five years, or both. In addition, civil penalties may be imposed for violators (Gatti et al., 1997).

Green Supply Chain Management

Supply chain activities often produce undesired byproducts such as CO_2 by using fossil-burning fuels for transportation, emitting toxic chemicals resulting from the manufacturing process, and dumping material waste. These undesired byproducts accelerate the speed of global warming that threatens the well-being of the human population. In an effort to tackle global warming issues, supply chain professionals should pay more attention to "green" supply chain management (GSCM). Generally, GSCM can be defined as the incorporation of environment-friendly initiatives into every aspect of supply chain activities encompassing sourcing, product design and development, manufacturing, transportation, storage, packaging, and post-sales services, including end-of-product life management. Herein, the examples of environment-friendly initiatives include the company-wide environmental guidelines/policy, compliance with environmental regulations (e.g., EPA rules) and standards (e.g., ISO 14000), supplier certification and selection based on its commitment to sustainability, use of renewable energy (e.g., sunlight, wind, rain, geothermal heat), use of bio-fuels, use of degradable or compostable packages, and environmental performance monitoring (Min and Kim, 2012). The successful implementation of GSCM requires the following ten steps (Colby and Fertal, 2007):

1. *Know where you stand*. Understanding your organization's culture, supply chain processes, and consumption patterns is naturally the first step because you cannot manage what you cannot see.
2. *Have a performance plan*. Create a set of goals and performance metrics that can be used to track progress.
3. *Have a single point of accountability*. Have a single point of accountability that is empowered to affect change.
4. *Market your progress internally and externally*. Be sure to communicate to all levels why environmental initiatives are being undertaken, what will be measured, and how an organization is going to get there.
5. *Incorporate "green" into your existing sourcing, manufacturing, and logistics processes*. Factoring green priorities into your existing supply chain processes is an effective way to drive environmental initiatives and goals.

6. *Communicate your environmental goals and standards to your supply chain partners.* By setting clear expectations to your supply chain partners during the supply chain processes and proactively monitoring compliance, you can quickly improve your sustainability performance.
7. *Stay up to date with environmental regulations.*
8. *Keep up with new materials, technologies, and processes.* Develop new approaches that can cost effectively address the challenges and opportunities that environmental initiatives present.
9. *Do the "easy stuff" first.* An overhaul of your supply chain is not necessary to obtain benefits from sustainability efforts.
10. *Get everyone involved.* To be effective, you need to involve every unit, including groups of sourcing, manufacturing, sales, marketing, logistics, and finance across the organization and the supply chain for sustainability efforts.

With these steps in mind, the following subsections will elaborate on systematic tools needed for the management of products from their "cradle to grave" and various environment-friendly initiatives essential for GSCM.

Life Cycle Assessment

Life Cycle Assessment (LCA) is a technique to assess the environmental aspects and the full range of potential environmental impacts associated with a product, process, or service (U.S. Environmental Protection Agency, 2010):

- By compiling an inventory of relevant energy and material inputs and environmental releases
- By evaluating the potential environmental impacts associated with identified inputs and releases
- By interpreting the results to help an organization make a more informed decision

The main goal of LCA is to compare the environmental performance of products and services throughout their life cycle and be able to choose the least burdensome one. It also can be used to enhance the environmental friendliness of a single product (eco-design) or to improve the overall environmental performance and public image of an organization (http://www.greenoptions.com/wiki/life-cycle-assessments). Thus, LCA is useful for preventing pollution and assessing the long-term impact of green product design by helping its users develop environmental profiles of proposed environmental initiatives and green product/service design (e.g., renewable versus petro-chemical materials). LCA is also helpful for identifying areas where environmental improvements can be made.

LCA is typically composed of three interrelated components: a life cycle inventory analysis, a life cycle impact analysis, and a life cycle improvement analysis. The details of these components are described as follows (Svoboda, 1995, p. 2):

- **Life cycle inventory**—An objective, data-based process of quantifying energy and raw material requirements, air emissions, waterborne effluents, solid waste, and other environmental releases incurred throughout the life cycle of a product, process, or activity.
- **Life cycle impact assessment**—An evaluative process of assessing the effects of the environmental findings identified in the inventory component. The impact assessment should address both ecological and human health impacts, as well as social, cultural, and economic impacts.
- **Life cycle improvement analysis**—An analysis of opportunities to reduce or mitigate the environmental impact throughout the whole life cycle of a product, process, or activity. This analysis may include both quantitative and qualitative measures of improvement, such as changes in product design, raw material usage, industrial processes, consumer use, and waste management.

Within the preceding LCA framework, LCA can be performed by following the five distinct phases displayed in Figure 11.2.

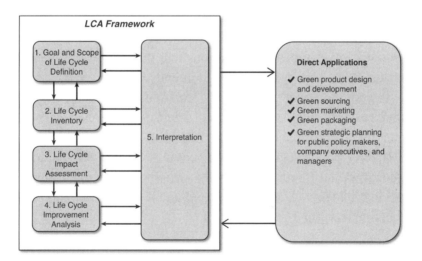

Figure 11.2. *The life cycle assessment phases (modified from Basset-Mens and van der Werf, 2007)*

Compliance with Environmental Regulations

Along with social well-being and economic benefits, regulatory compliance is one of the key drivers of GSCM. Therefore, it is imperative for supply chain professionals to understand the basic tenets of environmental regulations and keep up to date on changes in these regulations. Environmental regulations are policy guidelines or specifications that promote environmental safety by controlling human and business activities impacting our living environments (e.g., water, air, wildlife, habitats, plants, eco-systems). They aim to minimize the adverse impact of human and business activities on the environment and subsequently ensure the existing state of the environment without degradation. The rationale for environmental regulations includes environmental protection, improvement of social welfare/quality of life, inducement of cost-reducing innovation, and enhanced competitiveness in the marketplace (see Palmer et al., 1995; Shaw and Stroup, 1999). Despite controversy, compliance with environmental regulations is known to reduce the long-term cost of doing business (Hodges, 1997). For example, both Andersen Corporation and Ashland Chemical reported to reduce their costs of doing business by incorporating stringent environmental standards into supply chain activities such as purchasing, material handling, inventory control, and manufacturing (U.S. Environmental Protection Agency, 2000).

Environmental regulations can take a variety of forms: air regulation, water regulation, toxic waste regulation, chemical regulation, and transportation regulation. Some of the landmark environmental regulations that may have supply chain ramifications include the Clean Air Act, which intends to improve the nation's air quality and the stratospheric ozone layer; the Clean Water Act, which aims to restore and maintain chemical, physical, and biological integrity of the Nation's waters through prevention, reduction, and elimination of pollution; the Resource Conservation and Recovery Act (RCRA), which regulates the disposal of solid and hazardous waste; and the Hazardous Material Regulation (HMR), which governs the transportation of hazardous materials by highway, rail, vessel, and air. Detailed provisions of these regulations are summarized in Table 11.1.

Table 11.1. A Summary of Landmark Environmental Regulations

Regulations	Key Provisions
Clean Air Act (1955, 1970, 1997, 1990)	The Clean Air Act seeks to protect human health and the environment from emissions that pollute ambient or outdoor air.
	The Clean Air Act requires the Environmental Protection Agency to establish minimum national standards for air quality, and assigns primary responsibility to the states to assure compliance with the standards.
	The Clean Air Act establishes federal standards for mobile sources of air pollution, for sources of 188 hazardous air pollutants, and for the emissions that cause acid rain.
	The 1990 amendments of the Clean Air Act require that, in nonattainment areas, no federal permits or financial assistance may be granted for activities that do not "conform" to a State Implementation Plan (SIP). This requirement can cause a temporary suspension in funding for most highway and transit projects if an area fails to demonstrate that the emissions caused by such projects are consistent with attainment and maintenance of ambient air quality standards.
Clean Water Act (1977 and 1987)	The Clean Water Act authorizes federal financial assistance for municipal sewage treatment plant construction. Its regulatory requirements can also be applied to industrial and municipal dischargers.
	The Clean Water Act embodies the concept that all discharges into the nation's waters are unlawful, unless specifically authorized by a permit, which is the act's principal enforcement tool.
The Resource Conservation and Recovery Act (RCRA) of 1976 and 1986	The Resource Conservation and Recovery Act (RCRA) gives the EPA the authority to control hazardous waste from "the cradle to the grave." This includes the generation, transportation, treatment, storage, and disposal of hazardous waste. RCRA also sets forth a framework for the management of nonhazardous solid wastes. The 1986 amendments to RCRA enabled EPA to address environmental problems that could result from underground tanks storing petroleum and other hazardous substances.
	The objectives of the Resource Conservation and Recovery Act (RCRA) are to protect health and the environment and to conserve valuable material and energy resources by measures that include providing technical and financial assistance to state and local governments and interstate agencies for the development of solid waste management plans (including resource recovery and resource conservation) to promote improved solid waste management techniques; prohibiting open dumping on land; ensuring that hazardous waste is managed in a manner that protects human health and the environment; minimizing the generation and land disposal of hazardous waste by encouraging process substitution, materials recovery, properly conducted recycling and reuse, and treatment; and establishing a viable federal-state partnership to carry out the purposes of the Act. It is national policy that, wherever feasible, the generation of hazardous waste is to be reduced or eliminated as expeditiously as possible. Waste that is generated should be treated, stored, or disposed of as to minimize the present and future threat to human health and the environment.

Hazardous Material Regulation (HMR)	The HMR applies to each person who performs, or causes to be performed, functions related to the transportation of hazardous materials such as determination of, and compliance with, basic conditions for offering; filling packages; marking and labeling packages; preparing shipping papers; handling, loading, securing, and segregating packages within a transport vehicle, freight container, or cargo hold; and transporting hazardous materials.
	The HMR prescribes requirements for classification, packaging, hazard communication, incident reporting, handling, and transportation of hazardous materials.
	The HMR is enforced by RSPA and DOT's modal administrations: the FAA, the Federal Highway Administration (FHWA), the Federal Railroad Administration (FRA), and the United States Coast Guard (USCG). Federal law provides for civil penalties of not more than $25,000 and not less than $250 for each violation. An individual who willfully violates a provision of the HMR may be fined, under Title 18 U.S.C., up to $250,000, be imprisoned for not more than 5 years, or both; a business entity may be fined up to $500,000.

Data sources: Slightly modified from McCarthy, J.E. (2005), *Clean Air Act: A Summary of the Act and its Major Requirements,* CRS Report for Congress, Washington DC: Environmental Policy Resources, Science, and Industry Division.

Copeland, C. (2010), *Clean Water Act: A Summary of the Law*, CRS Report for Congress, Washington DC: Congressional Research Service

U.S. Environmental Protection Agency (1986), "Summary of the Resource Conservation and Recovery Act," http://www.epa.gov/lawsregs/laws/rcra.html, retrieved on May 25, 2011

Federal Wildlife and Related Law Handbook (1996), "Resource Conservation and Recovery Act of 1976," http://wildlifelaw.unm.edu/fedbook/rcra.html, retrieved on May 25, 2011

Oklahoma State University (2011), "An Overview of HAZMAT Transportation," http://ehs.okstate.edu/modules/dot/index.htm, retrieved on May 25, 2011.

ISO 14000 Standards

ISO 14000 is a series of voluntary environmental management standards developed by the International Organization for Standardization (ISO) that help the organization minimize the negative environmental impacts of its business activities by reducing its carbon footprint and waste. It also provides a set of guidelines that help the organization comply with environmental regulations/laws and continue to improve its environmental performances. Unlike ISO quality standards such as ISO 9000, ISO 14000 is considered a *generic* management system standard because it can be applied to any organization, large or small, whatever its product or service, in any sector of the industry, whether it is a private enterprise, public administration, or government organization. ISO 14000 can be subdivided into the following family of standards:

- Environmental Management Systems (14001, 14002, 14004)
- Environmental Auditing (14010, 14011, 14012)
- Evaluation of Environmental Performance (14031)
- Environmental Labeling (14020, 14021, 14022, 14023, 14024, 14025)
- Life Cycle Assessment (14040, 14041, 14042, 14043)

All but ISO 14001 standards are for guidance. ISO 14001 is a management standard that is characterized as follows (http://www.qsae.org/web_en/pdf/ISO14000Concepts.pdf):

- **Comprehensive**—It requires all members of the organization to participate in environmental protection. In other words, ISO 14001 considers all stakeholders and includes all processes that can help the organization identify all aspects of environmental impacts.
- **Proactive**—It focuses on forward thinking and action instead of reacting to command and control policies.
- **Systematic**—It stresses improving environmental protection by using a single environmental management system across all functions of the organization.

ISO 14001 can be implemented by following these five steps (http://14000store.com/ISO-14000-info-what-is-an-EMS.aspx):

1. Review your organization's impact on the environment ("Where are we today?").
   - Start an Initial Environmental Review as merely a baseline, but not an objective.
2. Define goals for environmental performances ("Where do you want to be tomorrow?").
   - Meet environmental regulations.
   - Define requirements for certification.
   - Meet customer requirements.
   - Reduce scrap or waste.
3. Create a plan to achieve the goals ("What are you going to do about it?")
   - Develop environmental management systems (EMS)
4. Monitor performance against goals ("Are we meeting the goals and realizing the benefits?").
   - If it isn't measured (and recorded), it can't be improved!
5. Report the results and then review/audit the results and continuously improve ("What corrective actions should be taken?").
   - If you meet your first goals, create more!

Once all the ISO 14000 standards are in place, the organization can reap various managerial benefits: savings in consumption of raw materials and energy, reduced use of resources, improved process efficiency, reduced waste generation and disposal costs, utilization of recoverable resources, reduced compliance costs, improved profits, and improved corporate image among regulators, customers, and general public (ISO, 2011).

Environmental Audits

An *environmental audit* generally refers to a systematic, documented, and objective assessment of an organization's environmental activities that helps the organization to determine whether or not it is in compliance with applicable environmental laws and regulations, to raise the internal awareness of the organization's environmental policy, to maintain credibility with the organization's stakeholders and the general public, and then to explore the future opportunities for the continuous improvement of environmental performances. The overall objective of an environmental audit is to help the organization safeguard the environment and minimize risks to human health. Specifically, its key objectives are as follows (http://www.ilo.org/safework_bookshelf/english?content&nd=857170621):

- To determine how well the environmental management systems and equipment are performing
- To verify compliance with the relevant national, local, or other environmental laws and regulations
- To minimize human exposure to risks from environmental, health, and safety problems.

The environmental audit process is composed of three steps, as summarized in Table 11.2.

Table 11.2. An Environmental Audit Process

Pre-Audit	Audit	Post-Audit
• Ensure support from both internal and external stakeholders. • Determine the scope of the audit. • Develop performance metrics and benchmarks. • Establish clear policy guidelines. • Select and authorize an audit team. • Collect necessary data.	• Review the level of compliance with environmental policies and procedures. • Check the current status and conditions of environmental activities for progress. • Perform appropriate tests. • Interview parties responsible for green supply chain activities.	• Summarize and document findings. • Make an action plan based on results for going forward. • Report any improvement plans to relevant internal and external stakeholders. • Make new plans for repeating the audit process in the future.

Data source: bcorporation.net (2006), *Conducting Environmental Audits and Reviews,* B Resource Guide, http://www.bcorporation.net/resources/bcorp/documents/B%20Resources%20-%20Environmental%20Audit.pdf, retrieved on May 28, 2011.

By following the preceding process, the environmental audit covers the following aspects (GTZ-APIIC Corporation, 2008):

- **Manufacturing process and mass balance of various materials used**—Process flow, optimal unit operations, materials used (raw materials, water, etc.), products and by-products, wastes (gaseous, liquid, solid) generated, and unit operations of poor performance
- **Emissions**—Identification of unit operations and processes where emissions are generated; type, characteristics, and quantity of emissions and their control measures; and compliances with norms, standards, and so forth under the Clean Air Act
- **Water balance**—Identification of unit operations and processes where water is used, where waste water is generated, its quantity and quality, pretreatment, final treatment, compliance with the norms, standards, and so forth under the Clean Water Act
- **Wastes**—Identification of unit operations and processes where wastes (hazardous and nonhazardous waste) are generated, their quantity, quality, storage, treatment, disposal, and compliances with the Hazardous Waste Management Rules
- **Pollution control measures**—Adequacy and performance
- **Potential impact of handling, transportation, and storage of materials used and products and byproducts generated by the industry on the surrounding environment**—Water bodies, ground water, surrounding population and agriculture, and so on
- **Compliances with environmental acts, rules, norms, and standards**—Clean Air Act, Clean Water Act, and Hazardous Waste Management as per the Environment Protection Act
- **Suggestive measures for environmental compliance**

Here are the key benefits of environmental audits:

- Identification of the area(s) of poor environmental performances needing improvements
- Reduction in the risk of litigation and fines arising from potential environmental breaches
- Better utilization of the organization's resources by reducing the use of virgin raw materials and the unnecessary waste, and identifying the opportunities for recycling, reusing, and remanufacturing
- Improvement of the organization's opportunity to gain organization-wide certifications such as ISO 14001 and Cradle to Grave, or product-specific certifications from Energy Star, LEED (Leadership in Energy and Environmental Design), the Forest Stewardship Council, Scientific Certification Services, and the Green Seal.

Chapter Summary

The following are key lessons learned from this chapter:

- Considering the profound impact of supply chain activities on everyone's daily lives (e.g., air pollution from emission of carbon dioxide through manufacturing/logistics activities, economic development through the hiring of labor forces), today's supply chain activities are geared toward both financial and nonfinancial goals. The nonfinancial goals include the awareness of corporate social responsibility (CSR) that helps improve the well-being of the community surrounding the company and enhance the quality of life of employees working for the company. The inclusion of these nonfinancial goals in the corporate strategy became the foundation for the so-called "triple bottom line."

- Because everyday business activities are often bound by various legal constraints and ethical dilemmas, supply chain professionals need to possess and utilize an in-depth knowledge of business laws and ethics that have supply chain implications. In particular, the globalization of supply chain activities increases the exposure of supply chain professionals to unfamiliar foreign business customs/laws and thus creates greater challenges for managing supply chains. For example, two different trading or supply chain partners may have disagreements over their supply chain roles and contracts. Then, the question may arise how those conflicts can be resolved in a manner that a relationship between those two parties can be sustained without serious damages. To answer such a question, supply chain professionals should look into four stages of conflict resolution: negotiation, mediation, arbitration, and litigation.

- The commercial laws are composed of the law of agency and the law of contracts. The law of agency sets limits on the authority of the agent. The law of contracts specifies provisions for legally binding contracts. To make the contract "legally binding," the contract requires *mutual assent* between two contracting parties, *sufficient consideration* by those parties, the *competency* of those parties, and a *legal purpose* of the contract terms.

- As an amendment of the Securities Exchange Act of 1934, the Foreign Corrupt Practice Act (FCPA) was enacted by the Carter Administration in 1977. This Act intended to curtail the involvement of U.S. multinational firms in foreign commercial bribery activities. This Act prohibits U.S. firms from making any form of payments (of substantial value) to foreign government officials with intent to influence their decision and corrupt them.

- Environment-friendly practices require the transformation of various supply chain practices such as supplier selection/evaluation, carrier/modal choice, new product design and development, packaging, labeling, material handling, waste disposal, and scrap disposition. This transformation, in turn, necessitates the careful formulation of green supply chain strategy in compliance with various

environmental standards and regulations. Some of the notable regulations include the Clean Air Act, the Clean Water Act, the Resource Conservation and Recovery Act, and the Hazardous Material Regulation.
- Life Cycle Assessment (LCA) allows the firm to identify the true cost of environmental initiatives throughout the entire life cycle of a product. As such, LCA is an important tool for assessing the long-term impacts of environmental initiatives and then providing opportunities to improve current environment-friendly practices.
- ISO 14000 is a series of international environmental standards that provide a framework for the development of an environmental management system (EMS), environmental audit programs, and the continuous improvement of environmental performance and positions. EMS follows a Plan-Do-Check-Act cycle in the following manner: Plan (develop environmental policy and plan environmental initiatives); Do (implement environmental initiatives); Check (monitor the performance of environmental initiatives); Act (take the necessary corrective actions).

Study Questions

The following are important thought starters for class discussions and exercises:
1. What comprises the "triple bottom line?" Why is there a need for the triple bottom line? How does the triple bottom line differ from traditional financial goals?
2. Define the term *corporate social responsibility* (CSR).
3. What is the best way to resolve contract disputes?
4. How do you differentiate between ethics and laws? Are ethics related to laws?
5. Categorize the U.S. laws.
6. What is the legal definition of bribery?
7. What are "questionable" payments? Provide examples of questionable payments.
8. What are "grease" payments? Are such payments legal?
9. What are key provisions of the Securities Exchange Act (SEC)?
10. What are the supply chain ramifications of the Sarbanes Oxley Act of 2002? How is it related to the Securities Exchange Act?
11. What is the rationale behind the Foreign Corrupt Practice Act (FCPA)? Why is FCPA so controversial?
12. How can environment-friendly practices transform supply chain practices? What are the biggest challenges for the implementation of green supply chain management?

13. Choose a product manufactured by the company you work for (or have worked for in the past) and then conduct its life cycle analysis from an environmental perspective.
14. What does ISO 14000 have to do with green supply chain management?
15. What advantages does the company have with ISO 14000 certification?

Case: Jumping Footwear, Inc.

Jumping Footwear (JF) has been manufacturing high-end premium comfort footwear, including work boots and sneakers, for people of all ages since 1954. Due to its commitment to high-quality and dedicated customer services, its business continued to grow and enjoyed a great run up until 2007. Its peak sales and earnings reached $235 million and $45 million in 2007, respectively, thanks in part to its innovative reflexology technology and unmatched "three-year guarantee." After the economic bubble burst in 2009, however, its sales began to get stagnant and profit declined substantially. For the last several years, the founder's eldest son and current CEO, Jimmy Stone, felt a lot of heat from disgruntled shareholders whose dividends shrank by one-third in value. To make it worse, Jimmy has just learned that a head of his subsidiary in China, Larry Chen, has been paying bribes to local government officials in Guangzhao, China. Although Jimmy does not want to believe that Larry has ever involved in the bribery scheme, he found traces of evidence that Larry paid bribes through customs brokers to facilitate the entry of JF's key materials from Indonesia to China without the necessary paperwork and proper inspection. The bribes were disguised as "miscellaneous loading and delivery expenses," as if those are part of legitimate charges. Also, Larry claimed that such practice has long been accepted by the Chinese government and was needed for him to maintain a good rapport with the local Chinese government officials essential for long-term business opportunities in China.

Results-Only Work Environments (ROWE)

Achieving a double-digit annual growth always has been Jimmy's overriding goal, no matter what the circumstances may be. The question was how such a goal could be attained. The culture at JF became a mirror image of Jimmy's character: ambitious, tenacious, demanding, and goal oriented. Once goals were set, Jimmy rarely accepted any excuses and shortfalls. There is no exception to this mindset. All of JF's managers and employees, including ones working for JF's overseas subsidiaries, are expected to follow this business philosophy. Such a philosophy seems to bring remarkable business success for a while. But things started to unravel in the late 2000s as growth slowed in

Asia and North America, where competition intensified and financial pressure mounted as a result of the global financial crisis. The Chinese subsidiary was the cash cow for JF's global business operations due to the rapid increase in Chinese middle-class consumers who yearned for premium foreign brands such as JF, Ralph Lauren, Adidas, and Nike. Unfortunately, for the last several years, the Chinese subsidiary was marginally profitable after its growth stalled. However, the unyielding growth targets remained the same. To cope with this environment, Larry was forced to take many shortcuts. First, to remove the unwanted inventory, Larry tried end-of-month promotion with deep discounts to JF's distributors and retailers, one after another. This practice diluted its brand by deluging its local market in China. Second, in January of 2013 and 2014, according to many internal sources, Larry Chen sent out lavish gifts, including vintage wine, perfume, dresses, and handbags (valued at $300 to $15,000 each) to the top managers of its retail stores and several local government officials a few days before the Chinese New Year's celebration. Third, there has been some speculation that JF's Guangzhao subsidiary—once the admired role model of their international business success—might have booked huge sales to Guangdong and Shenzhen region distributors, without actually shipping products to them. Instead, most of the goods might have been funneled into the gray and/or black market where profit margins are much lower. This means that some of the reported sales revenue in JF's Guangzhao subsidiary might be nothing more than bogus. If this rumor turns out to be true, JF will be subject to both a Securities Exchange Commission (SEC) investigation and a shareholder class action lawsuit accusing the company of misleading investors by falsely inflating sales and earnings.

Are these isolated problems? Some think so. According to Sam Perkins, JF's corporate lawyer, it is not unusual that when the company has operations scattered throughout the world, people do things they should not. More often than not, some questionable practices such as gift giving, exchange of favors, and extra fees for expedited services are all too common in other countries and therefore are accepted as typical business customs in those foreign countries. Although Sam does not mean to endorse and condone such practices, Sam reminded Jimmy that many expatriates are not aware of the seriousness of certain questionable practices (e.g., violation of the Foreign Corrupt Practice Act).

Painful Decisions in Waiting

Near the end of a highly acrimonious three-hour overseas phone call with Larry Chen, Jimmy asked him, "What will it take for you to resign?" After a stunned silence and sobbing, he stammered that he has been made a scapegoat and still intends to remain until he can consult with his attorney. A week later, Jimmy's old buddy tipped him off that a prominent business newspaper in the U.S. planned on publishing an investigative report on JF's Chinese operations and interviewing Larry as its anonymous source for the telltale story. Jimmy is aware that Larry has been an effective, talented manager who is well connected to the Chinese business community and the government. He not only can speak both English and Mandarin/Cantonese fluently, but also has expertise in international finance and logistics. Jimmy is in great pain and does not know exactly what he is supposed to do.

Discussion Questions
1. What is wrong with the JF's management practices? Which of JF's practices are considered unethical or illegal? (Be as specific as possible and explain why they are unethical or illegal.)
2. Do you think that Larry's dismissal will be justified? (If so, why? If not, why not?)
3. If you were JF's CEO, what actions (e.g., termination of customs brokers) should be taken as an immediate next step?
4. If you have to play the role of JF's CEO, how would you change the management practices, policies, and corporate culture of JF?
5. If you belonged to the top management of JF, what strategic changes would you like to recommend to Jimmy to prevent the recurrence of the current problems? (Justify your strategic plans.)

Bibliography

Basset-Mens, C. and van der Werf, H. (2007), "Life Cycle Assessment of Farming Systems," *The Encyclopedia of Earth*, July 10, http://www.eoearth.org/article/Life_cycle_assessment_of_farming_systems, retrieved on May 25, 2001.

Beer, J. (1997), *The Mediator's Handbook*, 3rd edition, Montpelier, VT: Capital City Press.

Bigelow, M.M. (1905), "Definition of Law," *Columbia Law Review*, 5(1), 1.

Bloomberg Businessweek (2014), "Japan Feels the Heat on Bribery," *Bloomberg Businessweek*, November 3–9, 17–18.

Burt, D.N., Dobler, D.W., and Starling, S.L. (2003), *World Class Supply Management: The Key to Supply Chain Management*, New York, NY: McGraw-Hill Higher Education.

Colby, K. and Fertal, D. (2007), "Ten Steps to a Green Supply Chain," *EETimes Supply Network*, October 30, 2007, http://www.eetimessupplynetwork.com/, retrieved on May 5, 2008.

Economist (2009), "Triple Bottom Line: It Consists of Three Ps," November 17, http://www.economist.com/node/14301663?story_id=14301663, retrieved on January 17, 2011.

Encyclopedia Britannica (2011), "Reciprocity," http://www.britannica.com/EBchecked/topic/493561/reciprocity, retrieved on May 20, 2011.

Fletcher, M.A. (2009), "Questionable Government Payments on Rise, Administration Says," *The Washington Post*, November 7, http://voices.washingtonpost.com/44/2009/11/questionable-government-paymen.html, retrieved on April 16, 2011.

Garner, B.A. (2006), *Black's Law Dictionary*, St. Paul, MN: West Publishing Company.

Gatti, M.M, Ogrady, C.R.G., and Morgan, O.F. (1997), "Foreign Corrupt Practices Act," *FindLaw for Legal Professionals*, http://library.findlaw.com/1997/Jan/1/126234.html, retrieved on May 23, 2011.

GTZ-APIIC Corporation (2008), "Concept Note on Environmental Audits," http://www.ecoindustrialparks.net/new/pdfs/Env%20audit.pdf, retrieved on May 28, 2011.

Hodges, H. (1997), "Falling Prices: Cost of Complying with Environmental Regulations Almost Always Less Than Advertised," Unpublished Briefing Paper, Washington, DC: Economic Policy Institute.

Honeyman, C. and Yawanarajah, N. (2003), "Mediation," BeyondIntractability.org, http://www.beyondintractability.org/essay/mediation/, retrieved on April 10, 2011.

ISO (2011), "Business Benefits of ISO 14000," http://www.iso.org/iso/iso_catalogue/management_standards/environmental_management/business_benefits_of_iso_14000.htm, retrieved on May 28, 2011.

Kerley, P.N., Hames, J.B., and Sukys, P.A. (2001), *Civil Litigation*, 3rd edition, Albany, NY: Delmar-Thomson Learning.

Koch, Jr., C.H. (2010), *Administrative Law and Practice*, 3rd edition, Egan, MN: West.

Kramer, H.S. (1998), *Alternate Dispute Resolution in the Work Place*, Brooklyn, NY: Law Journal Press.

Min, H. and Kim, I. (2012), "Green Supply Chain Research: Past, Present and Future," *Logistics Research*, 4(1), 39–47.

Moore, C.W. (1996), *The Mediation Process: Practical Strategies for Resolving Contracts*, 2nd edition, Boulder, CO; Jossey-Bass, Inc.

Palmer, K., Wallace, E.O. and Portney, P.R. (1997), "Tightening Environmental Standards: The Benefit-Cost or No-Cost Paradigm," *Journal of Economic Perspectives*, 9(4), 119–132.

Radack, D.V. (1994), "Understanding Confidentiality Agreements," *Journal of the Minerals, Metals, and Material Society*, 46(5), 68.

Scheuing, E.E. (1989), *Purchasing Management*, Englewood Cliffs, NJ: Prentice-Hall, Inc.

Shaw, J.S. and Stroup, R.L. (1999), "Do Environmental Regulations Increase Economic Efficiency," *Regulation 13*, 23(1), 1–2.

Stackhouse, D. (1993), "The Foreign Corrupt Practices Act: Bribery, Corruption, Recordkeeping and More," *Icemiller LLP: Legal Counsel*, http://www.icemiller.com/publication_detail/id/45/index.aspx, retrieved on May 23, 2011.

Svoboda, S. (1995), *Note on Life Cycle Analysis*, Unpublished Report, National Pollution Prevention Center, Ann Arbor, MI: University of Michigan.

Suranovic, S.M. (2010), *International Trade: Theory and Policy*, Irvington, NY: Flat World Knowledge Inc.

Tyworth, J.E., Cavinato, J.L., and Langley, Jr., C.J. (1987), *Traffic Management: Planning, Operations, and Control*, Prospect Heights, IL: Waveland Press, Inc.

U.S. Environmental Protection Agency (2000), *Enhancing Supply Chain Performance with Environment Cost Information: Examples from Commonwealth Edison, Andersen Corporation, and Ashland Chemical*, Unpublished Report, Washington DC: Office of Pollution Prevention and Toxics.

U.S. Environmental Protection Agency (2010), "Life Cycle Assessment (LCA)," *Life Cycle Assessment Research*, August 6, http://www.epa.gov/nrmrl/lcaccess/, retrieved on May 25, 2011.

12

Measuring the Supply Chain Performance

"If you aim at nothing, you'll hit it every time."
—Zig Ziglar

"An acre of performance is worth a whole world of promise."
—Red Auerbach

Learning Objectives

After reading this chapter, you should be able to:

- Grasp a big picture view of why supply chain performance measures are essential to business success.
- Identify key performance metrics, including key performance indicators (KPIs), in the supply chain management.
- Understand unique challenges of measuring cross-organizational supply chain performance in comparison to individual organizational performance.
- Understand the role of balanced scorecard approach in measuring the supply chain performance.
- Learn about the Supply Chain Operations Reference (SCOR) model and leverage it to measure the supply chain performance.
- Learn to improve supply chain performance continuously and deal with organizational resistance to changes in supply chain practices required for continuous performance improvement.

Supply Chain Performance and Its Impact on the Bottom Line

It is hard to imagine any basketball player scoring 101 points in an NBA game, a pitcher winning 512 games, or a sprinter running 9.58 seconds for a 100 meter dash. However, these seemingly "unassailable" records were set by remarkable human beings. Although these records seem to be unbreakable, we cannot rule out the possibility that these records will be broken by someone some day. In any competition, records are meant to be broken due to the competitive nature of human beings. Reflecting human nature, today's business world is characterized by never-ending competition and continuous improvement of organizational performance. In this fast-paced, increasingly competitive business world, an organization that cannot get the job done faster or better than its competitors is unlikely to survive. Therefore, the organizational performance can be directly translated into its competitiveness. Because we cannot improve organizational performance without measuring it, organizational performance measurement has been one of the most important management practices. Generally speaking, there are at least five different reasons why measuring organizational performance is essential for business success (Kaplan, 2009, pp. 4–5):

- **Improvement**—By tracking performance, an organization can identify and address potential problems, such as declining market share, flattening profits, and rising employee turnover.
- **Planning and forecasting**—Performance measurement can serve as a progress check or a monitoring tool that can help the organization determine whether it is meeting its performance goals and whether it needs to be revised to reflect a forecasted reality.
- **Competitive edge**—When the organization compares its performance against its rivals and industry benchmarks, it can identify its performance gaps and thus sharpen its competitive edge by reducing those gaps.
- **Reward**—By knowing how much any particular units or employees have excelled in achieving the organization's performance goals, managers can fairly distribute performance-based incentives (e.g., merit-based pay) and rewards to deserving units and employees.
- **Regulatory and standards compliance**—Performance measurement helps the organization to comply with government regulations (e.g., environmental and safety regulations) and international quality standards (e.g., ISO 9000 or 14000 guidelines).

As just specified, performance measurement is necessary for many reasons. However, without knowing its direct impact on the company's bottom line, it would be difficult to convince top management to commit themselves to performance measurement. As such, some efforts have been made to assess the impact of supply chain excellence on the company's bottom line. A 1997 Integrated Supply Chain Benchmarking Study revealed that best-in-class supply chain performance could give a firm 45% total supply chain cost advantage over its average rival (Stewart, 1997), the rationale being that cash flows are dramatically improved with enhanced supply chain performance due to reduction in total cycle time. Similarly, another study conducted by

Scott and Westbrook (1991) reported that improved supply chain performance could lead to lower costs, higher profit margin, enhanced cash flow, revenue growth, and a higher rate of return on assets. Indeed, companies participating in the MIT's Integrated Supply Chain Management Program reported a 17% revenue increase as a result of improved supply chain performance (Quinn, 1997). As these examples have illustrated, improved supply chain performance positively influences the company's bottom line. Regardless, the assessment of supply chain performance from a single company's perspective can be misleading given that the supply chain involves many different companies that cut across different industries with different roles in a given supply chain. In other words, supply chain performance and its impact on business success should be evaluated based on the performance of multiple companies involved in the supply chain as an integrated whole, rather than a collection of separate entities. This means that traditional performance metrics intended for an individual company or an individual business function are no longer relevant to the measurement of supply chain performance. Therefore, new performance metrics that can embrace the multidimensional, multifunctional, and multi-organizational aspects of the supply chain should be developed. Though such metrics are difficult to find, the next section will explore a variety of performance metrics that may be more relevant to the assessment of supply chain performance.

Supply Chain Performance Metrics

Given the importance of supply chain performance to business success, you would like to know how well your organization and its supply chain partners are performing in comparison to your competitors' supply chain performance. Once you discover that your organization and its supply chain partners' supply chain performance are lagging behind your competitors, you would like to know what the sources of the lagging supply chain performance are and how you can improve your organization and its partners' supply chain performance. In order to answer these critical questions, you should develop a yardstick that can determine both the effectiveness and efficiency of the organization, its supply chain partners, and its competitors. To have more meaningful measures of the supply chain performance, such a yardstick should have the following traits:

- **Multidimensional**—Because the supply chain spans many different boundaries of the business functions and organizations, its performance metric has to cover more than one aspect of business performance. For example, the company's effort to reduce inventory carrying costs will increase the possibility of stockouts/backorders and thus lower the order fill rate. That is to say, the use of a traditional single-dimensional measure such as inventory carrying cost may be irrelevant to the measurement of supply chain performance.

- **Specific**—Although the supply chain performance measure should reflect a big-picture performance of cross-functional and cross-organizational integrated activities, it should be specific enough to pinpoint exactly what you are measuring. For example, a broadly defined traditional measure such as capacity utilization, logistics productivity, and information accuracy may create confusion among the different organizations and their different units.
- **Universal**—A performance measure that can only be used for a particular company or industry will not be appropriate for a meaningful supply chain performance metric. For example, a traditional manufacturing measure such as the machine idle time, the number of defects, and cost per inspection may not be a meaningful supply chain performance metric because it cannot be universal across the entire supply chain, which can include non-manufacturing industry sectors as the supply chain partners.
- **Repeatable**—Because supply chain performance has to be monitored constantly over time, a one-time measure will not be appropriate for measuring the supply chain performance. For example, the size of the organization or the size of a particular facility such as a warehouse and a truck terminal at a certain point of time cannot be a proper supply chain performance metric.
- **Quantifiable**—Some customer service measures such as the level of customer satisfaction and the extent of customer loyalty are hard to quantify. However, those measures should be translated into numerical scales to provide a clear yardstick for objectively evaluating the comparative performance of a particular supply chain against others. Therefore, metrics that can be expressed in numerical terms (e.g., percentage, ratio, time) are more useful for measuring supply chain performance.
- **Traceable**—Performance measures that are based on confidential information, proprietary information, or untraceable data may not be appropriate for measuring supply chain performance. If the necessary performance data is hard to come by, metrics based on such data will be difficult to implement and therefore will be less meaningful.

Unfortunately, there has been no evidence to date that supply chain performance metrics that satisfy the preceding traits actually exist due in part to the complexity involved in measuring performance across the entire supply chain (Lambert and Pohlen, 2001). In most instances, metrics that management and academics refer to as "supply chain performance metrics" are limited to single-dimensional, function-specific measures, such as those listed in Table 12.1 (e.g., Lapide, 2000, p. 5). As shown in Table 12.1, a few exceptions are considered cross-functional measures, they but are still lacking specifics.

Table 12.1. A List of Commonly Used Supply Chain Performance Metrics

Customer Service Measures	Marketing-Related Measures	Financial Measures
• Order fill rate • Backorders/stockouts • Customer defection rate • Frequency of customer complaints/disputes • Order accuracy • Order cycle times • Product return rate	• Demand forecasting accuracy • Time to market • Market share • Sales revenue • Revenue growth	• Cost of goods sold • Cash flow • Profit margin • Return on investment • Debt ratio • Cash-to-cash cycle time
Purchasing-Related Measures	**Manufacturing-Related Measures**	**Logistics-Related Measures**
• Percentage supplier managed inventory • Supplier delivery reliability and time • Incoming material/part quality • Unit purchase costs • Frequency of expedited shipments • Supplier invoicing error	• Product defect rate • Costs of quality • Cost of production per unit • Setup/CHANGEOVER costs • Production stoppage • Percentage of scrap/rework • Overtime usage • Production cycle time • Wage rate • Labor productivity • Plant capacity • Plant utilization	• Total landed cost • Inventory turns • On-time delivery • Delivery consistency • Lines picked/hour • Damaged shipments including concealed damage • Inventory accuracy • Order picking accuracy • Shipment accuracy • Warehouse space utilization • Inventory shrinkage/pilferage • Cost of carrying inventory • Documentation accuracy • Transportation costs • Warehousing costs • Container utilization • Truck cube utilization

To overcome the shortcomings of the existing supply chain performance metrics listed in Table 12.1, efforts were made to simplify but capture the cross-functional aspects of supply chain performance. One of these efforts led to the development of the supply chain performance metrics summarized in Table 12.2 (Bolstroff and Rosenbaum, 2003, p. 70). These metrics became the foundation of the Supply Chain Operations Reference (SCOR) model that will be discussed later.

Table 12.2. Supply Chain Performance Metrics with a Cross-Functional Emphasis

Performance Category	Performance Metrics	Notes
Supply Chain Delivery Reliability: The performance of the supply chain in delivering the correct product, to the correct place, at the correct time, in the correct condition, in the correct quantity, with the correct documentation, to the correct customer.	Delivery performance	A measure of fulfilling the customer demand by the designated deadline.
	Fill rates (or line item fill rates)	A percentage of customer orders satisfied from inventory at hand.
	Perfect order fulfillment	A percentage of orders delivered on time and in full without quality failures or missing required documentation.
Supply Chain Responsiveness: The velocity (speed) at which a supply chain provides a product to the customer.	Order fulfillment lead time	A summation of order-processing and outbound transit time.
Supply Chain Flexibility: The agility of a supply chain in adapting to marketplace changes to gain or maintain competitive advantage.	Supply chain response time	A summation of new product development time, machine setup time, changeover time, customer response time, and delivery time.
	Production flexibility	A measure of how quickly production capacity can be adjusted to changing customer demand.
Supply Chain Cost: The costs associated with operating the supply chain.	Total supply chain costs	A summation of production cost, order management cost (e.g., customer service, field warehouse, outbound transportation) and material acquisition cost (e.g., purchasing, inbound transportation).
	Warranty/returns processing costs	An indirect measure of the rate at which the product is expected to fail.
Supply Chain Asset Management Efficiency: The ability of an organization in managing assets needed to meet end-customer demand. This includes the management of all assets such as facility/equipment and working capital.	Cash-to-cash cycle time	A measure of how long it takes to convert a dollar spent on materials, labor, and so forth to cash in hand.
	Inventory days of supply	A measure of how long the inventory level of a certain product will be sufficient to match the expected customer demand.
	Asset turns	The total amount of sales revenue generated for every dollar's worth of assets owned by the company.

Among the more refined supply chain performance metrics listed in Table 1.2, cash-to-cash cycle time has emerged as an increasingly popular measure because it can help the organization assess how lean its supply chain operations are with regard to its operating capital (cash at hand). Therefore, this measure reflects the organization's ability to get engaged in a greater number of value-adding activities with available cash that can be invested in those activities. The faster cash-to-cash cycle time has often led to improved earnings per share and the subsequent bottom line of the company (Ward, 2004).

Key Performance Indicators (KPIs)

Generally speaking, a key performance indicator (KPI) is a measure of how well an organization is making progress toward its strategic goals and can be used to define its critical success factor (CSF). Because every organization's goals and mission statements are different, the KPI will vary from one organization to another, one business function to another, and one industry to another. For example, the KPIs for a nonprofit organization such as a university may include student graduation rates and job placement rates, whereas the KPIs for a for-profit organization such as a company may include annual revenue growth and after-tax profits. The KPIs in the manufacturing sector may include an asset turn ratio, a percentage of rework during the scheduled period, and unscheduled downtime, whereas the KPIs in the logistics sector may include customer order-to-delivery time, total logistics cost as a percentage of sales, and order fill rate. Regardless of these differences, the KPIs can be categorized into three types (Kaplan, 2009):

- **Process KPI**—Measures the efficiency or productivity of a business process such as product repair cycle time and a number of rings before a customer phone call is answered
- **Input KPI**—Measures assets and resources invested in or used to generate business results such as dollars spent on research and development or budgets set aside for employee training
- **Output KPI**—Measures the financial and nonfinancial results of business activities such as return on investment (RPI), economic value-added (EVA), and market share.

EVA is a measure of a company's financial performance based on "residual wealth" ("economic rent" or "business value") calculated by deducting cost of capital from its operating profit (Investopedia, 2010). EVA is aimed at calculating the company's true economic cost. Put simply, EVA is the difference between return achieved on resources invested and the cost of resources. The higher the EVA, the better the level of resource unitization would be. It can be calculated as follows (Boehlje, 2010; Kaplan, 2009):

EVA = Earnings Before Interest and Taxes (EBIT) − Interest − Net Income − Cost of Equity Capital

or

EVA = Operating Profit Before Taxes − Federal, State, and County Taxes − (Total Capital Employed × Cost of Capital)

To understand and identify specific KPIs relevant to a particular organization or its units, one should ask the following questions:

- What are your company's strategic goals, mission statements, and value propositions? Which industry does your company belong to?
- Which business activities should be prioritized to maximize your customers' value?
- What are the best ways to communicate the progress of your company's performance improvement efforts to your company's customers and stakeholders?
- How can you visualize and simplify your performance improvement efforts?

A Balanced Scorecard for Supply Chain Performance Measurement

Over the years, financial measures such as profitability and annual revenue growth have been used in practice extensively to monitor a firm's performance. However, those measures have often led the firm focusing on its short-term performance, which does not necessarily reflect the firm's true operating efficiencies and therefore cannot tell its long-term competitive position in the marketplace. To overcome this shortcoming of the traditional finance-based measures, Kaplan and Norton (1992) introduced the concept of a balanced scorecard (BSC) to supplement traditional financial measures with criteria that measure business performance from three additional perspectives: customers, internal business processes, and innovation and learning, as shown in Figure 12.1. The details of these perspectives are summarized in Table 12.3, along with their potential measures. To develop a meaningful BSC that provides a balanced portfolio of the supply chain performance, these processes all have to be linked together and closely tied to supply chain activities. Therefore, the BSC framework displayed in Figure 12.1 should be aligned with the goals and performance measures of supply chain management, which are depicted in Figure 12.2. By incorporating the supply chain goals and performance measures into the BSC framework, we can develop a balanced scorecard framework intended for measuring supply chain performance, as graphically displayed in Figure 12.3.

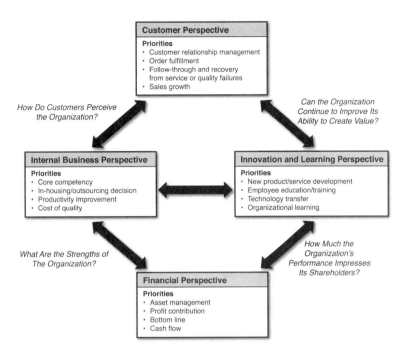

Figure 12.1. The balanced scorecard framework

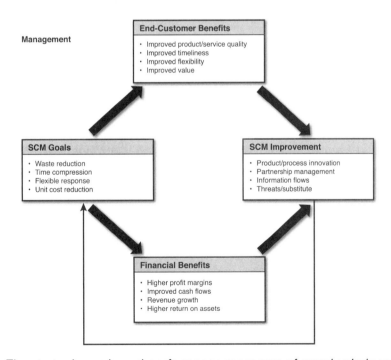

Figure 12.2. The strategic goals and performance measures of supply chain management (source: Brewer and Speh, 2000, p. 78)

Chapter 12: Measuring the Supply Chain Performance 427

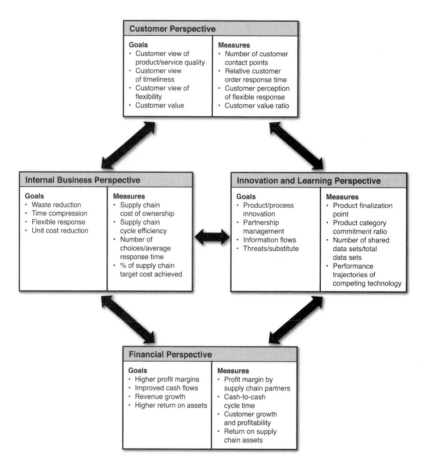

Figure 12.3. *A supply chain balanced scorecard framework (source: Brewer and Speh, 2000, p. 86)*

As illustrated in Figure 12.3, each perspective of the supply chain BSC framework has a different set of goals and performance measures. Some of these goals and their related performance measures may be relevant to a certain situation in a certain organization, but may be irrelevant to others. In other words, the supply chain BSC framework has to be tailored to specific needs and surrounding business environments of a particular organization and its supply chain partners. With that in mind, to better utilize the supply chain BSC framework, the following steps need to be taken:

1. Plot the organization's future competitive position and formulate a supply chain strategy that allows the organization and its supply chain partners to achieve that position.
2. Translate this strategy into specific goals/objectives and measures that can be classified within each of four perspectives of the BSC framework.

3. Set achievable targets for each measure.
4. Identify and develop courses of action (or initiatives) and then allocate the necessary resources to such courses of action to reach those targets.
5. Monitor the performance of these courses of action, identify performance gaps, and then communicate the outcomes of these courses of action to internal customers (e.g., employees) and external stakeholders (e.g., shareholders) within the organization and across the supply chain.
6. Develop specific plans to reduce performance gaps and then link the balanced scorecard to employees and supply chain partners (e.g., suppliers) to their performance evaluation and reward (e.g., merit pay, bonus, promotion, long-term contracts).
7. Readjust targets to changing customer demand and market positions, while continuously monitoring and improving the supply chain performance within each of the four perspectives of the BSC framework.

The Supply Chain Operations Reference (SCOR) Model

A Supply Chain Operations Reference (SCOR) model is an emerging management diagnostic tool that can be used to measure supply chain performance. It lays out the top-level supply chain processes in five key steps:

- **Plan**—This step balances aggregate demand and supply to develop a course of action that best meets sourcing, production, and delivery requirement.
- **Source**— This step procures goods and services to meet planned or actual demand.
- **Make**— This step transforms materials into finished products to meet planned or actual demand.
- **Deliver**—This step provides finished goods and services to meet planned or actual demand, typically including order management, transportation management, distribution management, and import/export management.
- **Return**—This step handles post-delivery customer support, including warranty administration, disposition, and replacement, as shown in Figure 12.4.

Unlike a traditional performance measurement tool, the SCOR model allows a company to communicate with its supply chain partners using common terminology and standard descriptions of the process elements that help explain the cross-functional aspects of the supply chain management processes and the best practices/strategies that yield the most desirable supply chain performance (Huang et al., 2005). The SCOR model can bring numerous managerial benefits. Examples of these benefits include the following (Davenport, 2005):

- Alcatel was reported to increase its on-time delivery from 10% to 50% in nine months and to reduce its material acquisition cost by third.
- Mitsubishi Motors reduced the number of their vehicles waiting in ports from 45,000 to zero, thus saving the company more than $100 million in supply chain costs.
- United Space Alliance improved its on-time delivery performances and logistic productivity after using the SCOR model.

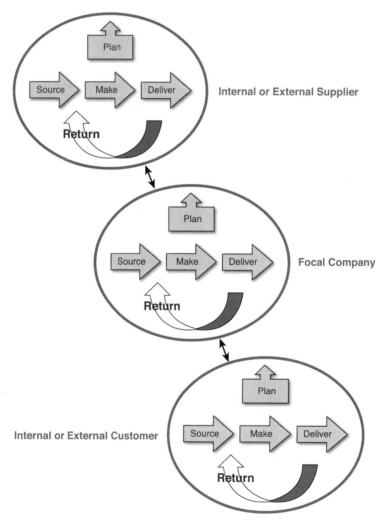

Figure 12.4. *The basic SCOR architecture within the supply chain*

In addition, Bolstorff and Rosenbaum (2003) observed an average of 3% improvement in total sales operating income after adopting the SCOR model. Hudson (2004) reported that the SCOR model could lead to an average of two to six times on return on investment. Similarly, Phelps (2006) discovered that SCOR users' annual revenues quadrupled those of non-SCOR users and generated profits eight times as much as non-SCOR users. In particular, the demand planning, supplier collaboration, joint manufacturing planning, and information technology support of delivery activities within the SCOR framework were found to have significant impact on the improvement of supply chain performances (Lockamy and McCormack, 2004). Despite these proven benefits, breadth and depth of the SCOR model, it should be noted that the SCOR model does not necessarily encompass every business or supply chain activity. For instance, it does not cover sales and marketing, finance, information technology, research and development, and new product development, whereas it heavily focuses on purchasing, production, and logistics functions (Larson et al., 2007). Another shortcoming of the SCOR model is its focus on performance measures such as cost of goods sold, fill rates, and order fulfillment lead time at the operational level as opposed to more tactical or strategic levels.

Given the aforementioned pros and cons of the SCOR model, the successful implementation of the SCOR model requires a careful step-by-step roadmap, as described next:

1. Analyze the basis of competition in terms of supply chain performances.
2. Perform SWOT analysis to identify the strengths and weaknesses of the current supply chain activities.
3. Develop a supply chain map to visualize the material and information flows throughout the supply chain and pinpoint the key sources of performance weaknesses in that map.
4. Develop specific courses of action that can improve supply chain performance and determine necessary changes in current supply chain practices that will help those courses of action be implemented successfully.
5. Clearly define milestones and deliverables to monitor the progress made by changes in supply chain practices.
6. Communicate the status of the progress to all supply chain partners and stakeholders on a regular basis.
7. Identify proven best practices
8. Continue to refine and improve those practices in response to dynamically changing business environments.

Other Performance Tools

In addition to the balanced scorecard and the SCOR model discussed earlier, the use of a dashboard for measuring supply chain performance has been increasing due in part to its simplicity and graphical diagnostic capability. A dashboard (or cockpit), complete with colorful graphical indicators and easy-to-read gauges, can help the company to monitor its progress and identify when it must change direction to improve its performance (Meyer, 1995). To elaborate, a dashboard is a measurement tool that combines the company's performance metrics, targets, and performance data into one online or printed document, such as a spreadsheet, on a regular basis to enable managers to easily digest the company's aggregated performance in cross-functional activities. The dashboard often uses a "traffic light" coding system—*red* indicating performances significantly below target; *yellow* indicating performances slightly below target; *green* indicating performances at or above target—to help managers evaluate performance on each metric as well as enable them to spot and correct problems promptly (Kaplan, 2009). One of the biggest advantages of the dashboard is its visual indicator that can display all the selected reports and charts showing the company's overall business performance, including revenue streams and sales figures in a single page, giving users a quick snapshot of the performance data that is most important to them. In addition, the dashboard immediately warns managers through alerts or warning signals when performance against any number of metrics deviates from the norm and thus prevents problems from being unnoticed or deteriorating beyond repair. Furthermore, it can be used in concert with a balanced scorecard with features complementary to the balanced scorecard, as summarized in Table 12.3 (e.g., Eckerson, 2006).

Table 12.3. The Comparative Features of a Dashboard and a Balanced Scorecard

Category	Dashboard	Scorecard
Purpose	Displays performance	Displays progress
Usage	Monitors performance improvement and then takes the necessary corrective actions	Shows performance milestones and identifies any room for (continuous) improvement
Update	On a real-time basis	For a periodic (e.g., monthly) snapshot
Data	Records events	Records summaries
Measures	Mainly based on metrics and gauges	Primarily based on key performance indicators (KPIs)
Context	Contains exceptions and alerts	Includes targets/goals and thresholds
Source	Linked to systems	Linked to (strategic) plans

Aside from balanced scorecards, SCOR models, and dashboards, other quantitative tools can be employed to measure supply chain performance. These tools include data envelopment analysis (DEA), analytic hierarchy process (AHP), and activity-based management. To elaborate, DEA is generally referred to as a linear programming (nonparametric) technique that converts multiple incommensurable inputs and outputs of each decision-making unit (DMU) into a scalar measure of operational efficiency, relative to its competing DMUs (Min and Joo, 2006). DMUs are collections of private firms, nonprofit organizations, departments, administrative units, and groups with the same (or similar) goals, functions, standards, and market segments. DEA is designed to identify the best-practice DMU without a priori knowledge of which inputs and outputs are most important in determining an efficiency measure (i.e., score) and assess the extent of inefficiency for all other DMUs that are not regarded as best-practice DMUs (Charnes et al., 1978). Because DEA provides a relative measure, it will only differentiate the least efficient DMU from the set of all DMUs. Therefore, the best-practice (most efficient) DMU is rated as an efficiency score of 1, whereas all other less efficient DMUs are scored somewhere between 0 and 1. To summarize, DEA determines the following (Sherman and Ladino, 1995):

- The best-practice DMU that uses the least resources to provide its products or services at or above the quality standard of other DMUs
- The less efficient DMUs compared to the best-practice DMU
- The amount of excess resources used by each of the less efficient DMUs
- The amount of excess capacity or ability to increase outputs for less efficient DMUs without requiring added resources

AHP introduced by Saaty (1980) also can be used to measure the supply chain performance. AHP is a scoring method that was designed to synthesize the overall performance of the organization, while enabling the managers to make tradeoffs among different performance criteria. It helps managers to assess the overall performance of the company or its supply chain partners relative to their principal competitors. Activity-based management is a method of evaluating business performance using activity-based costing (ABC) that helps management discern value-adding activities from non-value-adding activities and then improve customer value and efficiency (Forrest, 1996).

Chapter Summary

The following are key lessons learned from this chapter:

- Traditional performance measures such as order cycle time may be no longer suitable for measuring supply chain performance. Also, it is noted that financial measures are not the best reflection of supply chain efficiency and effectiveness due to their emphasis on short-term business outcomes.

- In order to continually improve supply chain efficiency and maintain supply chain excellence, an organization and its supply chain partners should uncover and adopt best-in-class supply chain practices on an ongoing basis. A key to the successful identification of those practices is the establishment of multifunctional, integrated performance measurement tools. In order for these tools to be meaningful and useful, they should correspond to the organization's strategic mission and be linked to its supply chain strategy. In other words, a key to developing supply chain performance measures is identifying and understanding the major elements (e.g., market share, customer value, business outcomes, revenue growth, and research and development) of the company's strategic goals.

- A supply chain performance measure is a means of providing feedback to the management throughout the supply chain that enables them to make informed decisions and take necessary corrective actions in a timely manner. It is also one of the ways to communicate with stakeholders of the organization and its supply chain partners regarding the efficiency and effectiveness of supply chain activities.

- The balanced scorecard is a set of financial ("current") and nonfinancial ("future") measures that reflect a balance between leading and lagging indicators of overall organizational performances. It aims to translate visions into strategic objectives/goals and actions. It also can show the cause-effect relationship and thus identify the sources of lagging performance. Typically, the balanced scorecard is composed of four components: perspectives, objectives/goals, measures, and initiatives/actions.

- The SCOR model is a cross-functional, cross-industry performance measurement tool that aims to capture the multidimensional aspects of supply chain activities. Such aspects include *reliability* (e.g., delivery performance, fill rates, perfect order fulfillment), *responsiveness* (e.g., order fulfillment lead time), *flexibility* (e.g., supply chain response time, production flexibility), *cost* (e.g., costs of goods sold, supply chain management cost, value-added productivity, warrant and return processing cost), and *assets* (cash-to-cash cycle time, inventory days of supply, asset turns). Although the SCOR model is gaining popularity as a legitimate supply chain performance measure, it is heavily geared toward the measurement of operational performance.

- A dashboard (or cockpit) provides a means of presenting performance information through multiple visual media, similar to the automobile's flash gauges. It can be considered part of business intelligence or an executive information system that stores, transforms, and updates performance data in the form of reports, graphs, and charts. Its usefulness often depends on data accuracy, timeliness, and relevance. It is primarily used as a performance monitoring tool rather than a performance management tool. That is to say, a dashboard intends to inform its users what they are doing rather than how well they are doing.

Study Questions

The following are important thought starters for class discussions and exercises:
1. How do you identify the "right" supply chain measures? How many of those measures should be considered? How do you combine different metrics?
2. Why is it so difficult to develop more integrated, inclusive supply chain measures?
3. What are the key performance indicators (KPIs) in the manufacturing, logistics, agricultural, energy, healthcare, pharmaceutical, and retail industries?
4. How does the balanced scorecard differ from the traditional measure?
5. How do you prioritize the four different perspectives of a balanced scorecard?
6. How do you collect, store, and update the necessary data for supply chain measures? How do you ensure data accuracy?
7. What are the biggest challenges for developing supply chain measures?
8. How would you measure nonfinancial dimensions such as customer loyalty, employee satisfaction, or R&D productivity?
9. How would you link your supply chain measures to your organization's business strategy?
10. What makes a dashboard more unique and powerful?
11. Compare and contrast various supply chain performance measurement tools (e.g., the balanced scorecard vs. the SCOR model).
12. Think of the organization you work for (or have worked for in the past) and then determine which supply chain performance tool is best suited to measure the supply chain performance of your organization and its supply chain partners.
13. Among various supply chain metrics, which metrics do you think matter most to your organization?
14. How would you discern value-adding activities from non-value-adding activities?

Case: La Bamba Bakeries

La Bamba Bakeries (LBB) traces their history back to the early 1990s, when Richardo La Bamba started selling homemade tortillas, sweet buns, muffins, and bagels mostly to Mexican immigrants in San Diego, California. As LBB's business grew rapidly, it acquired other local ethnic bakeries and has recently emerged as one of the largest family-owned bakeries in the United States, with annual net sales of $2.2 billion and a 17% share of the U.S. bakery market. In addition, LBB is a major producer of retail-branded pie crusts, pancake mixes, pizza crusts, brownies, croissants, and sweet desserts. LBB also produces custom-baked products for other marketers of branded food products. With a slogan of "Nobody does better than La Bamba" and its strong commitment to fresh-baked foods at a great value, LBB's business quadrupled in size over the last two decades, enabling it to deliver on its mission of serving America every meal, every day.

One of secrets behind its success may be LBB's direct store door (DSD) delivery practice. LBB distributes its branded bakery products through approximately 12,000 sales routes throughout the U.S. using its DSD distribution system. With this DSD distribution system, LBB delivers its products to individual convenience stores, grocery chain stores, and supermarkets directly via its suppliers or the suppliers' contracted carriers, such as FedEx and UPS. In many cases, DSD also includes stocking and appearance on retailer shelves. To increase DSD efficiency, LBB owns and operates two DSD distribution centers, and its sales employees (dubbed "route professionals") visit retail stores on average three times each week. During the visit, these employees review the retailer's on-hand inventory, help the retailer restock shelves, and ensure that the assigned shelf space generates sales revenue and profits sufficient enough for the retailer to allocate that much space to the product, while ensuring the correct inventory counts, proper arrangement, and free-standing display of LBB's products on store shelves. They also determine the product to be delivered to the retail store at a future time and communicate the future order to LBB's supplier's fulfillment center. This practice helps LBB with a high level of control over the stock level, future orders, and in-store presentation of its products. In particular, in-store presentations are important because LBB believes that purchases of bakery products are often impulse driven. Although the aforementioned process may sound simple, there can be complications.

To elaborate, LBB's DSD business model employs a so-called "guaranteed sale" program that does not require an individual retail store to carry any of LBB's products but rather to simply provide in-store display space to share the sales profits. In other words, LBB provides a complete turnkey merchandising solution by taking care of marketing, in-store display, logistics, and inventory control. As such, the retail store never has to touch LBB's products and worry about selling them. LBB is primarily responsible for rotating its products quickly if they are not selling and replacing them with newer items. If the product does not sell, LBB is not making money. Also, LBB manages its outbound logistics through a two-tier process, where product is transferred by truck from its two distribution centers, in Indianapolis, Indiana and Las Vegas, Nevada, to nearly 175 self-service warehouses, which in turn serve individual territories managed by route professionals. Each of LBB's route professionals is responsible for servicing the retail stores in his or her territory and creating solutions for optimal delivery routing and scheduling to minimize transportation costs.

Up until today, LBB has been using a third-party logistics provider (3PL) solution, but has experienced some delivery problems (e.g., missed time windows, shipping errors) because the 3PL's software was unable to adapt to the variables and scope of its increasingly complicated business practices. As an alternative, LBB considers physically delivering its products to the store by themselves, pulling pre-ordered products from its trucks and then participating in the delivery check process. This alternative may help LBB reduce fulfillment error, estimate its actual transportation costs more accurately, and figure out how many stops are being made per mile and how much time is being spent serving its customers. However, this alternative is likely to necessitate costly and time-consuming "peddle runs" (truck routes with frequent delivery stops) due to not knowing the exact volume of potential sales prior to delivery. To make matters more complicated, some large mass-merchant customers such as Walmart want to handle their own inbound transportation by themselves and thus pose difficulty creating economies of scale for LBB's own delivery option. Other chain stores also want LBB to ship products to their distribution centers directly and have their own people stock inventories, instead of leaving restocking to LBB's run professionals. On the other hand, most of LBB's smaller, independent customers are still happy with the LBB's DSD system, including a vendor managed inventory (VMI) approach, which takes the huge managerial/financial burden off of them (i.e., customers).

Despite the aforementioned complications, the DSD system is the wave of the future for food retailing. In today's food retail environment, with an increasing number of product choices and the inherent difficulty in managing store-specific assortments, DSD offers a unique opportunity for a retailer to deliver to grocery shoppers what is needed at the shelf where it counts most, when it is needed, on an individual store basis. Thus, DSD can build shopper loyalty, power sales growth, and improve cash flow. Given the needed collaboration between the supplier (e.g., LBB) and the retailer, DSD's success hinges on the supplier-retailer relationship and effective communication between the supplier and the retailer. Such communication, in turn, requires the use of wireless, real-time information and communication technology (ICT) such as DEX (Direct Exchange) EDI (Electronic Data Interchange), which not only can improve the visibility of product flows to retail stores, but can also reduce the waiting time and check-in process time of the delivery drivers. This technology should also immediately identify any order discrepancies in terms of item, price, and quantity, while allowing the retailer to make automated payments. Adoption of this technology, however, can create additional financial strain for LBB in the short run.

Another challenge stems from the customers' POS (point-of-sales) systems, which often transmit inaccurate demand information to LBB's mobile computer network. For example, LBB's route professionals have observed retail store cashiers scanning one bag for several different kinds of bakery products that are the same price (e.g., hotdog buns, hamburger buns, cinnamon buns, and cheese buns). This practice hurts LBB's inventory count in the computer because the sales record might show eight hotdog buns sold when the real sale was for two hotdog buns, two hamburger buns, two cinnamon buns, and two cheese buns. This means that the LBB's retail store's sales records do not match LBB's. It also means restocking is based on the misleading information if that retail store brought in LBB products through its own distribution network. Relying on this flawed information and the subsequently poor demand planning can create stockouts and overages/excess inventory—that is, none of one type of product and too much of another.

Given the various challenges just described, LBB's management is curious to know how well its current DSD system is doing and how much it contributes to the efficiency and effectiveness of its supply chain operations. To put it simply, LBB management would like to know how good they are in getting their information correctly and quickly, how accurate they are in making demand forecasts, and how fast they are making in their store deliveries. Without knowing how good LBB's current practice is, LBB management cannot figure out how competitive their company is and what it would take for them to become the leading company in the bakery industry. To further enhance its supply chain efficiency, LBB has begun to explore the possibility of downsizing plant production capacity, reducing overhead expenditures, and shrinking its sales force/route professionals. LBB also intends to continue to increase the efficiency and effectiveness of its supply chain operations by reducing overall logistics costs. For example, LBB plans to reduce transportation costs by handling its own outbound deliveries, further automating its transactions and payments, and upgrading its advanced ICT (such as electronic ordering systems and forecasting systems).

Key Discussion Questions

1. How can you evaluate the efficiency and effectiveness of the LBB's DSD distribution system? Why do you think the DSD distribution system works well for LBB?
2. What can be the key performance indicators (KPIs) in the bakery industry? (Be as specific as possible.)
3. What can be the right supply chain performance measurement tool for LBB and its supply chain partners (e.g., retailers)? (Be as specific as possible and explain why it is appropriate.)
4. What may be the most important challenges of developing LBB's supply chain performance metrics? (List at least three potential challenges and explain how to deal with these challenges.)
5. From the strategic perspective, what may be the best way to sustain LBB's competitive edge against its rivals?

Bibliography

Boehlje, M. (2010), *Economic Value Added*, Unpublished White Paper, Center for Food and Agricultural Business, West Lafayette: Purdue University.

Bolstorff, P. and Rosenbaum, R. (2003), *Supply Chain Excellence: A Handbook for Dramatic Improvement Using the SCOR Model*, New York, NY: AMACOM.

Charnes, A., Cooper, W.W., and Rhodes, E. (1978), "Measuring the Efficiency of Decision Making Units," *European Journal of Operational Research* 2, 429–444.

Davenport, T.H. (2005), "The Coming Commoditization of Processes," *Harvard Business Review*, 83(6), 1–8.

Forrest, E. (1996), *Activity-Based Management: A Comprehensive Implementation Guide*, New York, NY: McGraw-Hill.

Huang, S.H., Sheoren, S.K., and Keskar, S. (2005), "Computer-assisted Supply Chain Configuration Based on Supply Chain Operations Reference (SCOR) Model," *Computers and Industrial Engineering*, 48, 377–394.

Hudson, S. (2004), "The SCOR Model for Supply Chain Strategic Decisions," *Supply Chain Course Cooperatives*, October 27, http://scm.ncsu.edu/public/facts/facs041027.html, retrieved on December 30, 2010.

Investopedia (2010), "What Does Economic Value Added Mean," http://www.investopedia.com/terms/e/eva.asp, retrieved on December 24, 2010.

Kaplan, R.S. (2009), *Measuring Performance*, Pocket Mentor Series, Cambridge, MA: Harvard Business Press.

Lambert, D.M. and Pohlen, T. (2001), "Supply Chain Metrics," *International Journal of Logistics Management*, 12(1), 1–19.

Larson, P.D., Poist, R.F., and Halldorsson, A. (2007), "Perspectives on Logistics and SCM: A Survey of SCM Professionals," *Journal of Business Logistics*, 28(1), 1–24.

Lockamy III, A. and McCormack, K. (2004), "Linking SCOR Planning Practices to Supply Chain Performance: An Exploratory Study," *International Journal of Operations and Production Management*, 24(12), 1192–1218.

Meyer, C. (1995), "How the Right Measures Help Team Excel," in *Performance Measurement, Management, and Appraisal Sourcebook* edited by Shaw, D.G., Schneider, C.E., Beatty, R.W., and Baird, L.S., Amherst, MA: Human Development Press, Inc., pp. 118–126.

Min, H., and Joo, S-J. (2006), "Benchmarking the Operational Efficiency of Third Party Logistics Providers Using Data Envelopment Analysis," *Supply Chain Management: An International Journal*, 11(3), 259–265.

Phelps, T. (2006), "SCOR and Benefits of Using Process Reference Models," Presented at the Supply Chain Council Conference, January 12, http://www.gs1tw.org/twct/web/conference2006/01.pd, retrieved on December 30, 2010.

Quinn, F. (1996), "The Payoff," *Logistics Management*, 37(11), 56–62.

Saaty, T.L. (1980), *The Analytic Hierarchy Process*, New York, NY: McGraw-Hill.

Scott, C. and Westbrook, R. (1991), "New Strategic Tools for Supply Chain Management," *International Journal of Physical Distribution and Logistics Management*, 21(1), 23–33.

Sherman, H. David and Ladino, George (1995), "Managing Bank Productivity using Data Envelopment Analysis," *Interfaces* 25(2), 60–73.

Stewart, G. (1997), "Supply-chain Operations Reference Model (SCOR): The First Cross-industry Framework for Integrated Supply-chain Management," *Logistics Information Management*, 10(2), 62–67.

Ward, P. (2004), "Cash-to-Cash is What Counts," *Journal of Commerce*, February 16, http://www.hitachiconsulting.com/downloadPdf.cfm?ID=57, retrieved on December 24, 2010.

13

Emerging Technology in Supply Chain Management

"Information technology and business are becoming inextricably interwoven. I don't think anybody can talk meaningfully about one without talking about the other."
—Bill Gates, in his book titled *Business @ the Speed of Thought (1999)*

Learning Objectives

After reading this chapter, you should be able to:

- Understand how rapid advances in information technology have changed our daily lives and the way buyers and suppliers interact with each other to strengthen their supply chains.
- Describe the role and impact of information technology in the supply chain.
- Keep abreast of emerging technology available for effective and efficient supply chain operations.
- Appreciate the power of e-commerce for enhancing connectivity and visibility of the supply chain.
- Know how to make a better supply chain decision by leveraging information technology.
- Identify the various forms of information technology most commonly used for supply chain activities and understand their benefits and shortcomings.
- Learn to manage information technology projects and implement them successfully.
- Understand the evolution and future trends of information technology and learn to adapt to newly emerging information technology.
- Learn to deal with organizational resistance to adoption of new technologies.

The Emergence of E-commerce

The Internet may be considered one of the biggest inventions of human kind. It radically changed the way we buy and sell products/services and gather information about anything we are interested in knowing. Before the Internet became popular, could you imagine buying Christmas gifts online without even setting foot in a store? Could you imagine buying airline tickets online? Could you imagine ordering a pizza online? Could you imagine selling your old clothes and used furniture online? Could you imagine doing school homework online? Could you imagine shopping in your pajamas during the middle of the night? As these examples illustrate, the Internet has dramatically transformed our lives over the last couple of decades.

The Internet has also revolutionized the way we do business. In other words, physical processes (e.g., shopping in a store, paying money to the store clerk, and getting a sales receipt) often involved in business transactions have been replaced by electronic processes. Generally, electronic processes are faster and more accurate than traditional physical processes because they are performed at digital speed and are less prone to human error. With the availability of electronic transactions and interactions between the buyer and the seller, so-called "electronic commerce" (or "e-commerce") was born and quickly became entrenched in the fabric of our society. In fact, 77% of Americans currently have online access, and most of them have purchased something over the Internet. As of 2010, a total of nearly 2 billion people across the world used the Internet and joined the e-commerce-based economy (Internet World Stats, 2010). Even in a slumping economy, online retail sales in the U.S. increased 11% and reached an annual total of $155.2 billion, accounting for 7% of total retail sales in 2009 (Internet Marketing, 2009). Online retail sales in the U.S. are expected to grow from $263 billion in 2013 to $414 billion in 2018, a compound annual growth rate (CAGR) of 9.5%, according to the recent online retail sales forecast from Forrester Research Inc. (Enright, 2014). Put simply, e-commerce refers to conducting business on the Internet. Because e-commerce does not require any physical contact between the buyer and the seller at a retail store, many products ordered by the buyer over the Internet can be delivered directly to the buyer's home or office. The same idea can be applied to e-commerce between companies. This drastic change in buying/selling behaviors and delivery fulfillment of ordered products has gradually altered the business model many companies have adopted. Table 13.1 compares and contrasts the old business model (before the Internet) and the new "Internet age" business model.

Table 13.1. Old versus New Business Models

Paradigm	Business Model
Old (pre-Internet)	• Buy and sell off-line (at brick-and-mortar stores) • Hold product titles with investment risk • Sell from stock and build to stock for anticipated demand • Payments made via cash, check, or money order
New (Internet age)	• Buy and sell online • Source on order and build to order • Deliver directly to customer • Payments made via credit card or PayPal

A change in the business model also means the development of new business strategies (especially supply chain strategies) and the adaptation of emerging information technologies more relevant to ecommerce. In the next sections, such development and adaption are discussed in great detail.

Enterprise Resource Planning (ERP)

If people in different departments and companies within the supply chain all can share the same information, supply chain visibility will be enhanced and thus uncertainty/risk will be reduced. That is to say, the firm's ability to facilitate smooth information flow across its supply chain can dictate its supply chain excellence and subsequent business success. Typically, such ability is tied to the effectiveness and efficiency of the information technology (IT) the firm uses. Therefore, IT is a key enabler of supply chain management. IT solidifies connectivity among the supply chain partners by linking the point of production to the point of consumption seamlessly. Other roles of IT include the following (e.g., Davis and Spekman, 2004):

- Transformation of unstructured processes into automated, routine processes
- Making a free flow of information regardless of geography
- Reduction in the labor involvement of business processes
- Enabling parallel processing tasks
- Tracking physical flows of goods

Among the various IT tools, the one that is known to be effective in linking the entire organization and multiple partners/stages of the supply chain is enterprise resource planning (ERP). Generally, ERP is a cutting-edge software program that helps the firm coordinate and integrate company-wide business processes, including sales, marketing, manufacturing, logistics, purchasing, accounting, and human resources management, using a common database and shared management reporting tools (Brady et al, 2001). In a sense, ERP is a "dashboard" that provides some level of central oversight and controls needed to ensure that all of the company's resources are working together towards the same goal (Hwang and Min, 2013). Rather than separating inbound flows from outbound flows, ERP runs both flows together so that the company can react to its customers smoothly, correctly, and timely. A typical ERP architecture is graphically displayed in Figure 13.1. Its benefits may include the following (e.g., Davenport, 2000; ERPwire. com, 2010):

- **Lead time reduction**—By making the necessary parts/components always available for production, ERP can quickly react to customer demand and speed up the delivery schedule. For example, Autodesk reduced its delivery time to less than 24 hours from two weeks after implementing ERP.
- **Faster information transaction**—ERP has a master database as a central depository that links together all the related modules in the ERP software. The master database allows a company to maintain lists of all customers and suppliers, products the company sells, materials/parts the company procures, employees the company hires, and data about the equipment data the company owns. As such, it eliminates any duplicated entries by multiple departments and helps the company centrally control its data for easy/prompt access. In other words, ERP saves an enormous amount of time and effort in data entry. For instance, IBM's System Storage Division achieved a reduction in the time for entering pricing information from five days to five minutes, replacement part shipping went from 22 days to three, and credit checks (which previously took 20 minutes) were handled in three seconds after its ERP implementation.
- **Speedy payment**—Quick processing of customer information, less paperwork, and information queries through ERP can speed up the payment process; therefore, ERP improves cash flow.
- **Laying the groundwork for electronic commerce**—With the back-office business transactions structured and supported by ERP, ERP can facilitate the customers' web-based access to product ordering, shipment tracking, and delivery processes and subsequently smooth out e-commerce operations. For example, when a customer orders products online, ERP enables the company service representative to quickly access the customer profile information, including the customer's credit rating and order history from the finance module of ERP, the company's inventory levels from the material module, and the shipping dock's shipping schedule from the distribution module. As such, ERP helps the company handle online orders in a timely manner.

Although ERP can bring numerous benefits, it is not without its drawbacks. Some of those drawbacks include the following:

- **High implementation cost**—ERP software, hardware, and consulting costs can easily exceed $100 million (approximately $50 million to $500 million) for a big corporation such as a Fortune 500 company. Large companies can also spend $50 million to $100 million on ERP system upgrades. Full implementation of all modules can take years. Midsized companies (fewer than 1,000 employees) are likely to spend around $10 million to $20 million (Monk and Wagner, 2006).
- **Slow payback**—The exorbitantly expensive ERP implementation tends to lengthen payback periods (Hwang and Min, 2013). As such, the return on investment (ROI) in ERP can only be obtained when there is both a sufficient number of users and sufficient frequency of use (Monk and Wagner, 2009).
- **Challenges associated with change management**—ERP can change the way the company conducts business and subsequently the way its people work. A hasty implementation of ERP can lead to potential disasters similar to the ones experienced by Hershey, which suffered from a heavy loss in sales and profit after its ERP implementation. A successful ERP implementation often requires full integration of the company's business functions, seamless interfaces among different users, backup systems, customized software, compatible data formats, and extensive user training. Therefore, ERP necessitates a careful plan for changes in the company's strategy and business processes.

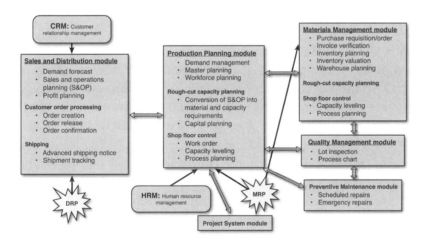

Figure 13.1. *ERP architecture*

Geographic Information System (GIS)

Since the 1970s, geographic information systems have been used to visualize spatial information associated with socio-economic planning, urban planning, market planning, transportation planning, emergency planning, environmental planning, land management, facility location analysis, crime analysis, and distribution network design. With the advent of the Web and advanced computer technology, the use of GIS has been widely spread to many different disciplines, including supply chain management. In general, GIS is a computer system that integrates hardware, software, and data for capturing, storing, managing, analyzing, and displaying all forms of spatial information, including locations and attributes of geospatial features (e.g., Chang, 2010; GIS.com, 2010). GIS allows us to view, understand, question, interpret, and visualize data in many ways that reveal relationships, patterns, and trends in the form of maps, globes, reports, and charts (GIS.com, 2010). For example, GIS enables us to identify specific locations in the U.S. that are earthquake prone by mapping seismic patterns and fault lines. As such, it allows us to monitor earthquake activities and then take necessary actions (e.g., early warning to residents living in earthquake-prone areas) to minimize earthquake victims. Likewise, GIS simplifies the data-display mechanism by separating data presentation from data storage and then allowing the supply chain manager to visualize how distinctive one geographic site is from another by superimposing spatial information, such as population density, racial makeup, and temperature/humidity, on an interactive map. Combined with a database management system (DBMS), GIS may also provide a friendly platform for the enhancement of dialogues among supply chain partners (Min and Zhou, 2002). In order to make GIS work, it requires the following components besides geospatial data (Change, 2010):

- **Hardware**—This includes computers such as PCs, workstations, and supercomputers with operating systems based on Windows, Linux, and UNIX platforms. Additional equipment may include monitors for display, digitizers and scanners for digitization of spatial data, global positioning system (GPS) receivers and mobile devices for fieldwork, and printers and plotters for hard-copy data display.
- **Software**—This includes the source code and the user interface. The code may be written in C++, Visual Basic, Java, or Python. Some notable GIS software includes GRASS GIS (originally developed by the U.S. Army Corps of Engineers), SAGA (System for Automated Geo-scientific Analyses) GIS, MapWindow GIS, and commercially available ArcGIS software and its related products—ArcINFO, ArcVIEW, and ArcLogistics—developed by the Environmental Systems Research Institute (ESRI).
- **Analyst**—GIS programs require years of training and experience. Therefore, GIS experts are often needed to enter and analyze GIS data.
- **Supporting infrastructure**—This refers to the necessary physical, organizational, administrative, and cultural environments (such as GIS laboratory, data standards, data repository, and data clearinghouses) that support GIS implementation and applications.

Once GIS is firmly in place, it can bring the following benefits:

- It can improve the way data is accessed, maintained, and analyzed by organizing geographic information into one seamless map.
- It can identify potential sources of problems (e.g., flooding, earthquakes, hurricanes, tornadoes, infectious diseases, water main breaks) on interactive maps.
- It can foster better communication and cooperation among the supply chain partners by sharing base map data.
- It can enhance logistics efficiency through improved workflows and scheduling based on GIS. For instance, Sears has seen dramatic improvements in its logistics operations after its use of GIS. Sears reduced the time it takes for dispatchers to create truck routes for their home delivery by about 75%. It also reduced the delivery costs with 12%–15% less driving time. Sears also improved customer service, reduced the number of return visits to the same site, and scheduled appointments more efficiently (GIS and Science, 2009).
- It can help develop strategic action plans in a timely manner. As explained earlier, GIS allows the decision maker to identify potential sources of problems well ahead of their occurrence. Thus, GIS allows the decision maker to prepare for future problems in advance. For example, with the help of GIS, supply chain professionals can map out potential supply chain risks (e.g., inclement weather delaying air express delivery, labor strikes at the port) and then develop action plans that may mitigate supply chain disruptions caused by such risks.

Despite great visualization and user interface capability, GIS can be easily misused. Its potential drawbacks include these two items (Wegener and Fotheringham, 2000):

- Garbage in, garbage out (GIGO), with the risk of using outdated and overly simplified geospatial data that can lead to misleading interpretation of the data and subsequent decision errors
- A lack of a temporal dimension for GIS programs, which hampers dynamic modeling and trend analysis.

Intelligent Transportation Systems

An intelligent transportation system (ITS) generally refers to the use of advanced information and communication technologies to improve transportation efficiency, mobility, and safety. Examples of such technologies encompass satellite navigation, wireless sensor networks, Bluetooth protocols, inter-vehicle communication protocols, broadband wireless communication devices, ultraviolet and infrared radiation tools, conformal imaging, highway advisory radios, in-vehicle en route guidance systems, intelligent cruise control systems, agent-based systems, smart cards, fiber optics, real-time simulation tools, mobile robots, interactive television, closed-circuit television (CCTV), personal digital assistants (PDAs), and global positioning systems (GPS). Also, there is a range of ITS-enabling technologies that enhance the effectiveness and efficiency of ITS. Exploiting these technologies, ITS enables vehicle drivers/operators to avoid traffic congestion, receive inclement weather warnings, get the real-time route guidance, detect road obstacles/hazards ahead, enhance night vision, avoid vehicle collisions, and pay tolls electronically. In addition to these benefits, ITS has received greater attention from policymakers and the general public after the enactment of The Intermodal Surface Transportation Efficiency Act of 1991 (ISTEA), which was designed to facilitate the research, development, and operations of ITS. Another rationale for increased ITS applications is that the development and improvement of the transportation infrastructure alone can no longer enhance transportation efficiency, flexibility, convenience, and safety. Although ITS applications so far have been heavily concentrated in automobile transportation (e.g., commercial vehicle operations technology for fleet and cargo management, truck scan systems) or urban transit systems (e.g., ramp meters, electronic transit fare payments), the potential for extended ITS applications is almost limitless.

The potential areas for further ITS applications may include seamless intermodal connections, terminal workforce scheduling, terminal congestion controls, yard management, emission reduction, noise abatement, and bidding on load bundles. In a broader sense, ITS application areas can be classified into eight different fields, as shown in Figure 13.2.

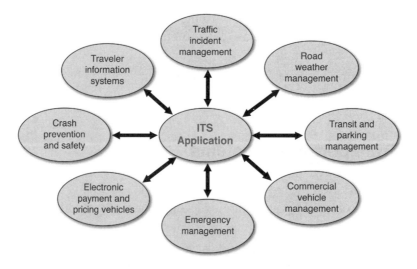

Figure 13.2. *The illustrated application areas of intelligent transportation systems*

Barcoding Systems

As a way to automatically read product information, barcode technology was introduced in the early 1950s and then first used for commercial purposes in the 1970s. Nowadays, a barcode is an essential part of our daily lives. If you look at the daily necessities you can find in your kitchen, such as a carton of milk, a box of cereal, a package of meat, and a can of soda—all of them have a series of vertical bar-like lines and numbers at the bottom of those lines. These are called "barcodes." A barcode refers to a standard symbol (marking) that consists of a series of vertical lines with various widths and spacing that contain product information regarding product supplier, product type, place of manufacture, and product price. This marking can be scanned and read electronically to identify an individual product. Typically, a barcode symbol has two parts: a machine-readable bar code and a human-readable 12-digit UPC (universal product code) number. The first six digits of the UPC number represent a product manufacturer's identification, whereas the next five digits represent the item number. The last digit of the UPC code is called a "check" digit, which lets the scanner determine whether or not it scanned the number correctly. Every single item that the manufacturer produces has its own number and its number differs depending on its package size. For example, a 12-ounce can of Coca-Cola has a different item number than a 16-ounce bottle of Coca-Cola (Brain, 2007).

The typical components of a barcode include barcode software, a barcode scanner, a barcode printer, and a barcode label. The barcode software can be also subclassified into three categories:

- **Barcode production software**—Generates the graphic images for barcodes in a Windows application or on the Internet.
- **Barcode data collection software**—Needed for data parsing, filtering, and formatting.
- **Barcode database and inventory software**—Helps create databases regarding customers, inventory, and sales. This software is often used for asset tracking by figuring out where the item is, where the item was, what the item is used for, and who used it.

The potential benefits of a barcode are as follows:

- **Improved supply chain efficiency**—Because a barcode scanner can read data five to seven times as fast as a skilled typist, barcode technology is far more effective in recording product information than the manual method. Also, barcode data entry has an error rate of one in 3 million, whereas human data entry has an error rate of one in 300; therefore, data error will be dramatically reduced with the use of a barcode system (Data ID Systems, 2006). The improved speed and accuracy of processing product information means greater

efficiency in tracking assets, including inventory and the subsequent supply chain efficiency.
- **Labor cost savings**—A barcode system minimizes clerical errors and the subsequent rework by automating many chores of inventory tracking and control. Thus, a barcode can enhance labor productivity, which can be translated into labor cost savings.
- **Improved customer services**—Inventory accuracy created by a barcode system reduces the chances of stockouts and backorders. Also, it reduces order fulfillment errors by collecting the shipping manifest information quickly and accurately. Therefore, a barcode system helps the company improve its customer services.

These benefit potentials may not be fully achieved unless a barcode system is right for a particular managerial setting and company. To ensure the suitability of a barcode application, its potential users need to check whether they can answer the following questions with a resounding "yes":

- Do you encounter problems or inefficiencies in tracking goods internally?
- Do you encounter problems or inefficiencies in handling and fulfilling customer orders?
- Are you part of a supply chain in which more efficient handling of goods is required?
- Do you feel insufficiently competitive in dealing with customer orders? Do you encounter problems in delivering orders?
- Do you plan on increasing SKUs (stock-keeping units) or diversifying product lines?
- Do you have frequent errors in transmitting or analyzing information?
- Is re-keying of data a common occurrence within the company? Do you have many keying errors?
- Do you plan on implementing WMS? Do you plan on incorporating barcoding technology into WMS?
- Are your company employees receptive of new technology?
- Is your top management willing to wait for 6–9 months to reap the benefit of barcoding?

Radio Frequency Identification (RFID)

Radio frequency identification (RFID) is one of the fastest-growing technologies that can revolutionize supply chain practices. RFID can be used just about anywhere unique identification is needed for tracking purposes, ranging from clothing tags to credit cards. RFID transmits signals containing product information using electromagnetic or electrostatic spectrums. Although its role is similar to that of a barcode, RFID eliminates the need for line-of-sight reading that the barcode requires. In other words, RFID can read product information from greater distances (up to 10 feet) than barcode scanning without direct physical contacts. This capability allows RFID to read data in harsh environments (e.g., oily or dirty surfaces) or chaotic circumstances (e.g., cluttered warehouse settings). Generally, RFID refers to a wireless device that carries data in transponders (i.e., "tags") attached permanently to an asset (e.g., item, container, case, pallet) to read data stored on a microchip using radio waves (Carr et al., 2010). Data stored in the RFID tag can provide unique identification for manufactured items, goods in transit, physical locations, vehicle identities, and humans (Seymour et al., 2008). Depending on the reading mechanism, the RFID tag can take several different forms (Moeeni, 2006):

- **Active tags**—These tags carry their own power source and broadcast their own signals to communicate with readers.
- **Passive tags**—These tags will be activated to broadcast data once they are within the electromagnetic field of the reader's antenna.
- **Semi-passive tags**—These tags contain a tiny battery for logging the data received from the connected sensors.

Depending on the electronic product code (EPC) structure, these tags can be further subdivided into five different levels (i.e., class 1 through class 5). These RFID tags can be read in two different ways:

- **Fixed readers**—Intended for large-scale deployments at fixed locations. Fixed readers can read high volumes of tags with one or more antennas. They detect any tagged object that passes near them—typically within a range of 10 to 20 feet for passive tags and up to 300 feet for active tags. Fixed readers require connection via cables to external antennas, a power source, and a middleware system that will manage the RFID read data (Greengard, 2007)
- **Mobile readers**—Usually deployed as peripheral devices on handheld or vehicle-mounted terminals that provide greater latitude over where, when, and how they are used and thus offer greater reading capabilities for reading tags on a few objects at a time.

Regardless of tag type and reading mechanism, RFID consists of three basic components, as graphically displayed in Figure 13.3.

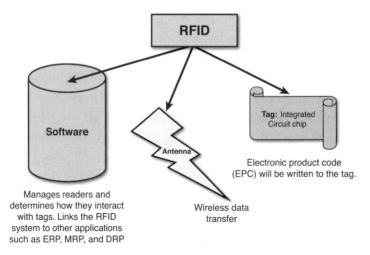

Figure 13.3. *Key components of the RFID*

The key advantages of RFID include the following:

- The ability to encode more information than a one-dimensional barcode
- The ability to read and write from a distance and in harsh environments
- Enhanced security due to difficulty in counterfeiting RF tags
- Prevention of theft/pilferage in warehouses through asset-tracking features
- Enhanced labor efficiency through improved data accuracy and less human involvement
- Better warehouse utilization through easier inventory control and planning
- Enhanced supply chain visibility through timely electronic transmission of necessary data

Despite many comparative advantages of RFID over traditional barcoding systems, as detailed in Table 13.2, RFID has the following drawbacks:

- It is far more (10 to 50 times more) expensive than one-dimensional barcodes.
- It requires a specific RF tag and transponder device, some of which are not widely available.
- There is a lack of uniform tag standards due to varying technology and data content standards.
- There is difficulty in reading liquid and metal products.
- Sunk costs are involved with old legacy systems.

After an organization weighs the pros and cons of RFID, the following questions should be addressed prior to making a full pledge to an RFID implementation:

- Do we have the right technology for all RFID initiatives?
- Can RFID talk to WMS? Can RFID be integrated into WMS or ERP?
- Can we absorb the cost of an RFID tag? Is RFID economical for less expensive items?
- Is our supply chain partner willing to get on board with RFID implementation?
- Is our supply chain partner willing to use the same or compatible RFID standards?
- Do we know where to tag (e.g., pallet, case, pack, item, tote, tray)?
- How does RFID affect the employee morale? (e.g., some labor jobs may be eliminated by RFID implementation.)

Table 13.2. A Comparison of RFID and Barcode

Barcode	RFID
One-dimensional information density via UPC code or two-dimensional information density via QR (Quick Response) code.	Multidimensional information density via EPC code.
A single identifier (reads one barcode at a time).	Multiple identifiers (reading multiple tags within the transmission field).
Obtains information by scanning data on laser or charge-coupled devices.	Obtains information from tags through a reader and an antenna (wireless).
Limited product information.	Extensive product information.
Reads data.	Reads and writes data.
Short distance reading requiring physical contact with the object.	Long-distance reading. Higher frequency tags typically have longer read distances and faster data transfer rates than lower frequency tags.

Artificial Intelligence

Artificial intelligence (AI) was introduced to develop and create "thinking machines" that are capable of mimicking, learning, and replacing human intelligence. Since the late 1970s, AI has shown great promise in improving human decision-making processes and the subsequent productivity in various business endeavors due to its ability to recognize business patterns, learn business phenomena, search information, and analyze data intelligently (Min, 2010). Due to such promise, AI can be used as an effective decision-aid tool for solving many supply chain problems. Generally speaking, artificial intelligence is the use of computers for reasoning, recognizing patterns, learning or understanding certain behaviors from experience, acquiring and retaining knowledge, and developing various forms of inferences to solve problems in decision-making situations where optimal or exact solutions are either too expensive or difficult to produce (Nilsson 1980; Russell and Norvig 1995; Luger 2002; Min, 2010). With advances in AI technology, AI encompasses a variety of tools:

- Artificial neural networks and rough set theory ("thinking like a human")
- Machine learning, expert systems, and genetic algorithm ("acting like a human")
- Fuzzy logic ("thinking rationally")
- Agent-based systems ("acting rationally")

The main objectives of AI are to understand the phenomenon of human intelligence and to design computer systems that can mimic human behavioral patterns and create knowledge relevant to problem solving. Thus, AI has the ability to learn and comprehend new concepts, learn from experience ("on their own"), perform reasoning, draw conclusions, impute meaning, and interpret symbols in context (Min, 2010). Due to such ability, AI has been successfully applied to game playing, semantic modeling, human performance modeling, robotics, machine learning, and data mining (see, e.g., Russell and Norvig 1995; Luger, 2002). Recently, the application of AI has been extended to various supply chain areas. These areas include demand planning, inventory control, order picking, transportation network design, make-or-buy decisions, location planning, and customer relationship management (see, e.g., Min 2010 for detailed applications of AI to supply chain areas).

Though still limited, an increasing application of AI to supply chain management decisions stem from the fact that supply chain management requires the comprehension of complex, interrelated decision-making processes. Because these processes often call for the creation of intelligent knowledge bases, AI fits the bill as a tool for managing supply chain knowledge. In particular, an agent-based system has emerged as one of the most popular AI tools for tackling various aspects of supply chain decision problems (Min, 2010).

Information Technology (IT) Project Management

Considering the numerous managerial benefits gained by IT, investment in IT projects is a business necessity. However, ill-advised IT investment decisions can cause more harm than good by draining limited financial resources and creating organizational resistance to changes in IT practices. Regardless of the extent of the IT project, ranging from the simple replacement of an old server to an enterprise-wide ERP implementation, the IT users should balance the three pillars of IT projects: budget (or cost), deadline (or time), and scope (or amount of work to be done). According to the IT consulting firm the Standish Group, 70% of IT projects are over budget and behind the schedule. In particular, the cost-overrun is so common that 52% of IT projects finish at 189% of the initial budget. Even some IT projects after years of time and million dollars of capital investment are never completed. For example, only 29% of the IT projects initiated in 2004 were completed successfully (www.standishgroup.com). Broadly speaking, the potential causes of IT project failures may include the following:

- The intended IT project is not clearly defined. In other words, its goals and expectations are not clearly set.
- The IT project team is not in sync and is therefore dysfunctional.
- The IT project draws limited support from top management.
- There are a lot of internal skeptics about the planned IT project and therefore project management responsibility rests on IT managers only rather than the direct involvement of potential IT users.
- The IT software/hardware vendors and the IT users do not communicate with each other on a regular basis.
- There is a lack of training and education for the IT project team and users before undertaking the project.
- There is no change management following the IT implementation.
- The return on investment (ROI) for the IT project is hard to estimate and is therefore difficult to claim.

To avoid IT project failure, the project should be broken down into a series of smaller, more manageable phases that can be completed sequentially, one at a time. These steps include (1) initiation, (2) goal setting, (3) planning and design, (3) executing, (4) monitoring and controlling, and (5) assessment and follow-up, as depicted in Figure 13.4. In particular, given that many IT projects fail at the beginning rather than at the end, the emphasis should be placed on the first two phases. At the initiation phase, the following questions should be answered:

- Do we have specific goals to achieve and deadlines to meet for the IT project?
- Is the IT implementation task unique?
- Does the IT implementation require interrelated tasks?
- Is IT implementation critical to the company?
- Does IT implementation cut across organizational lines?

At the planning and design phases, the ROI for the IT project needs to be calculated using one of the following three methods.

- **Pay-back period**—The investment will be "paid back" within "x" timeframe, "x" being typically 18–36 months (e.g., 6–18 months for WMS). For this method, the cost of capital is not considered.

- **Rate of return**—Requires the investment to return a specific rate calculated as a percentage of the annual savings divided by the total cost of the project. A return of 30% per year would roughly equate to a 3-year payback.
- **Net present value (NPV) / internal rate of return (IRR)**—Estimates the value of return over "x" periods (number of months or years) in terms of present dollar value (future returns expressed as current dollars).

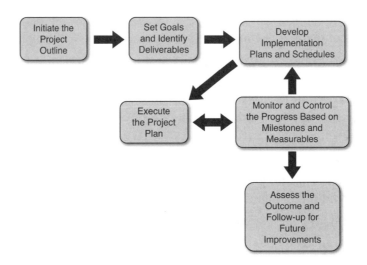

Figure 13.4. *IT project phases*

The execution phase of the IT project (involving the delivery of project plans) is followed by the monitoring and controlling phase of the IT project, which aims to help prevent potential failures or setbacks (e.g., delays or cost-overruns). At the monitoring and controlling phase, a cycle of measurement, correction, and evaluation, as shown in Figure 13.5, will continue until the progress report shows no issues to address. For example, a progress report may include the following details for monitoring and controlling purposes:

- Progress Made
  - Established and communicated expectations with the software vendor on 1/15/11.
  - Reviewed the electronic ordering system design with the help of the software vendor on 2/28/11.
- Progress Planned
  - Participated in the demonstration of the electronic ordering system by the vendor on 3/17/11.
  - Presented resource requirements to senior management on 4/20/11.

- Major Issues
  - Software vendor needed to formalize the delivery of EDI technology and documentation for quality control on 5/28/11.
- Budget Review
  - Original budget: $2 million. Spend to date: $980,000.
  - Estimate to complete: $2.25 million. Projected-over: ($250,000).

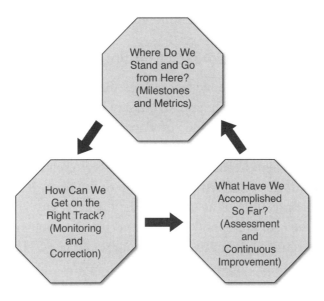

Figure 13.5. *The IT project monitoring and controlling cycle*

At the final phase of closing, the benefits of the IT project should be documented, and those benefits need to be compared to performance metrics developed at the monitoring and controlling phase. In particular, the benefits of the IT investment should be assessed with respect to customer satisfaction, the company's mission, and the company stakeholders' value propositions (e.g., contribution to profits and stock prices).

Future Trends of IT in Global Commerce

Only about 10 years ago, we could not imagine a small plastic card containing an embedded computer chip that allowed us to buy a meal at the school cafeteria, check out books from the library, watch a movie at the theatre, and call your friends at the airport telephone booth. As this example illustrates, new information technology (IT) changes our lives at a rapid pace. As the rapid advance in IT changes our lives, our behaviors will change as well. Changes in our behaviors may include our socialization habits, communication tools, and business transactions. Those changes will bring the dynamic transformation of business activities. To adapt to these dynamic changes in a timely fashion, we need to prognosticate which technology will be emerging in the future and how it will impact our business practices. Although it is very difficult for us to speculate over what the world of tomorrow will look like in the rapidly evolving digital era, we can notice several emerging trends from the world of IT. These trends are open collaboration through "cloud computing," "conversation economy" via online social networking media, mobile commerce in ubiquitous environments, and multitasking using smart objects. These trends are discussed in the following subsections in greater detail.

Open Collaboration through Cloud Computing

Cloud computing is a new leading-edge technology that allows the user to get access-hosted computing services over the Internet from any computer, anywhere, without installing the necessary infrastructure and software. As such, computing technology is hidden behind a "cloud" and invisible to its users, but can be accessed on demand through a client/server system. Cloud computing differs from traditional computing in that it is *used on demand,* typically by the minute or the hour. It is also *elastic*—a user can have as much or as little of a service as they want at any given time—and it is *inexpensive,* because the user needs nothing but a personal computer with Internet access (TechTarget, 2010). A simple example of cloud computing is a Yahoo! e-mail or Gmail system. According to a recent study conducted by the Aberdeen Group (2009), firms achieved on average an 18% reduction in their IT budget and a 16% reduction in data center power costs from cloud computing.

Depending on the needs of users, the hosting services provided by cloud computing are broadly divided into three segments (see, e.g., Rittinghouse and Ransome, 2010):

- Infrastructure-as-a-Service (IaaS)
- Platform-as-a-Service (PaaS)
- Software-as-a-Service (SaaS)

To elaborate, Infrastructure-as-a-Service, often called "utility computing," offers memory, data storage, computational capacity, and virtual servers as a pool of virtualized resources over the network that can be accessed by users on demand. This service is primarily used for supplemental and non-mission-critical computing. Platform-as-a-Service is a suite of cloud services that provides a set of software and product development tools hosted on the provider's infrastructure. This service allows its users to build their own applications that run on the provider's infrastructure and are delivered to the users via the Internet from the provider's servers. Prime examples of PaaS providers include NetSuite, Amazon AWS, Google App Engine, Microsoft, Heroku, AppFog, Engine Yard, and Coghead. Among these three service offerings, Software-as-a-Service may be the most popular, and a growing number of companies are exploiting it. SaaS provides its user with cost-effective solutions that can be quickly deployed with minimal upfront investment in software/server licensing and supporting hardware. For example, a typical installation of the WMS software can take at least 4–6 months. However, with SaaS, the WMS software can be deployed almost instantly and thus its users can reap its benefits immediately. Companies offering Software-as-a-Service include Salesforce.com (best known for its CRM and enterprise applications), Antenna Software (for mobile applications), Cloud9Analytics (for real-time sales forecasting), CVM Solutions (for supplier risk and performance management), and Workday (for human resource and ERP applications).

As discussed previously, cloud computing with on-demand service architectures shared by many users can encourage open collaboration among supply chain partners because it makes IT more affordable, accessible, and compatible, thereby making information sharing among the supply chain partners more reasonable.

Conversation Economy via Online Social Networking Media

Dissatisfied with traditional information portals and keyword search engines such as Yahoo!, Google, and AOL, a growing number of Internet users have tried to index and search for information in revolutionary ways. One of those ways may include the sharing of necessary information through social media and user-generated online content, including blogs. A rapid pace of advancement in this kind of communication media and technology facilitates quick exchange of information among customers, employees, and business partners located in different parts of the country and globe. For example, one of the online social media networks, Facebook, with approximately 550 million users across the globe, can help multinational firms (MNFs) develop an interactive communication platform that goes well beyond the interaction between back-office functions and client-facing functions through its web of friends, families, colleagues, and acquaintances. This platform can be a foundation for the MNF's customer relationship management (CRM), advertising tools, branding, business transactions, and information-sharing activities, and may reshape future supply chains.

Mobile Commerce in Ubiquitous Environments

As we have experienced through the use of our cell phones, mobile technology allows us to stay in touch with each other for virtually 24 hours a day, 7 days a week, from anywhere in the world. As such, this technology dramatically enhances connectivity among us. The extension of this mobile concept from "person-to-person" to "business-to-business" can revolutionize the way supply chain partners communicate and interact with each other. Generally, mobile commerce (or m-commerce) refers to the ability to conduct business transactions in ubiquitous environments using wireless handheld devices such as cell phones and personal digital assistants (PDAs). Based on the Wireless Application Protocol (WAP), mobile commerce can provide faster, more secure, and scalable e-commerce access without even needing a place to plug in mobile devices. Leveraging the technological advances in mobile devices, including their speech recognition capability, we can use mobile commerce to review bank accounts, order products/services, pay bills, and check financial news, traffic updates, and sports game scores on a handheld device, anywhere, at any time.

Multitasking Using Smartphones

Generally speaking, a smartphone is an electronic hand-held device that combines traditional voice services with built-in computing capabilities, including email, fax, pager, Internet access, camera, MP3 player, video player, television, and organizer. With such capabilities, a smartphone allows us to perform a variety of tasks: web browsing, personal information management, online banking, LAN connectivity, data entry, data storage, remote data transfer, remote control of home/business electronic systems, office document editing, and interactivity with unified messaging. Exploiting multitasking functions, approximately 45.4 million Americans are reported to own smartphones, which represent the fastest-growing segment of the mobile phone market, comprising 234 million subscribers in the U.S. alone as of 2010 (Gonsalves, 2010). Typical features of a smartphone include the following (Cassavoy, 2010):

- **Operating system**—A smartphone will be based on an operating system that allows it to run productivity applications, such as the Android, iOS, BlackBerry OS, Firefox OS, Palm OS, Symbian OS, Linux, BREW, and Window Mobile.
- **Software**—A smartphone will offer more than just an address book. It helps the user to create and edit Microsoft Office documents and manage finances/banking services.
- **Web access**—A smartphone allows for wireless Internet access using the 3G data network.
- **QWERTY keyboard**—A smartphone includes a QWERTY keyboard, so the keys are laid out in the same manner (not in alphabetical order) that they would be on your computer keyboard.
- **Email messaging**—All cell phones can send and receive text messages, but what sets a smartphone apart is its ability to send and receive email. Some smartphones can support multiple email accounts. Others include access to the popular instant messaging services, such as AOL's AIM and Yahoo! Messenger.

Despite these attractive features, a smartphone still has some constraints that a desktop computer does not. These constraints include the following (Buchanan, 2010):

- **Small screen**—A small-sized 3-to-4-inch screen of the typical smartphone does not offer high-resolution video capability.
- **Limited battery life**—A smartphone has to be recharged more frequently than a typical PC.
- **Computing speed**—Due to its small-sized CPU, a smartphone is still much slower than a typical PC.
- **A lack of memory**—The data storage capacity is constrained due to limited RAM and battery life.

Drone (Unmanned Aerial Vehicle) Technology

Drones are best known for their military applications during the Afghanistan war, but the technology has advanced a lot in recent years and has gotten more affordable for logistics applications. As a matter of fact, DHL started experimenting with the drone technology to ferry medicine to a sparsely populated island off the northwestern coast of Germany. Following suit, both Amazon and Google plan on exploiting drone technology to replace some of the traditional ways of delivering parcels to remote locations. What exactly is a drone? Generally speaking, a drone is an unmanned aerial vehicle typically guided by wireless GPS, cameras, and/or (inertial measurement) sensors. It is capable of delivering goods from point to point without a pilot. The buzz around drones has intensified after the FAA Reauthorization bill, which was recently signed into law for testing and licensing commercial drones by 2015. Although a lack of regulatory approval for the use of drones and their limited load capacity of 5 to 7 pounds can still be a challenge, drones can revolutionize air freight delivery services, particularly for reaching small, remote, sparsely populated areas where access to nearby airports is limited. If the current technological deficiencies associated with detection, safety, and flight stability can be overcome, the drone technology can be used to shuttle small packages (e.g., medical supplies, spare parts, books) from central depots or major airports to local pickup centers and residential homes. There is no doubt that drone technology is gaining momentum and will become the alternative mode of transportation in the near future.

Chapter Summary

The following are key lessons learned from this chapter:
- Information technology (IT) helps firms communicate with each other, anywhere, at any time, and thus helps them establish a solid business partnership, which in turn can facilitate information-sharing practices and subsequently enhance supply chain visibility.

- The emergence of e-commerce has revolutionized the way people conduct business. Because it helps businesses get closer to their end customers without the intervention of intermediaries, it accelerates the process of reducing the use of traditional channel intermediaries such as distributors and retailers. On the other hand, it often increases the burden of logistics and distribution responsibilities for manufacturers, who may want to sell and distribute their products directly to their end customers. Thus, logistics can be an important differentiator for firms competing in the ecommerce market.
- As collaboration has become the key focus of supply chain management, a growing number of firms have attempted to improve their ability to link, connect, and work effectively with their suppliers and customers. These attempts require the integration and harmonization of their company-wide business functions and channel-wide business processes. For such integration and harmonization, many firms began to recognize the usefulness of strategic information systems such as customer relationship management (CRM), supplier relationship management (SRM), and enterprise resource planning (ERP).
- Among new emerging technologies, RFID shows great promise in streamlining supply chain operations with its ability to enhance supply chain visibility. It not only helps improve customer services through improved inventory accuracy, but also helps improve supply chain security through the real-time tracking capability of moving inventory. Another benefit of RFID includes improved utilization of warehouse and retail store space through better management of inventory location.
- Although the investment in IT may be rewarding in the long run, many IT projects have not gone as well as expected. One of the biggest sources of failure is a lack of planning and poor project management. Therefore, before undertaking a new IT project, both IT users and providers must develop a step-by-step project plan and contingency plans in case the IT project suffers from cost-overruns and delays.

Study Questions

The following are important thought starters for class discussions and exercises:

1. How does IT transform a supply chain? How does IT change the business model and supply chain management practices?
2. How do you assess the impact of IT on each business function?
3. How does IT facilitate both vertical and horizontal integration of supply chain activities?

4. What can be the biggest challenges for implementing new, emerging IT such as leap motion devices, Google Glass with augmented reality features, eye-tracking devices, holographic heads-up displays, and speech recognition devices?
5. How do you estimate the return on investment (ROI) in IT?
6. What makes RFID different from barcoding technology? What does the future hold for barcoding technology with the emergence of RFID?
7. Is ERP a fad or a necessity? What are the most important prerequisites to ERP implementation?
8. What are the most important decision factors for selecting the right IT for your organization?
9. What is the best way to manage, control, and monitor IT projects?
10. Which technologies will have the biggest impact on supply chain performance?

Case: Red-Kitchen.com

This case was developed based on an actual situation, but was substantially modified to keep the company's identity confidential.

Selling Food Online

Red-Kitchen.com is a new breed of online grocer that markets, sells, and distributes groceries as well as precooked and prepackaged food for shoppers all across the Midwest. Red-Kitchen.com still publishes a high-quality catalog, which is sent to prospective and repeat customers who prefer to place their orders by mail or using a toll-free telephone number. Since Red-Kitchen.com's inception, its customer base has consisted principally of young busy professionals, college students, and outdoor campers. Nowadays, its customer base is extended to include some elderly people and foreign customers in Canada. Although the grocery/food industry is fiercely competitive and not very lucrative, the online business has recently started to pick up steam after experiencing tough times in the early 2000s. It is regarded as an appealing alternative for people who are "just too busy" to spend time shopping in grocery stores or supermarkets. Also, elderly and physically handicapped people who cannot drive to a grocery store seem to enjoy the convenience of ordering food online. The key benefit of online grocery shopping is that customers can browse the virtual shopping aisles 24 hours a day, 7 days a week, without leaving their home. They can also get instant answers to common ordering questions from an online assistant or through the online chat room.

E-logistics Challenges

During the last decade, some pioneering online grocers such as Webvan, Streamline, Homegrocer, and Shoplink failed to survive after investing hundreds of millions of dollars in finding their right business model and building infrastructure (e.g., warehouses). One of the main causes for their failure is attributed to their lack of understanding of e-logistics complexity. Typically, online grocers aim to deliver most items the day they are ordered, sometimes in as little as one hour. To maintain the same-day or one-day delivery commitment, many online grocers are forced to use an expedited shipping program and build bigger warehouses with more sophisticated software that allows "rapid-fire" order fulfillment and real-time inventory management. For example, Webvan once spent a staggering one billion dollars to build a number of state-of-art warehousing facilities throughout its targeted market areas and eventually ran out of money even before making break-even. Amazon.com, e-commerce giant and the mother company of the online grocer AmazonFresh, has also invested roughly $13.9 billion since 2010 to build 50 new state-of-art warehouses to help speed delivery from its boosted storage facilities.

Moving in the opposite direction, Red-Kitchen.com is currently exploring the possibility of downsizing its warehouse network and outsourcing its last-mile delivery services through contract shoppers who can drive their own cars to customers. In the past, Red-Kitchen.com experimented with the idea of letting its customers pick up their ordered groceries at a determined time in a predetermined brick-and-mortar store that had a profit-sharing partnership with Red-Kitchen.com. However, this self-service idea has not been well received by some customers who still did not like the hassle of traveling to the grocery store. As a better alternative, Red-Kitchen.com has decided to push the envelope on delivery options to further enhance online shopping convenience. It is currently piloting the use of "collection lockers," which allow the customer to retrieve their package using a unique pickup code outside of partnering grocery stores so that they do not need to visit the store to pick up their "click-and-collect" orders before the store closes. Also, under consideration are collection hubs in convenience stores, business parks, universities, mass transit stations, and park-and-ride stations.

The ultimate goal of Red-Kitchen.com is to become the industry leader in terms of brand recognition, product assortment, product quality, and post-sale customer service. In particular, its recent customer survey indicates that approximately 40% of its customers prefer a high-profile, well-known brand over others. The company realizes that there is no room for any managerial mistakes and the subsequent damage to the company's reputation. Therefore, it is willing to trade short-term losses for a better word-of-mouth reputation, a greater market share, and sustained profits later. Critical elements of customer service at Red-Kitchen.com are that orders are received, packed, and shipped in less than two hours, and procedures for returning unwanted products such as dented cans, spoiled milk, and expiring breads are "risk free" and "customer friendly." Although Red-Kitchen.com succeeds in accommodating product returns with little or no hassle to customers, this practice is getting more expensive and of growing concern to top management. In particular, nearly 10% of goods sold online were returned for refunds or product exchanges, which began to take a toll on Red-Kitchen.com's operating expenditures and the already thin profit margin. Red-Kitchen.com does not produce any of the food/grocery products it sells and distributes. Instead, it contracts with local farmers and food manufacturers in the Midwest to meet the needs of its largely perishable and seasonal product lines. Container loads of labeled and pre-tagged food/grocery items are shipped by express carriers to the company's centralized distribution center in Chicago, Illinois.

Technology as the Selling Point

Technology remains a crucial element of the online grocery business, both in retailer-customer interactions and back-end retail capabilities. Red-Kitchen.com starts experimenting with cutting-edge customer interface technologies to increase its sales volume and customer base. It uses an intuitive, interactive online experience that takes advantage of the latest interface technologies whose features allow customers to upload photos of grocery/food products on virtual dinner tables and review different recipes of those products without visiting a store. In addition to this customer interface technology, Red-Kitchen.com is now considering adopting new technology such as the Kiva System to enhance its order-fulfillment capabilities by automating and streamlining warehousing operations. With the paramount importance of technology to online grocery business success, executives at Red-Kitchen.com view themselves as being in the "logistics/technology business" in that the company's profitability hinges on e-logistics efficiency, advanced shipping notice, and shipment-tracking services. They feel that the company's e-logistics capabilities are the key to its excellent reputation in the online grocery industry. An area of nagging concern to those supply chain managers, however, is that consumer tastes (e.g., healthier, low-calorie, GMO-free products) and preferences (e.g., two-hour delivery window) for the company's products tend to change very quickly. As a result, the continued ability to forecast changing customer behavior well in advance and then react quickly to changing customer needs separates the market leaders from the rest. In addition, the company's ability to embrace state-of-art technology associated with e-fulfillment and logistics operations may give it an enormous competitive edge over its competitors.

Key Discussion Questions

1. Do you support Red-Kitchen.com's strategic option of outsourcing its entire distribution operations? (Do you think that it should consider developing its own brand? If so, why?)
2. Do you think that Red-Kitchen.com will be better off by outsourcing its reverse logistics operations for handling returned products? If so, why?
3. What return policy will be most appropriate for Red-Kitchen.com?
4. What do you think will be the core competency of Red-Kitchen.com?
5. What are the hidden perils of last-mile delivery services? (Be as specific as possible.)
6. What is unique about the Red-Kitchen.com's current business model (especially in comparison to traditional catalog business)?
7. What would you suggest to Red-Kitchen.com to tap the growing market in Canada?
8. Do you think that Red-Kitchen.com should continue to focus on online grocery/food sales? Otherwise, how should it expand its product line? (What kind of product/service does it need to add to its offering?)
9. From a supply chain perspective, how should Red-Kitchen.com protect its business from the onslaught of competitive offerings from spinoffs of e-commerce giants such as AmazonFresh and mass merchants such as Walmart.com?
10. Can you justify additional capital investment in new technology such as the Kiva System, which uses robots to sort and pack ordered items?

Bibliography

Aberdeen Group (2009), "Business Adoption of Cloud Computing," September 30, http://www.aberdeen.com/Aberdeen-Library/6220/RA-cloud-computing-sustainability.aspx, retrieved on September 29, 2010.

Brady, J.A., Monk, E.F., and Wagner, B.J. (2001), *Concepts in Enterprise Resource Planning*, Boston, MA: Course Technology-Division of Thomson Learning.

Brain, M. (2007), "How UPC Bar Codes Work," http://electronics.howstuffworks.com/gadgets/high-tech-gadgets/upc.htm, retrieved on September 2, 2010.

Buchanan, M. (2010), "How Multi-tasking Works on a Phone," *MSNBC.com*, May 4, http://www.msnbc.msn.com/id/36922756/, retrieved on October 6, 2010.

Carr, A.S., Zhang, M., Klopping, I., and Min, H. (2010), "RFID Technology: Implications for Healthcare Organization," *American Journal of Business*, 25(2), 1–16.

Cassavoy, L. (2010), "What Makes a Smartphone Smart," About.com, http://cellphones.about.com/od/smartphonebasics/a/what_is_smart.htm, retrieved on October 6, 2010.

Chang, K. (2010), *Introduction to Geographic Information Systems*, 5th edition, New York, NY: McGraw-Hill.

Data ID Systems (2006), "What's a Bar Code." http://www.dataid.com/benefitsbc.htm, retrieved on September 4, 2010.

Davis, E.W. and Spekman, R.E. (2004), *The Extended Enterprise: Gaining Competitive Advantage through Collaborative Supply Chains*, Upper Saddle River, NJ: Prentice-Hall.

Davenport, T.H. (2000), *Mission Critical: Realizing the Promise of Enterprise Systems*, Boston, MA: Harvard Business School Press.

Enright, A. (2014), "U.S. Online Retail Sales Will Grow 57% by 2018," *Internet Retailer*, May 12, 2014, https://www.internetretailer.com/2014/05/12/us-online-retail-sales-will-grow-57-2018, retrieved on March 23, 2015.

ERPwire.com (2010), "Advantages and Disadvantages of ERP," http://www.erpwire.com/erp-articles/erp-advantages-disadvantages.htm, retrieved on August 28, 2010.

GIS.com (2010), "What is GIS," The Guide to Geographic Information Systems, http://www.gis.com/content/what-gis, retrieved on August 30, 2010.

GIS and Science (2009), "Top Five Benefits of GIS," *Applied Geography*, http://gisandscience.com/2009/09/14/top-five-benefits-of-gis/, retrieved on August 30, 2010.

Gonsalves, A. (2010), "Android Phones Steal Market Share," *Information Week*, April 7, http://www.informationweek.com/news/mobility/smart_phones/showArticle.jhtml?articleID=224201881, retrieved on October 6, 2010.

Greengard, S. (2007), "How to Choose between Fixed and Mobile Interrogators," *RFID Journal*, July 30, http://www.rfidjournal.com/article/purchase/3349, retrieved on September 23, 2010.

Hwang, W. and Min, H. (2013), "Assessing the Impact of ERP on Supplier Performance," *Industrial Management and Data Systems*, 113(7), 1025–1047.

Internet Marketing (2009), "Forrester: Online Sales to Continue Growth in 2009," http://www.hotelmarketing.com/index.php/article/forrester_online_sales_to_continue_growth_in_2009/, retrieved on August 26, 2010.

Internet World Stats (2010), "Top 20 Countries with the Highest Number of Internet Users," http://www.internetworldstats.com/top20.htm, retrieved on August 27, 2010.

Luger, G.F. (2002), *Artificial Intelligence: Structures and Strategies for Complex Problem Solving*, 4th edition, Essex, England: Pearson Education Limited

McQueen, B. and McQueen, J. (1999), *Intelligent Transportation Systems Architectures*, Norwood, MA: Artech House, Inc.

Min, H. (2010), "Artificial Intelligence in Supply Chain Management: Theory and Applications," *International Journal of Logistics: Research and Applications*, 13(1), 13–39.

Min, H. and Zhou, G. (2002), "Supply Chain Modeling: Past, Present and Future," *Computers and Industrial Engineering*, 43(1), 231–249.

Moeeni, F. (2006), "From Light Frequency Identification (LFID) to Radio Frequency Identification (RFID) in the Supply Chain," *Decision Line*, 37(3), 8–13.

Monk, E.F., and Wagner, B.J. (2006), *Concepts in Enterprise Resource Planning*, 2nd edition, Boston, MA: Course Technology-Division of Thomson Learning.

Nilsson, N.J. (1980), *Principles of Artificial Intelligence*, Los Altos, CA: Morgan Kaufmann Publishers.

Rittinghouse, J.W. and Ransome, J.F. (2010), *Cloud Computing: Implementation, Management, and Security*, Boca Raton, FL: CRC Press.

Russell, S. and Norvig, P. (1995), *Artificial Intelligence: A Modern Approach*, Upper Saddle River, NJ: Prentice-Hall.

Seymour, L., Lambert-Porter, E., and Willuweit, L. (2008), "Towards a Framework for RFID Adoption into the Container Supply Chain," *Journal of Information Science Technology*, 5(1), 31–46.

TechTarget (2010), "What is Cloud Computing," April 5, http://searchcloudcomputing.techtarget.com/sDefinition/0,,sid201_gci1287881,00.html, retrieved on September 29, 2010.

Wegener, M. and Fotheringham, A.S. (2000), "New Spatial Models: Achievements and Challenges," *Spatial Models and GIS: New Potential and New Models* edited by A.S. Fotheringham and M. Wegener, London, United Kingdom: Taylor & Francis.

Wikipedia (2010), *Project Management*, September 24, http://en.wikipedia.org/wiki/Project_management#The_traditional_triple_constraints, retrieved on September 26, 2010.

Index

Numerics

3PL (third-party logistics provider), 325-326
   asset-based, 330
   market trends, 329, 332-334
   non-asset-based, 330
   performance metrics, 332-333
4PL (fourth-party logistics provider), 327
15-Day rules, 402

A

ABC (activity-based costing, 16, 273-276
ABC inventory analysis, 131-134
acculturation, 377
acquiring customers, 69
acquisition costs, 273
active RFID tags, 451
adjustable leases, 181
administrative expenses for warehousing, 182
administrative law, 392
advanced forecasting techniques, 103
advantages
   of ABC, 276
   of overseas distributors, 322
   of private warehousing, 178
   of public warehousing, 179
affective commitment, 70
agents, 293
agile supply chain strategy, 49-50
AHP (analytic hierarchy process), 286, 433
AI (artificial intelligence), 453-454
Air Cargo Deregulation Act of 1977, 229
air carriers, 244-245
air freight carriers, 244
Airline Deregulation Act of 1978, 229
airline industry, deregulation of, 229
air passenger carriers, 244
air taxi carriers, 245
anticipation inventory, 129
antitrust law, 393
arbitration, 395
asset-based 3PLs, 330
asset management for warehousing, 190
asset utilization, 15
attributes of total systems approach, 7
auctions, 305-308
auditing, supply chain strategies
   external supply chain, 52-53
   internal supply chain, 51-52
Aurora Jewelers case study, 380-384
automation, warehousing, 196
avoiding
   failed projects, 455
   traps in change management, 23

B

bait and switch, 398
Bank Export Services Act of 1982, 325
barcoding systems, 449
   benefits of, 449
   categories, 449
   comparing with RFID, 453
   ensuring suitability of, 450
barter, 374
batch picking, 193
bearer bill of lading, 247
benchmarks for customer service, 78-79
benefits
   of ABC, 276
   of barcoding systems, 449
   of DRP, 159
   of ERP, 444
   of GIS, 447

of intermodal transportation, 246
of JIT, 162-163
of MRP, 155
of private warehousing, 178
of public warehousing, 179
of supply chain integration, 4
of TMS, 253
birdyback transportation, 246
B/L (bill of lading), 247-248, 368
blue ocean strategy, 39-40
BOM (bill of materials), 148
bonded warehouses, 178
bottleneck products, 42
boundary-spanning activities, roles of, 8
Box-Jenkins method, 103
bribery, 398
   grease payments, 397
   questionable payments, 397
BSC (balanced scorecard), 426-428, 432
buffer inventory, 129
Buffet, Warren, 389
building trust, 18
bulk storage warehouses, 178
bullwhip effect, 5, 113-116
business doctrine, 38
business models, Internet age versus pre-Internet, 442-443
business unit strategy, 38
buy-back, 374

C

CAGR (compound annual growth rate), 442
calculating
   coefficient of variation, 147
   moving average, 98-100
   MRP, 149-150
   on-hand inventory, 150
   ROI for IT projects, 455
   total annual inventory cost, 144
   total inventory cost, 138, 141
cargo agents, 322

carrier management, 235-238, 247-248
case studies
   Aurora Jewelers, 380-384
   bad apple, 211-215
   Dell, Inc., 54-58
   Falcon Supply Chain Solutions, 336-339
   Jumping Footwear, 414-416
   La Bamba Bakeries, 436-439
   Louis Cab On Demand, 259-262
   Lucas Construction, Inc., 311-313
   Red-Kitchen.com, 463-466
   Sandusky Winery, 170-172
   Shiny Glass, Inc., 83-85
   Seven Star Electronics, 119-122
   Zara, 27-31
cash flow
   economic exposure risks, 356
   foreign exchange fluctuations, 356-357
   translation exposure risks, 356
cash-to-cash cycle time metric, 425
categorical method of supplier selection, 285
categories of barcode systems, 449
categorizing customers, 67-69
centralization strategy, 184-185
Central Place Theory, 224-225
certification programs for suppliers, 290
change management, 22-24
characteristics
   of global strategic alliances, 351
   of performance metrics, 421-422
   of supply chain links, 14
check digits, 449
Class I railroads, 242
class rates, 249
Clean Air Act, 407
Clean Water Act, 407
cloud computing
   IaaS, 458
   PaaS, 459
   SaaS, 459

CMAT (Customer Management Assessment Tool), 73
CO (certificate of origin), 368
coefficient of variation, calculating, 147
cognitive planning processes, 41-42
cold storage warehouses, 178
collaborative commerce, 107-109
   CPFR, 110-113
   ECR, 110
co-managed services, 268
combined new business creation, 113
commercial air carriers, 244
commercial invoices, 368
commercial law, 394
   arbitration, 395
   litigation, 396
   mediation, 395
commodities, 42
commodity rates, 250
common carriers, 234
common law, 392
common markets, 346
communicating with customers, 71
commuter air carriers, 245
comparing
   barcoding systems and RFID, 453
   competitive bidding and negotiation, 299-301
   DRP and MRP, 157
   EOQ and MRP, 149
   foreign market entry modes, 350-351
   forward and reverse logistics, 206-207
   JIT
      and MRP, 162
      and push systems, 161-162
   perpetual and periodic systems, 136
   pre-Internet and Internet age business models, 442-443
   push and pull strategies, 48
   red and blue ocean strategies, 40
   supply chain strategies, 50

competitive bidding versus negotiation, 299-301
compliant organizations, 299
components
   of inventory management, 128
   requirements for GIS, 446
concealed damage, 401-402
confidence tricks, 399
confidentiality agreements, 400
conflicts of interest, 399
consequences of forecasting errors, 5
constraints, 20
consular invoice, 368
container leasing companies, 323
continuance commitment, 70
contract carriers, 234
contract disputes, resolving
   arbitration, 395
   cross-cultural negotiations, 376-378
   litigation, 396
   mediation, 395
contract manufacturing, 350
controlling
   dependent demand inventory, MRP, 147-149
      benefits of, 155
      calculating, 149-150
      example of, 151, 154
   independent demand inventory
      economic order, 135-140
      quantity and, 143-146
      quantity discount, 141-143
      risk pooling, 146
controlling phase of IT projects, 456-457
cooperative supplier partnership, 291-293
coordinated strategies, 113
copyright, 393
COQ (cost of quality), 16
corporate philosophy, 38
corporate strategy, 38
cost of ERP implementation, 445

Index   471

cost analysis for sourcing. *See also* value analysis
  ABC, 273-276
  LCC, 276
  performing, 270
  TCO, 272-273
  traditional costing, 271
cost behavior, 15
cost ratio method of supplier selection, 286-288
cost strategy, 38
cost value, 278
costs
  of service pricing, 250
  of warehousing, 181
counter-purchase, 374
countertrade, 373-374
CPFR (collaborative planning, forecasting, and replenishment), 110-113
creating customer service strategy, 79-81
critical products, 42
critical value analysis, 131
CRM (customer relationship management), 67
  customers
    acquiring, 69
    communicating with, 71
    retaining, 69-70
    service strategy, formulating, 79-81
    valued customers, selecting, 71-72
  ROI, measuring, 72-73
  service benchmarks, 78-79
  service gaps, 76-78
  service performance measurement, 74
  service quality, 75-76
cross-cultural negotiations, 376-378
CSF (critical success factor), 425
CSI (Container Security Initiative), 360
C-TPAT (Customs-Trade Partnership against Terrorism), 360
customers
  acquiring, 69
  communicating with, 71
  retaining, 69-70
  ROI, measuring, 72-73
  segmenting, 67-69
  valued customers, selecting, 71-72
customer satisfaction, 15
customer service, 62
  elements of, 63
  initiatives, 15
  post-transaction elements, 66-67
  pre-transaction elements, 63
  service benchmarks, 78-79
  standards, 75
  strategy, creating, 79-81
  transaction elements, 64-66
customs (house) brokers, 324
cycle counting, 164
cycle stock, 129

D

DAP (Delivered at Place), 369
dashboard, 432
DAT (Delivered at Terminal), 369
DBR (drum-buffer-rope) logic, 21
DEA (data envelope analysis), 433
"deadheading," 199
decentralization strategy, 185
decoupling inventory, 129
defendants, 396
Dell, Inc., 54-58
demand chain, 8-10
demand forecasting, 94
  advanced techniques, 103
  bullwhip effect, 113-116
  exponential smoothing, 97-98
  method, selecting, 104
  moving average, 98-100
  trend analysis, 100-102
demand management, 90
  communicating demand, 91-92
  influencing demand, 92-93

planning demand, 91
prioritizing demand, 94
demand volatility risk, 295
dependent demand inventory, MRP, 147-149
  benefits of, 155
  calculating, 149-150
  example of, 151, 154
deregulatory acts, impact on transportation industry, 229-231
designing warehouse networks
  centralization strategy, 184-185
  decentralization strategy, 185
  profile analysis, performing, 185
design patents, 393
destinations for global outsourcing, selecting, 355
DHS (Department of Homeland Security), 359
differentiation strategy, 38
direct costs, 271
direct distribution channels, 364
direct exporting, 349
disadvantages
  of ABC, 276
  of ERP, 444-445
  of GIS, 447
  of JIT, 163-166
  of overseas distributors, 322
discrete picking, 192
dispute resolution
  arbitration, 395
  litigation, 396
  mediation, 395
disruptions in supply chain as threat to revenue, 5
distribution channels, 364
  grey markets, 365
  selecting, 365
diversity programs for suppliers, 290-291
DMUs (decision-making units), 433
documentation, B/L, 247-248

DOT (U.S. Department of Transportation), 226
drawbacks of global sourcing, 304-305
drivers of supply chain linkages
  customer service, 15
  information/knowledge, 16
  monetary value, 15
  risk elements, 16-17
drone technology, 461
DRP (distribution resource planning)
  benefits of, 159
  comparing with MRP, 157
  example of, 157
  functions of, 159
  objectives, 155-156
DSD (direct store door) delivery, 436-438

E

e-auctions, 307-308
EC (electronic commerce), 305
  e-auctions, 307-308
  EDI, 306
  electronic integration, 306
  XML, 306
e-commerce, 442
  collaborative commerce, 107-109
    CPFR, 110-113
    ECR, 110
economic exposure risks in global supply operations, 356
economic order quantity model, 135-140
ECR (efficient consumer response), 110
EDI (Electronic Data Interchange), 305-306
efficient supply chain strategy, 50
EFTA (European Free Trade Association), 345
electronic integration, 306
elements of customer service
  post-transaction elements, 66-67
  pre-transaction elements, 63
  transaction elements, 64-66

Elkins Act of 1903, 228
EMCs (export management companies), 324, 350
employment law, 393
ensuring suitability of barcoding systems, 450
environmental audits, 410
environmental law, 394
environmental regulation compliance, 406-408
EOQ (economic order quantity), 184
e-purchasing, 305
    e-auctions, 307-308
    EDI, 306
    electronic integration, 306
    XML, 306
equations
    exponential smoothing, 97-98
    moving average, 98-100
    total annual inventory cost, calculating, 144
    total inventory cost, calculating, 138, 141
    trend analysis, 100-102
ERP (enterprise resource planning), 443
    benefits of, 444
    disadvantages of, 444-445
errors in forecasting, consequences of, 5
esteem value, 278
estimating customer profitability, 72
ETC (export trading company), 325
ethics, 396
    confidentiality agreements, 400
    unethical supply chain behavior, 396
        bribery, 397-398
        concealed damage, 401-402
        conflicts of, 399
        fraud, 398-399
        questionable, 402-403
        reciprocity, 400-401
EU (European Union), 346
EVA (economic value-added), 425
evaluating suppliers, 283-284
evolution of supply chain management, 6

examples
    of DRP, 157
    of global strategic alliances, 352
    of MRP, 151, 154
excess inventory, 128
exchange value, 278
execution phase of IT projects, 456-457
exempt carriers, 234
explicit knowledge, 18
exponential smoothing, 97-98
Export Administration Regulations, 321
export brokers, 294
exporting, 349, 367-368
export packers, 324
external risk, 295
external supply chain strategies, auditing, 52-53

F

FAA Reauthorization bill, 461
failed IT projects
    avoiding, 455
    reasons for, 455
failure rate of global strategic alliances, 353
false advertising, 399
FAST (Free and Secure Trade), 361
FDI (foreign direct investment), 347
features
    of dashboard and BSC, 432
    of smartphones, 461
    of WMS, 200-203
finance-based 3PLs, 330
financial benefits of supply chain integration, 4
financial risk, 295
finished goods, 130
fishyback transportation, 246
fixed costs, 272
fixed RFID readers, 451
flat rental leases, 180
focus strategy, 38

forecasting
  advanced techniques, 103
  Box-Jenkins method, 103
  bullwhip effect, 113-116
  errors in, consequences, 5
  exponential smoothing, 97-98
  method, selecting, 104
  moving average, 98-100
  trend analysis, 100-102
Foreign Corrupt Practice Act of 1977, 402-403
foreign exchange fluctuations as risk in global supply chain management, 356-357
foreign market entry strategies of MNFs, 348
  comparing, 350-351
  contract, 350
  exporting, 349
  foreign market, 349
  franchising, 350
  join venture, 350
  licensing, 350
formulas
  exponential smoothing, 97-98
  moving average, 98-100
  trend analysis, 100-102
formulating
  customer service strategy, 79-81
  transportation strategy, 223-224
forward logistics, 206-207
franchising, 350
fraud, 398
  bait and switch, 398
  confidence tricks, 399
  false advertising, 399
  identity theft, 399
free trade associations, 346
free trade movements
  GATT, 345
  impact on global supply chain, 344, 347
  NAFTA, 346
free trade zones, 366

freight
  Incoterms, 369-373
  rates, 249
    cost of service pricing, 250
    negotiating, 250-252
    published tariffs, 249
    value of service pricing, 250
freight forwarders, 321-322
FTZ (foreign trade zones), 366-367
full outsourcing, 268
functional strategy, 38
functions
  of DRP, 159
  of inventory, 129
  of warehouses, 176
future trends in IT
  cloud computing, 458-459
  drone technology, 461
  mobile technology, 460
  smartphones, 460-461
  social networking, 459

G

gaps in service quality, 76-78
Gartners S&OP maturity model, 105-107
gatekeeping, 208
GATT (General Agreement on Tariffs and Trade), 345
general carriers, 245
general merchandise warehouses, 178
GIS (geographic information systems), 445
  benefits of, 447
  disadvantages of, 447
  required components, 446
global market penetration strategies of MNFs, 348
  comparing, 350-351
  contract, 350
  exporting, 349
  foreign, 349
  franchising, 350

Index   475

joint venture, 350
licensing, 350
global outsourcing, 354-355
global sourcing, 302-305
global strategic alliances, 351
characteristics of, 351
examples of, 352
failure rate among, 353
motivating factors for entering, 352-353
global supply chain operations
countertrade, 373-374
cross-cultural negotiations, 376-378
distribution channels, 364-365
documentation, 367-368
free trade zones, 366
FTZ, 366-367
impact of free trade, 344, 347
Incoterms, 369-373
international trade, growth, 344
risks
economic exposure risks, 356
foreign exchange, 356-357
maritime piracy, 361-363
natural disasters, 363-364
terrorism, 358
transaction exposure risks, 356
translation exposure risks, 356
supply chain security, 359-361
transfer pricing, 375-376
goals
setting, 14
of strategic alliances, 11
grease payments, 397, 402
grey markets, 365
gross leases, 180
gross requirements, 150
growth of international trade, 344
GSCM (green supply chain management), 403
environmental audits, 406-410
implementing, 403-404

ISO 14000 standards, 408-409
LCA, 404-405

H

handling returned products, 205-207
Hepburn Act of 1906, 228
hidden costs for warehousing, 183
high context culture, 377
high power distance culture, 378
historical background of transportation regulations, 227-229
HMR (Hazardous Material Regulation), 406-408
hockey stick phenomenon, 116
horizontal alliances, 11-13
HOS (hours of service) regulations, 232
household goods warehouses, 178
hybrid distribution channels, 365

I

IaaS (Infrastructure-as-a-Service), 458
IATA (International Air Transport Association), cargo, 322
ICC (Interstate Commerce Commission), 227
identifying
potential supply sources, 280
internal sources, 282
international sources, 282
professional sources, 282
published industry, 281
suppliers, 282-288
problem suppliers, 289
identity theft, 399
impact of free trade movement on global supply chain, 344, 347
imperatives of the value chain, 9
implementing
GSCM, 403-404
environmenal audits, 410
environmental regulation compliance, 406-408

ISO 14000 standards, 408-409
LCA, 404-405
SCOR model, 431
importing, documentation, 367-368
Incoterms (International commercial terms), 369-371, 373
independent demand inventory
  economic order quantity model, 135-140
  quantity and freight, 143-146
  quantity discount model, 141-143
  risk pooling model, 146
indirect costs, 272
indirect distribution channels, 364
indirect exporting, 349
influencing demand, 92-93
information transactions, 16-17
in-housing, 266-267
initiatives, customer service, 15
input KPIs, 425
inspection certificates, 368
integrated systems, 2
integrating supply chain processes, 4, 42
intellectual property law, 393-394
interfunctional coordination, 19
intermediaries
  3PL, 325-326
    asset-based, 330
    market trends, 329, 332-334
    non-asset-based transportation, 330
    performance metrics, 332-333
  4PL, 327
  container leasing companies, 323
  customs house brokers, 324
  EMC, 324
  ETC, 325
  export packers, 324
  freight forwarders, 321-322
  NVOCC, 323
  overseas distributors, 322
    advantages of, 322
    disadvantages of, 322
  potential challenges for using, 327-328
  role of, 320-321
  shipping agents, 323
  for sourcing, 293-294
intermodalism, 245
  benefits of, 246
  TOFC, 246
internal risk, 295
internal sources, 280-282
internal supply chain strategies, auditing, 52
international sources, 281-282
international trade
  countertrade, 373-374
  rapid growth of, 344
international trade documentation, 367-368
Internet, e-commerce, 442
Internet age business model, 442-443
interorganizational coordination, 19
Interstate Commerce Act, 227
inventory
  loss of
    preventing, 195
    reasons for, 195-196
  WMS, 199-205
inventory management, 128
  components of, 128
  cycle counting, 164
  dependend demand inventory, 147-151, 154-155
  DRP
    benefits of, 159
    comparing with MRP, 157
    example of, 157
    functions of, 159
    objectives, 155-156
  functions of inventory, 129
  independent demand inventory, 143-146
    economic, 135-140
    quantity, 141-143
    risk, 146
  inventory classification, 131-134

JIT
  benefits of, 162-163
  comparing with MRP, 162
  comparing with push system, 161-162
  disadvantages of, 163-166
  principles, 160
  waste, 160
on-hand inventory, calculating, 150
the Pareto principle, 131
total annual inventory cost, 144
total inventory cost, calculating, 138, 141
types of inventory, 130
VMI, 167-168
inverted tariff relief, 366
ISO 14000 standards, 408-409
ISPS (International Ship and Port Security) Code, 360
ISTEA (Intermodal Surface Transportation Efficiency Act of 1991), 448
IT (information technology)
  AI, 453-454
  barcoding systems
    benefits of, 449
    categories, 449
    comparing with RFID, 453
    ensuring suitability of, 450
  ERP, 443
    benefits of, 444
    disadvantages of, 444-445
  future trends
    cloud computing, 458-459
    drone technology, 461
    mobile technology, 459
    smartphones, 460-461
    social networking, 459
  GIS
    benefits of, 447
    disadvantages of, 447
    required components, 446
  ITS, 447-448

project management
  failures, reasons for, 455
  project phases, 456-457
  ROI, calculating, 455
RFID
  benefits of, 452
  disadvantages of, 452
  EPC, 451
  readers, 451
roles in supply chain, 443
ITS (intelligent transportation systems), 447-448

J

JIT, 143
  benefits of, 162-163
  comparing with MRP, 162
  comparing with push system, 161-162
  disadvantages of, 163-166
  principles, 160
  sourcing, 43
  waste, 160
joint ventures, 350
Jumping Footwear case study, 414-416

K

kanban, 161
knowledge, 18
knowledge transactions, 16
KPIs (key performance indicators), 425-426

L

L4L (lot-for-lot) rule, 151
La Bamba Bakeries case study, 436-438
laws, 391. *See also* **ethics; legislation, 397**
  administrative law, 392
  antitrust law, 393
  commercial law, 394-395
    arbitration, 395
    litigation, 396
    mediation, 395

common law, 392
employment law, 393
environmental law, 394
intellectual property law, 393-394
with supply chain implications, 392
written law, 391
layouts for warehouses, 186
modular spine design, 189
multistory design, 189
straight-thru design, 188
U-shaped flow, 187
LCA (Life Cycle Assessment), 404-405
LCC (life cycle costing), 276
leagile supply chain strategy, 49-50
lean supply chain strategy, 49-50
leases for warehouses, 180
legislation
Bank Export Services Act of 1982, 325
FAA Reauthorization bill, 461
ISTEA, 448
licensing, 350
linkages among supply chain activities, characteristics of, 14
litigation, 396
local line-haul railroads, 242
logistics, 20. *See also* **logistics intermediaries**
maritime piracy, effects on, 361-363
natural disasters, effects on, 363-364
sourcing strategies, 43-44
supply chain security initiatives, 359
  CSI, 360
  C-TPAT, 360
  FAST, 361
  ISPS, 360
strategies, 43
terrorism, effects on, 358
logistics intermediaries
3PL, 325-326
  asset-based, 330
  market trends, 329, 332-334
  non-asset-based, 330
  performance metrics, 332-333
4PL, 327
container leasing companies, 323
customs (house) brokers, 324
EMC, 324
ETC, 325
export packers, 324
freight forwarders, 321-322
NVOCC, 323
overseas, 322
overseas distributors, 322
potential challenges for using, 327-328
role of, 320-321
shipping agents, 323
long-term forecasting, trend analysis, 100-102
loss of inventory, reasons for, 195-196
lot-size inventory, 129
Louis Cab On Demand case study, 259-262
low context culture, 377
low power distance culture, 378
LTL carriers, 238
Lucas Construction, Inc. case study, 311

M

maintaining productive relationships with outsourcers, 269-270
managed business process links, 14
managed services, 268
managing
outsourcing, 269-270
returns, 207
Mann-Elkins Act of 1910, 228
maritime piracy as risk in global supply operations, 361-363
mass theft, 195
material handling, 190-192
order picking, 192-193
principles, 191
material management, 3
MAUT (multiple attribute utility theory) of supplier, 287

maximizing supply chain benefits, 9
measuring
   customer ROI, 72-73
   customer service, service benchmarks, 78-79
   organizational performance, 420-421
   performance
      BSC, 426-428
      dashboard, 432
      DEA, 433
      KPIs, 425-426
      SCOR model, 429-431
   service delivery performance, 74
mediated sourcing, 293-294
mediation, 395
merchants, 293-294
metrics, 421-422
   cash-to-cash cycle time, 425
   SCOR, 423-425
minority suppliers, 291
mistakes made in change management, 23-24
mitigating risk, 16-17
MNFs (multinational firms)
   countertrade, 373-374
   cross-cultural negotiations, 376-378
   distribution channels, 364
      grey markets, 365
      selecting, 365
   foreign market entry strategies, 348
      comparing, 350-351
      contract, 350
      exporting, 349
      foreign market, 349
      franchising, 350
      joint venture, 350
      licensing, 350
   free trade zones, 366
   FTZ, 366-367
   global outsourcing, 354-355
   Incoterms, 369-373
   international trade documentation, 367-368
   risks in global supply chain operations
      foreign, 356-357
      maritime, 361-363
      natural, 363-364
      terrorism, 358
   strategic alliances, 351
      characteristics of, 351
      failure rate among, 353
      motivating factors for entering, 352-353
   supply chain security initiatives, 359
      CSI, 360
      C-TPAT, 360
      FAST, 361
      ISPS, 360
   transfer pricing, 375-376
mobile RFID readers, 451
mobile technology, 460
modular-spine design (warehouses), 189
monetary value, 15
monitored business process links, 14
monitoring phase of IT projects, 456-457
motivating factors for entering global strategic alliances, 352-353
Motor Carrier Act of 1980, 230
moving average, calculating, 98-100
MRP (material requirements planning), 147-149
   benefits of, 155
   calculating, 149-150
   comparing
      with DRP, 157
      with EOQ, 149
      with JIT, 162
   example of, 151, 154
multistory warehouse design, 189

N

NAFTA (North American Free Trade Agreements), 346
natural disasters as risk in global supply operations, 363-364
negative reciprocity, 401

Negotiated Rates Act of 1993, 250
negotiating
   cross-cultural negotiations, 376-378
   freight rates, 250-252
   versus competitive bidding, 299-301
net leases, 180
net requirements, 150
netting, 149
non-asset-based transportation management 3PLs, 330
nondiscrimination, 345
non-economic regulations, 225-226
non-member business links, 14
normative commitment, 70
nuisance products, 42
NVOCC (non-vessel-operating common carrier), 323

O

objectives
   of AI, 454
   of DRP, 155-156
obsolete inventory, 128
Ocean Shipping Reform Act, 230
OEM (original equipment manufacturer), 3
offset, 374
on-hand inventory, calculating, 150
on-hand tracking, 164
online social networking media, 459
operational learning, 18
operational risk, 295
order bill of lading, 247
order picking, 192-193
organizational learning, 17
   operational learning, 18
   strategic learning, 18
organizational performance
   measuring, 420-421
     DEA, 433
     KPIs, 425-426
     SCOR model, 429-431
   metrics, 421-425

organizational planning processes, 41-42
output KPIs, 425
outsourcing, 266-267
   co-managed services, 268
   full outsourcing, 268
   global outsourcing, 354-355
   managed services, 268
   managing, 269-270
   out-tasking, 268
outsourcing return management, 208
out-tasking, 268
overseas-based merchants, 294
overseas distributors, 322
ownership costs, 273

P

PaaS (Platform-as-a-Service), 458-459
parcel carriers, 238
Pareto, Villefredo, 131
the Pareto principle, 131
partnerships
   primary partners, 12
   supporting partners, 12
   trust, building, 18
   VOP, 13
passive RFID tags, 451
patents, 393
percentage leases, 181
performance
   KPIs, 425-426
   measuring, 420-421
     BSC, 426-428
     dashboard, 432
     DEA, 433
     SCOR model, 429-431
   metrics, 421-422
     for 3PLs, 332-333
     cash-to-cash cycle time, 425
     SCOR, 423-425

performing
  cost analysis for sourcing, 270
    ABC, 273-276
    LCC, 276
    TCO, 272-273
    traditional costing, 271
  profile analysis for warehouse network design, 185
periodic inventory systems, 135
perpetual inventory systems, 135
phases of IT projects, 456-457
physical distribution, 3
piggyback transportation, 246
pilferage, 195
pipeline inventory, 129
plaintiffs, 396
planned order receipts, 151
planned order releases, 151
planning demand, 91
planning processes, 41-42
plant patents, 393
political planning processes, 41-42
pooled negotiation power, 113
positive reciprocity, 401
positive transportation regulation, 228
post-ownership costs, 273
post-transaction elements of customer service, 66-67
potential challenges for using logistics intermediaries, 327-328
potential supply sources, identifying, 280
  internal sources, 282
  international, 282
  professional sources, 282
  published industry sources, 281
  suppliers, 282-283
    evaluating, 283-284
    selecting, 285, 288
power of expertise, 301
pre-compliant organizations, 299
preemptive risk mitigation initiatives, 296-297

pre-Internet business model, 442-443
preparing for risk, 298-299
pre-transaction elements of customer service, 63
preventing inventory loss, 195
pricing, transfer pricing, 375-376
primary partners, 12
principles
  of material handling, 191
  of GATT, 345
  of JIT, 160
prioritizing demand, 94
private carriers, 233
private warehousing, advantages of, 178
problem suppliers, identifying, 289
process KPIs, 425
product availability, 15
productive relationships with outsourcers, maintaining, 269-270
productivity, warehousing, 194
products, types of, 42
professional sources, 281-282
profile analysis, performing for warehouse network design, 185
profitability of customers, estimating, 72
project management
  failed projects, avoiding, 455
  project phases, 456-457
  ROI, calculating, 455
public warehousing, advantages of, 179
published industry sources, 280-281
published tariffs, 249
pull strategy, 46-49
purchasing, 279
  e-auctions, 307-308
  e-purchasing, 305
    EDI, 306
    electronic integration, 306
    XML, 306
purchasing portfolio tool, 280
push strategy, 46-49, 161-162

Q

QR (quick response), ECR, 110
quality, gaps in service quality, 75-78
quantity and freight discount model, 143-146
quantity discount model, 141-143
questionable payments, 397, 402-403

R

Railroad Revitalization and Regulatory Reform Act, 229
rail transportation, 241
rate tariffs, 249
raw materials, 130
RCRA (Resource Conservation and Recovery Act), 406
readers (RFID), 451
Reciprocal Tax Agreements Act of 1934, 303
reciprocity, 345, 400-401
red ocean strategy, 39-40
refreezing, 22
refrigerated warehouses, 178
regional railroads, 242
regulations, 225, 392
   economic regulations, 225-226
   environmental regulation compliance, 406-408
   transportation regulations
      agencies, 226-227
      deregulatory acts, 229-231
      historical, 227-229
      hours of service, 232
relationship management, 291-293
relaxation of transportation regulation, 229
reorder point, 135
resilient organizations, 299
resolving contract disputes, 395
   arbitration, 395
   litigation, 396
   mediation, 395
response time, 15
responsive supply chain strategy, 50
restrictive transportation regulation, 228

retaining customers, 69-70
returned products, handling, 205-207
revenue management, 15, 252-253
reverse logistics, 206-207
RFID (radio frequency identification), 450
   benefits of, 452
   comparing with barcoding systems, 453
   disadvantages of, 452
   EPC, 451
   readers, 451
risk
   demand volatility risk, 295
   external risk, 295
   financial risk, 295
   in global supply operations
      economic exposure risks, 356
      foreign exchange, 356-357
      maritime piracy, 361-363
      natural disasters, 363-364
      supply chain security, 359-361
      terrorism, 358
      transaction exposure, 356
      translation exposure, 356
   internal risk, 295
   mitigating, 16-17
   operational risk, 295
   preemptive mitigation initiatives, 296-297
   preparedness, 298-299
   sourcing risk, 295
risk-hedging supply chain strategy, 50
risk pooling model, 146
ROI (return on investment), 15
   for IT projects, calculating, 455
   measuring, 72-73
roles
   of boundary spanning activities, 8
   of intermediaries, 320-321
   of IT in supply chain management, 443
      AI, 453-454
      barcoding systems, 449
      ERP, 443-445

GIS, 446-447
ITS, 447-448
RFID, 450-452
of transportation agencies, 226-227

S

Saas (Software-as-a-Service), 459
safety stock, 129
Sandusky Winery case study, 170-172
scheduled receipts, 150
SCOR (Supply Chain Operations Reference) model, 423-425, 429-431
secure organizations, 299
security
  maritime piracy, effect on global logistics, 361-363
  natural disasters, effect on global logistics, 363-364
  supply chain security initiatives, 359
    CSI, 360
    C-TPAT, 360
    FAST, 361
    ISPS, 360
  terrorism, effect on global logistics, 358
  warehousing, theft, 195
segmenting customers, 67-69
selecting
  distribution channels, 365
  forecasting method, 104
  global outsourcing destinations, 355
  suppliers, 285, 288
  supply chain strategy, 38-39
  transportation services, 235-238
  valued customers, 71-72
semi-passive RFID tags, 451
semivariable costs, 272
service benchmarks, 78-79
service performance measurement, 74
service quality, 75-78
setting goals, 14
Seven Star Electronics case study, 119-122
shared know-how, 112

shared tangible resources, 113
Sherman Antitrust Act of 1890, 393
shipment consolidators, 323
Shipping Act of 1984, 230
shipping agents, 323
short-term forecasting, exponential smoothing, 97-98
SKUs (stock keeping units), 130
smartphones, 461
smoothing constant, 97
social networking, 459
social planning processes, 41-42
S&OP (sales and operational planning), 105-107
sources of service gaps, 78
sourcing
  and logistics, 43-44
  competitive bidding, 299-301
  cost analysis
    ABC, 273-276
    LCC, 276
    performing, 270
    TCO, 272-273
    traditional costing, 271
  global sourcing, 302-305
  identifying potential supply sources, 280
    internal, 282
    professional, 282
    published, 281
    supplier, 282-285, 288
  in-housing, 266-267
  intermediaries, 293-294
  JIT, 43
  outsourcing, 266-267
    co-managed services, 268
    full outsourcing, 268
    maintaining productive relationships, 269-270
    managed services, 268
    out-tasking, 268
  purchasing, 279
  risk preparedness, 298-299

splitting orders, 288
suppliers
   certification, 290
   diversity, 290-291
   relationship management, 291-293
supply base reduction, 289
value analysis, 277-279

sourcing risk, 295
spatial utility of transportation, 222
special commodity warehouses, 178
splitting orders, 288
square-root rule, 184
Staggers Rail Act of 1980, 230
stakeholders, 3
standards of customer service, 75
state trading agencies, 294
straight bill of lading, 247
straight-thru design (warehouses), 188
strategic alliances, 10
   global strategic, 351
   global strategic alliances, 351
      failure, 353
      motivating, 352-353
   goals of, 11
   horizontal alliances, 11
   organizational learning, 17
   partnerships, building trust, 18
   primary partners, 12
   supporting partners, 12
   vertical alliances, 11
   VOP, 13

strategic learning, 18
strategies
   blue ocean strategy, 39-40
   business unit strategy, 38
   corporate strategy, 38
   customer service strategy, creating, 79-81
   external supply chain strategies, auditing, 52-53
   functional strategy, 38

   internal supply chain strategies, auditing, 51-52
   planning processes, 41-42
   pull strategy, 46-49
   push strategy, 46-49
   red ocean strategy, 39-40
   sourcing and logistics, 43-44
   supply chain strategy, selecting, 38-39
   transportation strategy, formulating, 223-224
   victory model, 44-45

structural dimensions of the supply chain, 13
S&T (switching and terminal) railroads, 242
subsidiaries, 350
suitability of barcoding systems, ensuring, 450
suppliers
   certification, 290
   diversity programs, 290-291
   evaluating, 283-284
   mediated sourcing, 293-294
   prequalification checklist, 282-283
   relationship management, 291-293
   selecting, 285, 288
   splitting orders, 288-289

supplies, 130
supply base reduction, 289
supply chain. *See also* **supply chain management**
   comparing with value and demand chain, 10
   OEM, 3

supply chain management
   change management, 22-24
   ethics, 396
      confidentiality agreements, 400
      unethical supply chain behavior, 396-403
   evolution of, 6
   linkages
      characteristics of, 14
      drivers, 15-17
   material management, 3
   physical distribution, 3

risk
- *demand volatility risk, 295*
- *external risk, 295*
- *financial risk, 295*
- *internal risk, 295*
- *operational risk, 295*
- *preemptive mitigation initiatives, 296-297*
- *sourcing risk, 295*

stakeholders, 3
strategies, 50
- *comparing, 50*
- *external strategies, auditing, 52-53*
- *internal strategies, auditing, 51-52*
- *pull strategy, 46-49*
- *push strategy, 46-49*
- *selecting, 38-39*
- *victory model, 44-45*

structural dimensions, 13
TOC, 20-22
total systems approach, 6-7
VOP, 13

supporting partners, 12
supporting tariffs, 249
surface transportation, 239
- rail, 241
- trucking industry, 240-241

surrender bill of lading, 248
sustainability, triple bottom line, 390
switch trading, 374
symptoms of bullwhip effect, 5

T

tacit knowledge, 18
tags (RFID), 451
tariffs
- free trade zones, 366
- FTZ, 366-367
- inverted tariff relief, 366

TCO (total cost of ownership, 272-273
terminal operations, 254-255

terrorism
- as risk in global supply operations, 358
- supply chain security initatives, 359
  - *CSI, 360*
  - *C-TPAT, 360*
  - *FAST, 361*
  - *ISPS, 360*

theft, 195
time series methods
- exponential smoothing, 97-98
- moving average, 98-100
- trend analysis, 100-102

time utility of transportation, 222
TL carriers, 238
TMS (transport management system), 253
TOC (theory of constraints), 20-22
TOFC (Trailers On Flat Cars), 246
total annual inventory cost, 144
total inventory cost, calculating, 138, 141
total systems approach, 6-7
trademarks, 394
traditional business paradigm, 6
traditional costing, 271, 275
traditional supplier relationship, 291-293
transactional tracking, 165
transaction elements of customer service, 64-66
transaction exposure risks in global supply operations, 356
transfer pricing, 375-376
translation exposure risks in global supply operations, 356
transparency, 345
transportation, 222
- air carriers, 244-245
- carrier management, 235-238
- Central Place Theory, 224-225
- common carriers, 234
- contract carriers, 234
- documentation, B/L, 247-248
- exempt carriers, 234
- intermodalism, 245-246

pricing, 249
   cost of service pricing, 250
   freigh rate negotiation, 250-252
   published tariffs, 249
   value of service pricing, 250
private carriers, 233
regulations, 226
   agencies, 226-227
   deregulatory acts, impact on, 229-231
   historical backgrounds, 227-229
   hours of service regulations, 232
revenue management, 252-253
spatial utility of, 222
strategies, formulating, 223-224
surface transportation, 239
   rail, 241
   trucking industry, 240-241
terminal operations, 254-255
time utility of, 222
TMS, 253
water carriers, 242-244
traps, avoiding in change management, 23
treaties, GATT, 345
trend analysis, 100-102
trends in IT
   cloud computing, 458-459
   drone technology, 461
   mobile technology, 459
   smartphones, 460-461
   social networking, 459
triple bottom line, 389
trucking industry, 240-241
trust, building, 18
TSA (Transportation Security Administration), 359
types of inventory, 130
types of supply chain strategies, 50

U

UCC (Uniform Commercial Code), 394
unethical supply chain behavior, 396
   bribery, 397-398

   concealed damage, 401-402
   conflicts of interest, 399
   fraud
      bait and switch, 398
      confidence tricks, 399
      false advertising, 399
      identity theft, 399
   questionable payments, 402-403
   reciprocity, 400-401
unfreezing, 22-24
Uniform Freight Classification, 249
unmanaged business process links, 14
unmanned aerial vehicles, 461
UPC (universal product code), 449
use value, 278
U-shaped layouts (warehouses), 187
utility patents, 393

V

value-adding strategy, 44-45
value analysis, 277-279
value chain, 8-10
valued customers
   ROI, measuring, 72-73
   selecting, 71-72
value of service pricing, 250
VANs (value-added networks), 305
variable costs, 272
vertical alliances, 11
vertical integration, 113
vertical structures, 13
victory model strategy, 44-45
VMI (vendor managed inventory), 167-168
VOP (value-offering point), 13

W

WAP (Wireless Application Protocol), 460
warehousing
   adminstrative expenses, 182
   asset management, 190
   automation, 196

costs of, 181
functions, 176
handling returned products, 205-207
hidden costs, 183
inventory loss, reasons for, 195-196
layouts, 186
   modular-spine design, 189
   multistory design, 189
   straight-thru design, 188
   U-shaped flow, 187
leases, 180
material handling, 190-193
network design, 184-185
private warehousing, advantages of, 178
productivity, 194
public warehousing, advantages of, 179
theft, 195
WMS, 198-205
workforce planning, 197

waste, 160
water carriers, 242-244
wave picking, 193
weighted-point method of supplier selection, 285-287
Windom Committee, 227
winning model strategy, 44-45
WIP (work in progress), 130
WMS (warehouse management systems), 198-205
workforce planning, warehousing, 197
written law, 391
WTO (World Trade Organization), 345

X-Y-Z

XML, 306

yield management, 252-253

Zara, 27-31
zone picking, 193